Springer Series in

OPTICAL SCIENCES 120

founded by H.K.V. Lotsch

Springer Series in
OPTICAL SCIENCES

The Springer Series in Optical Sciences, under the leadership of Editor-in-Chief *William T. Rhodes*, Georgia Institute of Technology, USA, provides an expanding selection of research monographs in all major areas of optics: lasers and quantum optics, ultrafast phenomena, optical spectroscopy techniques, optoelectronics, quantum information, information optics, applied laser technology, industrial applications, and other topics of contemporary interest.
With this broad coverage of topics, the series is of use to all research scientists and engineers who need up-to-date reference books.

The editors encourage prospective authors to correspond with them in advance of submitting a manuscript. Submission of manuscripts should be made to the Editor-in-Chief or one of the Editors. See also www.springer.com/series/624

Hamlet Karo Avetissian
Editor

Relativistic Nonlinear Electrodynamics

Interaction of Charged Particles with Strong and Super Strong Laser Fields

With 23 Illustrations

Springer

Prof. Dr. Hamlet Avetissian
1 Alex Manoukyan St.
Department of Quantum Electronics
Yerevan State University
375025 Yerevan, Armenia
avetissian@ysu.am

Library of Congress Control Number: 2005935209

ISBN-10: 0-387-30069-4 e-ISBN: 0-387-30070-8

ISBN-13: 978-0387-30069-6

Printed on acid-free paper.

Printed in the United States of America. (EB)

9 8 7 6 5 4 3 2 1

springer.com

Preface

With the appearance of lasers have come real possibilities of revealing numerous nonlinear phenomena of diverse nature resulting from the interaction of strong electromagnetic field either with matter or with free charged particles. First attempts of investigators, especially experimentalists, were directed toward studying the processes of interaction of laser radiation with matter, which led to the rapid formation of a new field — Nonlinear Optics. The numerous published monographs on this subject are evidence of that. The situation regarding the processes of interaction of laser radiation with free charged particles (free–free transitions) is different. Whereas the experimental results on atomic systems frequently had preceded the theoretical ones, the experimental investigations on free electrons began gathering power only recently. It is enough to mention that the first experiments on the observation of multiphoton exchange between free electrons and laser radiation started in 1975 (the Cherenkov and bremsstrahlung processes) whereas, due to the progress of Nonlinear Optics, the precision laser spectroscopy of superhigh resolution on atomic systems had already been established. This situation is explained by two objective factors. Whereas the experiments on atoms require only laser devices in common laboratories, the experiments on free electron beams require accelerators of charged particles and laser laboratories, i.e., this field is a synthesis of Accelerator and Laser Physics. The second major factor is the smallness of the photon–electron interaction cross section in comparison with the photon–atom one; revealing nonlinear phenomena on free electrons thus requires laser fields of relativistic intensities (e.g., even the observation of the second harmonic in nonlinear Compton scattering). Such superpower femtosecond laser sources have appeared only recently. Hence, the time for experimental development of this branch of Nonlinear Electrodynamics — interaction of charged particles with laser fields of relativistic intensities — has come. In presenting the current state-of-the-art in this field and gathering up-to-date theoretical material in this book we have pursued the goal of stimulating the laser driven experiments on relativistic electron beams and comprehensive theoretical investigations of nonlinear electromagnetic processes in currently available coherent radiation fields of relativistic intensities.

Increasing interest in free–free transitions is connected with the realization of the two most important problems of modern physics, namely, the creation of shortwave coherent radiation sources — X-ray and γ-ray lasers — and high energy laser accelerators of charged particles. It is noteworthy that a great deal of the works on free–free transitions is related to the Free Electron Laser (FEL) problem, i.e., to the discussion of concrete schemes of relativistic electron beam radiation amplification in coherent systems, such as the undulator, and to the search for their optimization. A small number of monographs and large number of reviews are devoted to this problem in the linear regime of amplification. However, particularly for the implementation of X-ray lasers, the most promising candidate of which at the present time are FEL devices, the need for nonlinear mechanisms of generation of coherent radiation due to induced interaction of electron beam with strong laser fields may be crucial, compared with the current undulator-based FELs in the linear regime of amplification. On the other hand, the present FELs operate in the classical regime where the electron wave packet size over the interaction length is less than a wavelength of radiation. This means that the photon frequency shift due to the electron quantum recoil must be less than the gain bandwidth. This condition is satisfied for current FELs typically operating at optical or smaller frequencies. For the X-ray photons in expected X-ray FELs, the downshift in frequency as well as other quantum effects become important. Thus, because of the absence of mirrors (resonator) or other drivers operable at these wavelengths, FEL systems currently under consideration for X-ray sources, operate in the so-called Self-Amplified Spontaneous Emission (SASE) regime in which the initial shot noise on an electron beam is amplified over the course of propagation through a long wiggler. In turn, large pulse-to-pulse variations arise in both output power and radiation spectrum, and quantum effects on the start-up from noise will be important. Finally, the absence of resonators at X-ray wavelengths requires a single-pass high-gain FEL, which in the linear regime will have an extremely large size. Hence, to reach the required gain on distances much smaller than the coherent length in the linear regime of amplification, which would reduce greatly the present size of projected X-ray lasers (several kilometers), nonlinear quantum mechanisms of generation due to laser induced coherent interaction becomes of prime importance. On the other hand, the inverse problem of laser induced nonlinear FEL schemes is the problem of creation of novel accelerators of charged particles of superhigh energies — laser accelerators. Therefore, the nonlinear interaction of charged particles with strong laser fields will be considered in general aspects from the point of view of both nonlinear quantum FEL schemes and classical laser accelerator problems. At the same time, we will not overload the material of this book, the subject of which is nonlinear electromagnetic processes, with the consideration of linear schemes of FELs taking also into account the existence of well-known monographs by T. Marshall (1987), C. Brau (1990), H. Freund and T. Antonsen (1996), and E. Saldin, E. Schneidmiller, and M. Yurkov (1999) devoted especially to this problem.

Besides the mentioned problems there is a third important problem concerning the quantum electrodynamic vacuum in superstrong laser fields. With the appearance of superpower lasers of relativistic intensities in recent years, for which the energy of an electron acquired at a wavelength of laser radiation exceeds the electron rest energy, multiphoton excitation of the Dirac vacuum via nonlinear channels becomes real and, consequently, electron–positron pair production becomes available. It is a strongly nonlinear process in superintense laser fields, which occurs inevitably in all processes where the conservation laws for the pair production are permitted. Thus, while considering such nonlinear processes we will give special consideration to the multiphoton electron–positron pair production from superintense laser fields.

Among the considered processes and, in general, stimulated processes with the charged particles the coherent processes like Cherenkov, Compton, and undulator essentially differ due to a peculiarity, which fundamentally changes the common picture of electromagnetic processes in dielectric media, and in vacuum — the presence of a second wave or an undulator. Because of the coherent character of the corresponding spontaneous radiation process (the existence of coherence condition for radiation) in the presence of an external electromagnetic wave a critical value of the wave field exists above which a plane wave becomes a potential barrier or well for a particle and specific threshold nonlinear phenomena arise. The latter open new possibilities for laser acceleration and FEL, since in these regimes the induced process proceeds only in one direction: the inverse concurrent process of radiation in acceleration regime, and absorption process for the FEL regime are absent. Therefore, we expect that this book will help to direct the attention of experimentalists to nonlinear phenomena of "reflection" and capture of charged particles by a plane electromagnetic wave in Cherenkov, Compton, and undulator processes, which have been left in the shadows for more than three decades. This especially relates to the experiments on the induced Cherenkov process made at SLAC by R. Pantell and collaborators since 1975 where the laser intensities were left below the critical value for the induced nonlinear Cherenkov process. It was necessary to increase the laser intensity a bit to reveal the existence of critical intensity and electron acceleration due to the "reflection" phenomenon, proving thereby the peculiarity of the induced Cherenkov process with its nonlinear threshold nature.

It is worth emphasizing another threshold phenomenon of nonlinear cyclotron resonance in an arbitrary medium (dielectric or plasma). That is so-called electron hysteresis, which can serve as an actual mechanism for laser acceleration of charged particle beams in plasma media where the use of superpower laser fields is not restricted and significant acceleration may be reached.

As is known, the spontaneous radiation of relativistic electrons and positrons channeled in a crystal is of great interest due to two major factors: the radiation is in the X-ray and γ-ray domains, and its spectral intensity noticeably exceeds that of other radiation sources in the short-wave range. Thus, induced channeling radiation in the presence of an external wave field becomes important as a potential source for short-wave coherent radiation. On the other hand, due to the induced channeling effect the inverse process — absorption of the wave photons by the particles — will also take place reducing the particles' acceleration and other coherent classical and quantum effects. As a periodic system with high coherency and having the same character of a particle motion, the crystal channel may be compared with an undulator — it is a "micro-undulator" with the space period much smaller than the undulator one. We thus give consideration to the induced channeling process in general aspects of coherent interaction of relativistic electrons and positrons with a plane electromagnetic wave in a crystal.

Concerning the consideration of induced noncoherent processes, please note that in the present book we included only induced processes related to plasma media where they provide actual energy conversion between the particles and transverse electromagnetic wave and, due to the nonlinear interaction, one can reach the effective outgrowth from the point of view of the above-mentioned problems. In particular, Stimulated Bremsstrahlung (SB) is of interest in plasma in the presence of an electromagnetic radiation field, since bremsstrahlung is one of the major electrodynamic processes in plasma, and is the actual mechanism for plasma heating (a scattering center performs the role of a third body for actual absorption/radiation of the wave photons by a charged particle). Besides, the role of SB is significant in the process of particle acceleration with plasma/laser fields, as well as in the process of high harmonics generation in atomic/ionic systems through the continuum states in strong laser fields as an alternative means for implementation of coherent X-ray sources, which has witnessed significant experimental advancement in recent years. However, the consideration of these processes is beyond the scope of this book. We will consider here the relativistic SB in strong and superstrong radiation fields in regard to general aspects with nonlinear effects (nonrelativistic SB in various approximations has been considered in many monographs). We will also consider the coherent SB in crystals, which is of relativistic nature in itself, having in mind consideration of a high-gain X-ray FEL scheme based on coherent bremsstrahlung in a crystal.

A separate chapter has been devoted to the so-called induced nonstationary transition effect based on the spontaneous transition radiation effect in a medium at the abrupt variation of its properties, to describe the nonlinear particle–strong wave interaction processes in plasma. Such a situation takes place inevitably at the interaction of superintense femtosecond laser pulses with any medium, which instantly turns into plasma. It is thus of certain interest to study the nonlinear processes at the formation of laser plasma. This process may also be of great interest in astrophysics related to conversion of electromagnetic radiation frequencies in nonstationary plasma, in particular formation of hard γ-quanta of relativistic energies, electron–positron pair production, and other nonlinear processes at the abrupt variation of the matter properties in cosmic objects.

In order not to overload the reader, the references on a given subject are presented separately in each chapter. My apologies go to all authors whose works are not covered in this book. I included only the ones that are most directly related to this monograph.

Indeed, the problems discussed in this monograph do not exhaust the frame of induced nonlinear phenomena at the interaction of charged particles with strong electromagnetic radiation. By considering a certain class of induced processes, we have aimed at revealing principal features of nonlinear behavior of a particle–strong wave interaction in coherent and noncoherent induced processes, which are of primary importance for the implementation of contemporary problems of FEL, laser accelerators, and electron–positron pair production from superintense laser fields. And if the consideration of these nonlinear processes based on relativistic classical and quantum theories and the presentation of the main results are helpful to specialists in this field, then the publication of this monograph will be justified.

In closing, I would like to thank Dr. G. Mkrtchian for assistance in preparation of the manuscript, Dr. H. Koelsch, physics editor Springer-Verlag New York, and associate editor V. Lipscy, for their patience and encouragement in the writing and publishing of the book.

Yerevan, Armenia
June 2005

Hamlet K. Avetissian

Contents

1 Interaction of a Charged Particle with Strong Plane Electromagnetic Wave in Vacuum

What can we expect from particle–strong wave interaction in vacuum?

It is well known that the radiation or absorption of photons by a free electron in vacuum is forbidden by the energy and momentum conservation laws, which means that the real energy exchange between a free electron and plane monochromatic wave in vacuum is impossible, isn't it?

Then, is it worth considering the interaction of a free electron with strong monochromatic wave in vacuum? In other words, what can we expect from the strong wave fields in nonlinear theory with respect to the weak ones described by the linear theory?

For example, what are the changes in cross section of the major electrodynamic process of electron–photon interaction, that is, Compton effect (which in the one-photon approximation within quantum electrodynamics is described by the Klein–Nishen formula) at a high density of incident photons?

Lastly, how strong should a wave field be for revelation of nonlinear effects in vacuum? What are the criteria of the strong field?

To answer these questions one must first study the dynamics of a charged particle in the field of a plane electromagnetic wave of arbitrary high intensity in vacuum on the basis of the classical and quantum equations of motion. Then, with the help of the classical trajectory of the particle and dynamic wave function in the quantum description, the nonlinear radiation in the scope of the classical and quantum theories — the Compton effect in the field of electromagnetic waves of arbitrary high intensity — will be treated.

We will start from the relativistic equations, because in the field of a strong wave even a particle initially at rest becomes relativistic. Then, the amplitude of a strong wave will be assumed invariable, i.e., the radiation effects do not influence the magnitude of a given strong wave field.

1.1 Classical Dynamics of a Particle in the Field of Strong Plane Electromagnetic Wave

Let a particle with a mass m and a charge e (let $e > 0$) interact with a plane electromagnetic (EM) wave of arbitrary form and intensity propagating in vacuum along a direction ν_0 ($|\nu_0| = 1$). Then, for the electric ($\mathbf{E}$) and magnetic ($\mathbf{H}$) field strengths we have

$$\mathbf{E}(t,\mathbf{r}) = \mathbf{E}(t - \nu_0\mathbf{r}/c); \quad \mathbf{H}(t,\mathbf{r}) = \mathbf{H}(t - \nu_0\mathbf{r}/c); \quad \mathbf{H} = [\nu_0\mathbf{E}] . \tag{1.1}$$

Relativistic classical equation of motion of the particle in the field (1.1) will be written in the form

$$\frac{d\mathbf{p}}{dt} = e\mathbf{E} + \frac{e}{c}[\mathbf{vH}] , \tag{1.2}$$

where $\mathbf{p}$ and $\mathbf{v}$ are the particle momentum and velocity in the field and c is the speed of light in vacuum.

For the integration of the equation of motion (1.2) the latter should be written in components:

$$\nu_0 \frac{d\mathbf{p}}{dt} = \frac{e}{c}(\mathbf{vE}), \tag{1.3}$$

$$\frac{d\mathbf{p}_\perp}{dt} = e\left(1 - \frac{\mathbf{v}\nu_0}{c}\right)\mathbf{E}. \tag{1.4}$$

Then the integration of Eqs. (1.4) is very simple if one takes into account that $\mathbf{E}$ is the function of the variable $\tau = t - \nu_0\mathbf{r}/c$ and passes on the left-hand side of (1.4) from the variable t to τ . So, for the transversal components of the particle momentum we will have

$$\mathbf{p}_\perp = \mathbf{p}_{0\perp} + e\int_{\tau_0}^{\tau} \mathbf{E}(\tau)d\tau, \tag{1.5}$$

where $\mathbf{p}_{0\perp}$ is the particle initial transversal momentum at $\tau = \tau_0$ when $\mathbf{E}(\tau)\mid_{\tau=\tau_0} = \mathbf{H}(\tau)\mid_{\tau=\tau_0} = 0$ corresponding to the free particle state before the interaction. Such definition of the particle free state at the finite moment τ_0 at the interaction with the EM wave is justified when we consider the general case of a plane wave of arbitrary form, which actually corresponds to wave pulses of finite duration, let here $\tau_f - \tau_0$. Then, the interaction will be automatically turned on at $\tau = \tau_0$ and turned off at $\tau = \tau_f$, when $\mathbf{E}(\tau)\mid_{\tau=\tau_f} = \mathbf{H}(\tau)\mid_{\tau=\tau_f} = 0$ too, and the free particle states before the interaction will correspond to $\tau \leq \tau_0$ and after the interaction to $\tau \geq \tau_f$. Such approach also allows passing from the wave pulses of finite duration to quasi-monochromatic or monochromatic waves by extending $\tau_0 \to -\infty$ and $\tau_f \to +\infty$.

The expressions (1.5) can be written in a simpler form through the vector potential ($\mathbf{A}$) of the field according to known relations with the electric and magnetic field strengths for radiation field in the Lorentz gauge

$$\mathbf{E} = -\frac{1}{c}\frac{\partial \mathbf{A}}{\partial t}; \qquad \mathbf{H} = \mathrm{rot}\mathbf{A}; \quad \mathrm{div}\mathbf{A} = 0, \tag{1.6}$$

consequently

$$\mathbf{A}(\tau) = -c \int_{\tau_0}^{\tau} \mathbf{E}(\tau) d\tau. \tag{1.7}$$

The condition $\text{div}\mathbf{A} = 0$ in Eq. (1.6) is the condition of transversality of a plane wave: $\nu_0 \mathbf{A}(\tau) = 0$.

So, the particle transversal momentum (1.5) can be represented in the form

$$\mathbf{p}_{\perp} = \mathbf{p}_{0\perp} - \frac{e}{c}\mathbf{A}(\tau), \tag{1.8}$$

where $\mathbf{A}(\tau) \mid_{\tau=\tau_0} = 0$ according to Eq. (1.7) ($\mathbf{A}(\tau) \mid_{\tau=\tau_f} = 0$ as well because of $\mathbf{E}(\tau) \mid_{\tau=\tau_f} = \mathbf{H}(\tau) \mid_{\tau=\tau_f} = 0$).

Note that Eq. (1.8) may be written without integration of the equation of motion taking into account the space properties in this issue. Thus, the existence of a plane wave does not violate the homogeneity of the space in the plane of the wave polarization. Consequently, the corresponding transversal components of generalized momentum are conserved: $\mathbf{p}_{\perp} + (e/c)\mathbf{A}(\tau) = \text{const}$ and we come at once to Eq. (1.8).

For the integration of Eq. (1.3) for the longitudinal component of the particle momentum we will use the additional equation for the particle energy variation in the field

$$\frac{d\mathcal{E}}{dt} = e\,(\mathbf{vE})\,. \tag{1.9}$$

From Eqs. (1.3) and (1.9) follows the integral of motion for the charged particle in the field of a plane EM wave:

$$\mathcal{E} - c\mathbf{p}\nu_0 = \text{const} \equiv \Lambda. \tag{1.10}$$

Now we can define the particle momentum and energy in the field with the help of Eqs. (1.8) and (1.10), utilizing the dispersion law of the particle energy-momentum as well:

$$\mathcal{E}^2 = \mathbf{p}^2 c^2 + m^2 c^4. \tag{1.11}$$

The following formulas in the field of a plane EM wave of arbitrary form and polarization are obtained:

$$\mathbf{p} = \mathbf{p}_0 - \frac{e}{c}\mathbf{A}(\tau) + \nu_0 \frac{e^2 A^2(\tau) - 2ec\,(\mathbf{p}_0 \mathbf{A}(\tau))}{2c(\mathcal{E}_0 - c\mathbf{p}_0 \nu_0)}, \tag{1.12}$$

$$\mathcal{E} = \mathcal{E}_0 + \frac{e^2 A^2(\tau) - 2ec\,(\mathbf{p}_0 \mathbf{A}(\tau))}{2(\mathcal{E}_0 - c\mathbf{p}_0 \nu_0)}, \tag{1.13}$$

where $\mathbf{p}_0$ and $\mathcal{E}_0$ are the initial momentum and energy of a free particle ($\Lambda = \mathcal{E}_0 - c\mathbf{p}_0\nu_0$).

Then, to obtain the law of the particle motion $\mathbf{r} = \mathbf{r}(t)$ one must integrate the equation

$$\frac{d\mathbf{r}(t)}{dt} = \mathbf{v}(t) = \frac{c^2 \mathbf{p}(t)}{\mathcal{E}(t)}\,. \tag{1.14}$$

However, since the general expressions of particle momentum and energy in the field of a plane EM wave depend only on retarding time τ, the last equation allows exact analytical solution in the parametric form $\mathbf{r} = \mathbf{r}(\tau)$. Thus, passing in Eq. (1.14) from the variable t to τ and taking into account the integral of motion (1.10) we obtain

$$\frac{d\mathbf{r}(\tau)}{d\tau} = \frac{c^2 \mathbf{p}(\tau)}{\mathcal{E}_0 - c\mathbf{p}_0 \nu_0}. \tag{1.15}$$

Integration of Eq. (1.15) with the help of Eq. (1.12) gives

$$\mathbf{r}(\tau) = \mathbf{r}_0 + \frac{c^2 \mathbf{p}_0}{(\mathcal{E}_0 - c\mathbf{p}_0 \nu_0)}(\tau - \tau_0) + \frac{c}{(\mathcal{E}_0 - c\mathbf{p}_0 \nu_0)}$$

$$\times \int_{\tau_0}^{\tau} \left\{ \frac{\nu_0}{2(\mathcal{E}_0 - c\mathbf{p}_0 \nu_0)} \left(e^2 A^2(\tau') - 2ec\mathbf{p}_0 \mathbf{A}(\tau')\right) - e\mathbf{A}(\tau') \right\} d\tau', \tag{1.16}$$

where $\mathbf{r}_0(x_0, y_0, z_0)$ is the particle initial position at $t = t_0$ ($\tau = \tau_0$).

1.2 Intensity Effect. Mass Renormalization

Equations (1.12), (1.13), and (1.16) describe the particle motion in the field of a strong plane EM wave of arbitrary form and polarization. They show that after the interaction ($\tau \geq \tau_f$) $\mathbf{p} = \mathbf{p}_0$, $\mathcal{E} = \mathcal{E}_0$, i.e., the particle remains with the initial energy-momentum, which means that real energy exchange between a free charged particle and a plane EM wave in vacuum is impossible. This result is in congruence with the fact that the real absorption or emission of photons by a free electron in vacuum is forbidden by the energy and momentum conservation laws, which will be discassed in regard to the quantum consideration of this process. Nevertheless, in vacuum the wave intensity effect in the field exists, for revealing of which it should be taken into account the oscillating character of periodic wave field, for which $\overline{\mathbf{A}}(\tau) = 0$.

Then, averaging the expressions in Eqs. (1.12) and (1.13) over time we obtain the following formulas for the particle average momentum and energy in the field:

$$\overline{\mathbf{p}} = \mathbf{p}_0 + \nu_0 \frac{e^2 \overline{\mathbf{A}^2}(\tau)}{2c(\mathcal{E}_0 - c\mathbf{p}_0\nu_0)}; \quad \overline{\mathcal{E}} = \mathcal{E}_0 + \frac{e^2 \overline{\mathbf{A}^2}(\tau)}{2(\mathcal{E}_0 - c\mathbf{p}_0\nu_0)}. \tag{1.17}$$

Taking into account the dispersion law of the particle energy-momentum (1.11) for these average values we can introduce the "effective mass" of the particle due to the intensity effect of strong wave:

$$m^* = m\sqrt{1 + \overline{\xi^2}(\tau)}. \tag{1.18}$$

This formula describes the renormalization of the particle mass in the field. Here we introduced a relativistic invariant dimensionless parameter of a plane EM wave intensity

$$\xi^2(\tau) = \left(\frac{e\mathbf{A}(\tau)}{mc^2}\right)^2. \tag{1.19}$$

The parameter ξ is the basic characteristic of a strong radiation field at the interaction with the charged particles, which represents the work of the field on the one wavelength in the units of the particle rest energy, i.e., it is the energy (normalized) acquired by the particle on a wavelength of a coherent radiation field.

As strong radiation fields actually relate to laser sources of high coherency, we will consider the case of quasi-monochromatic or monochromatic wave fields (we look aside from the actual intensity profiles of laser beams over space coordinates — deviation from a plane wave because of their finite sizes).

Let us consider the case of a monochromatic wave. Without loss of generality we will direct vector ν_0 along the OX axis of a Cartesian coordinate system: $\nu_0 = \{1, 0, 0\}$, then retarding wave coordinate: $\tau = t - x/c$. In the general case of elliptic polarization the vector potential of a monochromatic wave with a frequency ω_0 and amplitude A_0 may be presented in the form

$$\mathbf{A}(\tau) = \{0, A_0 \cos(\omega_0\tau), gA_0 \sin \omega_0\tau\}, \tag{1.20}$$

where g is the parameter of ellipticity; $g = 0$ corresponds to a linear polarization, while $g = \pm 1$ describes a wave of a circular polarization (right or left). Let $g = 1$ and the initial velocity of the particle is parallel to the wave propagation direction ($\mathrm{v}_0 = \mathrm{v}_{0x}$). In such geometry and circular polarization of the wave the intensity effect becomes apparent (only the latter exists with invariable magnitude, because $\mathbf{p}_0\mathbf{A}(\tau) = 0$). In the future we will mainly consider this case of interaction at which the energy and longitudinal velocity of the particle in the field are invariable, which allows, first, a more simple picture of a particle–wave nonlinear interaction, and second, exact solutions in many processes where the existence of the particle initial transverse momentum prevents obtaining exact analytical solutions.

Concerning the definition of the particle initial and final free states at the interaction with a monochromatic wave of infinite duration we will assume an arbitrarily small damping for the amplitude A_0 to switch on adiabatically the wave at $\tau = -\infty$ and switch off at $\tau = +\infty$, i.e., $\mathbf{A}(\tau)\mid_{\tau=\pm\infty}= 0$ (according to the above-mentioned conditions for a plane wave of finite duration $\tau_f - \tau_0$ it should be extended to $\tau_0 \to -\infty$ and $\tau_f \to +\infty$). For a quasi-monochromatic wave (spectral width $\Delta\omega \ll \omega_0$) it should be $A_0 \Rightarrow A_0(\tau)$, where $A_0(\tau)$ is a slowly varying amplitude with respect to the phase oscillations over the $\omega_0\tau$ and the conditions of adiabatic switching on and switching off will take place automatically.

Hence from Eqs.(1.12) and (1.13) we have simple formulas for the particle momentum and energy in the field of a monochromatic wave of circular polarization:

$$p_x = p_0\left[1 + \frac{1}{2}\frac{c}{\mathrm{v}_0}\left(1 + \frac{\mathrm{v}_0}{c}\right)\xi_0^2\right], \tag{1.21}$$

$$p_y = -mc\xi_0 \cos\omega_0\tau, \tag{1.22}$$

$$p_z = -mc\xi_0 \sin\omega_0\tau, \tag{1.23}$$

$$\mathcal{E} = \mathcal{E}_0\left[1 + \frac{1}{2}\left(1 + \frac{\mathrm{v}_0}{c}\right)\xi_0^2\right], \tag{1.24}$$

where the relativistic parameter of the wave intensity (1.19) $\xi^2(\tau) = \xi_0^2 =$ const and, consequently, one can represent it by the amplitude of the vector potential A_0 or electric field strength E_0:

$$\xi_0 = \frac{eA_0}{mc^2} = \frac{eE_0}{mc\omega_0}. \tag{1.25}$$

Equation (1.24) shows that for the significant energy change of a particle in the field of a plane wave in vacuum the superpower laser beams of relativistic intensities $\xi_0 >> 1$ are necessary. Such intensities corresponding to gigantic femtosecond laser pulses became available in recent years.

To elucidate the law of particle motion in the field of a monochromatic wave we will choose the frame of reference for the free particle initial position, in which the coordinates $\mathbf{r}_0$ at the moment $t = t_0$ correspond to $\mathbf{r}_0 = \mathbf{v}_0 t_0$. By that we exclude the infinities in the expression $\mathbf{r} = \mathbf{r}(\tau)$ connected with the initial infinity values of the parameters t_0 and $\mathbf{r}_0$, which have no physical meaning. Then one can extend $t_0 \to -\infty$ and, consequently, $\tau_0 = (1 - \mathrm{v}_{0x}/c)t_0 \to -\infty$ in Eq. (1.16) providing the particle free state before the interaction ($t_0 \to -\infty$) at infinity ($\mathbf{r}_0 \to -\infty$) with the adiabatic switching on the monochromatic (quasi-monochromatic) wave due to $A_0(-\infty) = 0$. Hence, from Eq. (1.16) follows the particle law of motion in the field (1.20) in parametric form. However, considering special cases it is analytically available to represent directly the law of motion $\mathbf{r} = \mathbf{r}(t)$ because of the invariability of longitudinal velocity of the particle in the field

$$\mathrm{v}_x = \mathrm{v}_0 \frac{1 + \frac{1}{2}\frac{c}{\mathrm{v}_0}\left(1 + \frac{\mathrm{v}_0}{c}\right)\xi_0^2}{1 + \frac{1}{2}\left(1 + \frac{\mathrm{v}_0}{c}\right)\xi_0^2}, \tag{1.26}$$

which is exposed only to permanent renormalization due to the intensity effect of the strong wave. Then, with the help of Eq. (1.26) we have the following formulas for the particle law of motion:

$$x(t) = \mathrm{v}_x t, \tag{1.27}$$

$$y(t) = -\frac{mc^3\xi_0}{\mathcal{E}_0\omega_0(1 - \frac{\mathrm{v}_0}{c})} \sin\omega_0(1 - \frac{\mathrm{v}_x}{c})t, \tag{1.28}$$

$$z(t) = \frac{mc^3\xi_0}{\mathcal{E}_0\omega_0(1 - \frac{\mathrm{v}_0}{c})} \cos\omega_0(1 - \frac{\mathrm{v}_x}{c})t. \tag{1.29}$$

Equations (1.27)–(1.29) show that the particle performs circular motion

$$y^2(t) + z^2(t) = \mathrm{const} \tag{1.30}$$

in the plane of the wave polarization (yz) with the radius

$$\rho_\perp = \frac{mc^3\xi_0}{\mathcal{E}_0\omega_0(1 - \frac{\mathrm{v}_0}{c})} \tag{1.31}$$

and translational uniform motion along the wave propagation direction (OX axis), i.e., performs a helical motion (Fig. 1.1). Consider now the case of linear polarization of the wave

$$\mathbf{A}(\tau) = \{0, A_0\cos(\omega_0\tau), 0\}. \tag{1.32}$$

From Eqs. (1.12) and (1.13) for the particle momentum and energy in the field (1.32) we have

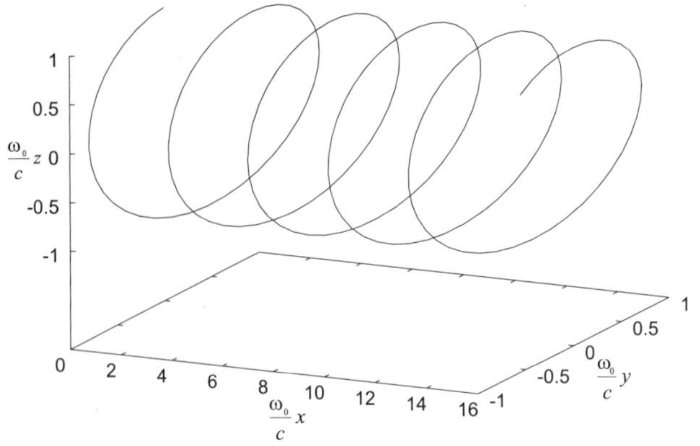

Fig. 1.1. Trajectory of the particle (initially at rest) in the field of circularly polarized EM wave. The relativistic parameter of intensity is taken to be $\xi_0 = 1$.

$$p_x = p_0 \left[1 + \frac{1}{2} \frac{c}{\mathrm{v}_0} \left(1 + \frac{\mathrm{v}_0}{c} \right) \xi_0^2 \cos^2 (\omega_0 \tau) \right], \tag{1.33}$$

$$p_y = -mc\xi_0 \cos \omega_0 \tau, \tag{1.34}$$

$$p_z = 0, \tag{1.35}$$

$$\mathcal{E} = \mathcal{E}_0 \left[1 + \frac{1}{2} \left(1 + \frac{\mathrm{v}_0}{c} \right) \xi_0^2 \cos^2 (\omega_0 \tau) \right]. \tag{1.36}$$

In contrast to the case of circular polarization, in the field of linearly polarized wave the intensity effect has the oscillating character (at the second harmonic $2\omega_0$, as follows from Eqs. (1.33) and (1.36)) and the representation of the particle trajectory analytically is unavailable. The latter may be performed in parametric form with the help of the particle law of motion $\mathbf{r} = \mathbf{r}(\tau)$, which in the field (1.32) has the following form

$$x(\tau) = \left[1 + \frac{1}{4} \frac{c}{\mathrm{v}_0} \left(1 + \frac{\mathrm{v}_0}{c} \right) \xi_0^2 \right] \frac{\mathrm{v}_0 \tau}{(1 - \frac{\mathrm{v}_0}{c})} + \rho_{\shortparallel} \sin(2\omega_0 \tau), \tag{1.37}$$

$$y(\tau) = -\rho_\perp \sin(\omega_0 \tau), \tag{1.38}$$

$$z = 0, \tag{1.39}$$

where

$$\rho_{\shortparallel} = \frac{1}{8} \frac{c}{\omega_0} \frac{1 + \frac{\mathrm{v}_0}{c}}{1 - \frac{\mathrm{v}_0}{c}} \xi_0^2 \tag{1.40}$$

is the amplitude of longitudinal oscillations of the particle along the wave propagation direction.

To determine the particle trajectory we pass to an inertial system of coordinates connected with the uniform motion of the particle along the axis OX with the velocity

$$V = \mathrm{v}_0 \frac{1 + \frac{1}{4}\frac{c}{\mathrm{v}_0}\left(1 + \frac{\mathrm{v}_0}{c}\right)\xi_0^2}{1 + \frac{1}{4}\left(1 + \frac{\mathrm{v}_0}{c}\right)\xi_0^2}, \tag{1.41}$$

to exclude the uniform part of translational movement in the direction of the wave propagation. After the Lorentz transformations for coordinates and wave frequency we have the following law of motion in this system:

$$x'(\tau') = \frac{1}{8}\frac{c}{\omega'}\frac{\xi_0^2}{1 + \frac{\xi_0^2}{2}} \sin(2\omega'\tau'), \tag{1.42}$$

$$y'(\tau') = y(\tau) = -\frac{c}{\omega'}\frac{\xi_0}{\sqrt{1 + \frac{\xi_0^2}{2}}} \sin(\omega'\tau'), \tag{1.43}$$

$$z' = 0, \tag{1.44}$$

where

$$\omega' = \frac{\omega_0}{\sqrt{1 + \frac{\xi_0^2}{2}}}\sqrt{\frac{1 - \frac{\mathrm{v}_0}{c}}{1 + \frac{\mathrm{v}_0}{c}}} \tag{1.45}$$

is the Doppler-shifted frequency of the wave in the system moving with the velocity (1.41).

Now from Eqs. (1.42) and (1.43) one can obtain the trajectory of the particle in the plane XY

$$\left(\frac{x'}{2\rho'_{\shortparallel}}\right)^2 = \left(\frac{y'}{\rho_\perp}\right)^2 - \left(\frac{y'}{\rho_\perp}\right)^4 \tag{1.46}$$

with the parameters $\rho'_{\shortparallel}$ and $\rho_\perp$:

$$\rho'_{\shortparallel} = \frac{c}{8\omega'}\frac{\xi_0^2}{1 + \frac{\xi_0^2}{2}}; \quad \rho'_\perp = \rho_\perp = \frac{c}{\omega'}\frac{\xi_0}{\sqrt{1 + \frac{\xi_0^2}{2}}}. \tag{1.47}$$

Equation (1.46) performs a symmetric 8-form figure with the longitudinal axis along the OY (Fig. 1.2).

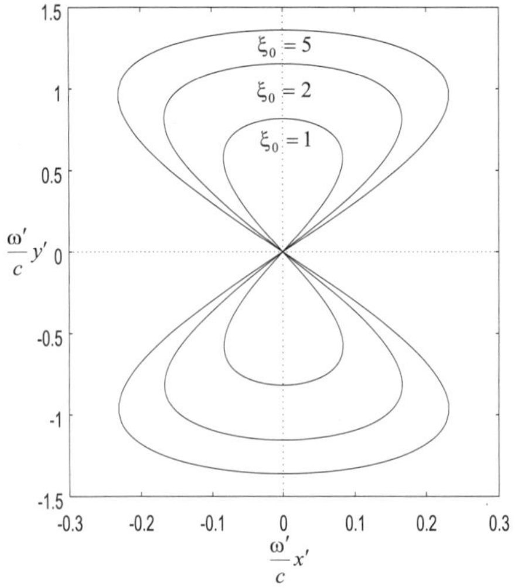

Fig. 1.2. Trajectory of the particle in the field of linearly polarized EM wave (excluding the uniform part of translational movement in the direction of the wave propagation) for the various ξ_0.

1.3 Radiation of a Particle in the Field of Strong Monochromatic Wave

Let us now consider the radiation of a charged particle in the specified wave field (1.20) of arbitrary high intensity in the scope of the classical theory. In the strong wave field the radiation of a particle is of nonlinear nature — radiation of high harmonics — which in quantum terminology means that the multiphoton absorption by the particle from the incident wave takes place with subsequent radiation of the corresponding photon. Taking into account certain dependence of harmonics radiation on the direction of particle motion with respect to the initial strong wave propagation and its polarization we will consider the general case of a particle–wave interaction geometry and arbitrary polarization of monochromatic wave (elliptic)

$$\mathbf{A}(\tau) = A_0\{\mathbf{e}_1 \cos\omega_0\tau + \mathbf{e}_2 g \sin\omega_0\tau\}; \tag{1.48}$$

$$\tau = t - \frac{\nu_0\mathbf{r}}{c}; \quad \mathbf{e}_1\nu_0 = \mathbf{e}_2\nu_0 = \mathbf{e}_1\mathbf{e}_2 = 0,$$

where $\mathbf{e}_{1,2}$ are the unit polarization vectors.

The energy radiated by a charged particle in the domain of solid angle dO and interval of frequencies $d\omega$ in the direction of the wave vector $\mathbf{k}$ (summed by all possible polarizations) is given by the formula

$$d\varepsilon_{\mathbf{k}} = \frac{e^2}{4\pi^2 c} \left| \int_{-\infty}^{\infty} [\mathbf{kv}]\, e^{i(\mathbf{kr}-\omega t)} dt \right|^2 d\omega dO, \tag{1.49}$$

where $\mathbf{v} = \mathbf{v}(t)$ and $\mathbf{r} = \mathbf{r}(t)$ are the particle velocity and law of motion in the wave field (1.20), which are determined by Eqs. (1.12), (1.13), and (1.16) in parametric form. The latter requires passing in Eq. (1.49) from the variable t to the wave coordinate τ. Then the equation for the radiation energy will be written in the form

$$d\varepsilon_{\mathbf{k}} = \frac{e^2 c^3}{4\pi^2 \Lambda^2} \left| \int_{-\infty}^{\infty} [\mathbf{kp}\,(\tau)]\, e^{i\psi(\tau)} d\tau \right|^2 d\omega dO, \tag{1.50}$$

where

$$\psi(\tau) = \omega\tau + k(\nu_0 - \nu)\mathbf{r}(\tau) \tag{1.51}$$

is the phase of radiated wave $(\mathbf{kr} - \omega t)$ as a function of the incident strong wave coordinate τ and the unit vector ν in Eq. (1.49) is $\nu = \mathbf{k}/k$.

Using Eqs. (1.12), (1.13) and introducing the functions

$$G_0 = \int_{-\infty}^{\infty} e^{i\psi(\tau)} d\tau,$$

$$\mathbf{G}_1 = \int_{-\infty}^{\infty} \mathbf{A}(\tau) e^{i\psi(\tau)} d\tau, \tag{1.52}$$

$$G_2 = \int_{-\infty}^{\infty} \mathbf{A}^2(\tau) e^{i\psi(\tau)} d\tau,$$

after the long but straightforward transformations for the radiation energy we obtain

$$d\varepsilon_{\mathbf{k}} = \frac{e^2 m^2 c^3 \omega^2}{4\pi^2 \Lambda^2} \left(\frac{e^2}{m^2 c^4} \left(|\mathbf{G}_1|^2 - Re\,(G_0 G_2^*) \right) - |G_0|^2 \right) d\omega dO. \tag{1.53}$$

This is the general formula of the spectral-angular distribution of radiation energy for the arbitrary plane EM wave field. Considering the case of monochromatic wave (1.48) with the corresponding law of motion (1.16) for the phase of radiated wave (1.51), which determines the functions (1.52) and, consequently, the energy of radiation (1.53), we have

$$\psi(\tau) = \left(\frac{\overline{\mathcal{E}} - c\nu\overline{\mathbf{p}}}{\Lambda}\right)\omega\tau + \alpha\sin(\omega_0\tau - \varphi) - \beta\sin 2\omega_0\tau, \tag{1.54}$$

where the parameters α, β, and φ are

$$\alpha = \rho_\perp k\sqrt{\left(\nu\mathbf{e}_1 + (\nu\nu_0 - 1)\frac{c\mathbf{p}_0\mathbf{e}_1}{\Lambda}\right)^2 + g^2\left(\nu\mathbf{e}_2 + (\nu\nu_0 - 1)\frac{c\mathbf{p}_0\mathbf{e}_2}{\Lambda}\right)^2},$$

$$\beta = (\nu\nu_0 - 1)\rho_{\shortparallel}k, \tag{1.55}$$

$$\tan\varphi = \frac{g\left(\nu\mathbf{e}_2 + (\nu\nu_0 - 1)\frac{c\mathbf{p}_0\mathbf{e}_2}{\Lambda}\right)}{\nu\mathbf{e}_1 + (\nu\nu_0 - 1)\frac{c\mathbf{p}_0\mathbf{e}_1}{\Lambda}}.$$

In these expressions the quantities $\rho_\perp$ and $\rho_{\shortparallel}$ are determined by Eqs. (1.31) and (1.40). Here we have omitted the terms with $\mathbf{r}_0$ and τ_0 as these terms (constant phase factor) do not contribute to the single-particle radiation energy. All functions in Eq. (1.53) can be expressed by the series of Bessel function production using the following expansion:

$$e^{i\alpha\sin(\omega_0\tau - \varphi) - i\beta\sin 2\omega_0\tau} = \sum_{n,k=-\infty}^{\infty} J_n(\alpha)J_k(\beta)e^{-in\varphi}e^{i(n-2k)\omega_0\tau}.$$

The latter in turn can be expressed by the so-called generalized Bessel function $G_s(\alpha, \beta, \varphi)$:

$$G_s(\alpha, \beta, \varphi) = \sum_{k=-\infty}^{\infty} J_{2k-s}(\alpha)J_k(\beta)e^{i(s-2k)\varphi}. \tag{1.56}$$

Then the functions (1.52) will be written by the function $G_s(\alpha, \beta, \varphi)$ as follows:

$$G_0 = 2\pi\sum_{s=-\infty}^{\infty} G_s(\alpha, \beta, \varphi)\delta\left(\frac{\overline{\mathcal{E}} - c\nu\overline{\mathbf{p}}}{\Lambda}\omega - s\omega_0\right),$$

$$\mathbf{G}_1 = \pi A_0\sum_{s=-\infty}^{\infty}\{\mathbf{e}_1\left(G_{s-1}(\alpha, \beta, \varphi) + G_{s+1}(\alpha, \beta, \varphi)\right)$$

$$+\mathbf{e}_2 ig\left(G_{s-1}(\alpha,\beta,\varphi)-G_{s+1}(\alpha,\beta,\varphi)\right)\}\,\delta\left(\frac{\overline{\mathcal{E}}-c\nu\overline{\mathbf{p}}}{\Lambda}\omega-s\omega_0\right), \tag{1.57}$$

$$G_2=\frac{A_0^2}{2}(1+g^2)G_0+\pi A_0^2(1-g^2)$$

$$\times\sum_{s=-\infty}^{\infty}\left(G_{s-2}(\alpha,\beta,\varphi)+G_{s+2}(\alpha,\beta,\varphi)\right)\delta\left(\frac{\overline{\mathcal{E}}-c\nu\overline{\mathbf{p}}}{\Lambda}\omega-s\omega_0\right).$$

The function $\delta(x)$ in Eqs. (1.57) is the Dirac δ-function expressing the resonance condition between the particle oscillation frequency in the incident strong wave field and radiation frequency (conservation law of the Compton effect in quantum terminology). According to Eqs. (1.57) the radiation energy (1.53) is proportional to the δ^2-function, which should be represented via particle–strong wave interaction time Δt (in the wave coordinate $\Delta\tau=\Delta t\Lambda/\overline{\mathcal{E}}$)

$$\delta\left(\frac{\overline{\mathcal{E}}-c\nu\overline{\mathbf{p}}}{\Lambda}\omega-s\omega_0\right)\delta\left(\frac{\overline{\mathcal{E}}-c\nu\overline{\mathbf{p}}}{\Lambda}\omega-s'\omega_0\right)$$

$$=\begin{bmatrix}0,\ \text{if } s\neq s',\\ \frac{\Delta\tau}{2\pi}\delta\left(\frac{\overline{\mathcal{E}}-c\nu\overline{\mathbf{p}}}{\Lambda}\omega-s\omega_0\right),\ \text{if } s=s'.\end{bmatrix} \tag{1.58}$$

Then instead of the radiation energy (1.53) one can determine the radiation power

$$dP_{\mathbf{k}}=\frac{d\varepsilon_{\mathbf{k}}}{\Delta t}.$$

Substituting Eqs. (1.57) into Eq. (1.53) taking into account Eq. (1.58) for the radiation power we obtain (from $\omega>0$ follows $s>0$)

$$dP_{\mathbf{k}}=\frac{e^2m^2c^3\omega^2}{2\pi\Lambda\overline{\mathcal{E}}}\sum_{s=1}^{\infty}\left\{\frac{\xi_0^2}{4}\left[(1+g^2)\left(|G_{s-1}|^2+|G_{s+1}|^2\right)\right.\right.$$

$$\left.+2(1-g^2)Re\left(G_{s-1}^*G_{s+1}-\frac{1}{2}G_s^*\left(G_{s-2}+G_{s+2}\right)\right)\right]$$

$$\left.-\left(1+\frac{\xi_0^2}{2}(1+g^2)\right)|G_s|^2\right\}\delta\left(\frac{\overline{\mathcal{E}}-c\nu\overline{\mathbf{p}}}{\Lambda}\omega-s\omega_0\right)d\omega dO. \tag{1.59}$$

In the case of the circular polarization of an incident strong wave ($g = \pm 1$) the second argument of the generalized Bessel function $G_s(\alpha, \beta, \varphi)$ is zero and $|G_s|^2 = J_s^2(\alpha)$, so that for the radiation power we have

$$dP_{\mathbf{k}} = \frac{e^2 m^2 c^3 \omega^2}{2\pi \Lambda \overline{\mathcal{E}}} \sum_{s=1}^{\infty} \left[\frac{\xi_0^2}{2} \left(J_{s-1}^2(\alpha) + J_{s+1}^2(\alpha) \right) - \left(1 + \xi_0^2\right) J_s^2(\alpha) \right]$$

$$\times \delta \left(\frac{\overline{\mathcal{E}} - c\nu\overline{\mathbf{p}}}{\Lambda} \omega - s\omega_0 \right) d\omega dO. \tag{1.60}$$

Using the known recurrent relations for the Bessel functions

$$J_{s-1}(\alpha) + J_{s+1}(\alpha) = \frac{2s}{\alpha} J_s(\alpha),$$

$$J_{s-1}(\alpha) - J_{s+1}(\alpha) = 2J_s'(\alpha),$$

Eq. (1.60) can be represented in the following form:

$$dP_{\mathbf{k}} = \frac{e^2 m^2 c^3 \omega^2}{2\pi \overline{\mathcal{E}} (\overline{\mathcal{E}} - c\nu\overline{\mathbf{p}})} \xi_0^2 \sum_{s=1}^{\infty} \left[\left(\frac{s^2}{\alpha^2} - 1 - \xi_0^{-2} \right) J_s^2(\alpha) + J_s'^2(\alpha) \right]$$

$$\times \delta \left(\omega - \frac{s\omega_0 (\overline{\mathcal{E}} - c\nu_0 \overline{\mathbf{p}})}{\overline{\mathcal{E}} - c\nu\overline{\mathbf{p}}} \right) d\omega dO. \tag{1.61}$$

For the linear polarization of an incident strong wave ($g = 0$) the third argument of the generalized Bessel function $G_s(\alpha, \beta, \varphi)$ is zero and G_s functions become real. Then for the radiation power in this case we have

$$dP_{\mathbf{k}} = \frac{e^2 m^2 c^3 \omega^2}{2\pi \overline{\mathcal{E}} (\overline{\mathcal{E}} - c\nu\overline{\mathbf{p}})} \sum_{s=1}^{\infty} \left[\frac{\xi_0^2}{4} \left((G_{s-1} + G_{s+1})^2 - G_s \left(G_{s-2} + G_{s+2} \right) \right) \right.$$

$$\left. - \left(1 + \frac{\xi_0^2}{2} \right) G_s^2 \right] \delta \left(\omega - \frac{s\omega_0 (\overline{\mathcal{E}} - c\nu_0 \overline{\mathbf{p}})}{\overline{\mathcal{E}} - c\nu\overline{\mathbf{p}}} \right) d\omega dO. \tag{1.62}$$

1.4 Nonlinear Radiation Effects in Superstrong Wave Fields

Equations (1.59)–(1.62) for the radiation power of a charged particle show that as a result of the particle–strong wave nonlinear interaction in vacuum, numerous harmonics in the radiation spectrum arise, i.e., the radiation process is also nonlinear. In quantum terminology this means that due to multiphoton absorption by a particle from the strong wave the nonlinear Compton effect takes place. The power of harmonics radiation nonlinearly depends on incident strong wave intensity and for its considerable value, laser fields must have relativistic intensities $\xi > 1$.

Up to the last decade such intensities practically were unachievable (even then the strongest laser fields were $\xi < 1$) and to expect to reach high harmonics radiation via nonlinear Compton channels in vacuum with laser fields of intensities $\xi < 1$ (or any other nonlinear effect at the charge particle–EM wave interaction in vacuum, particularly laser acceleration), as will be shown below, was unreal. For this reason, actual interest in the nonlinear Compton effect until recently was only theoretical. However, the rapid development of laser technology in the last decade made available laser sources of supershort duration — femtosecond pulses, the intensity of which today much exceeds its relativistic value in the optical domain: $I_{rel} \sim 10^{18}$ W/cm^2 ($\xi \sim 1$), laser fields with $\xi >> 1$ became available. The latter has provided the necessary intensities for actual radiation of high harmonics in the Compton process. Therefore, we will analyze the process of high harmonics radiation in the nonlinear interaction of a charged particle with superstrong laser fields ($\xi >> 1$) on the basis of Eqs. (1.59)–(1.62).

We will analyze the cases of circular and linear polarizations of the incident wave taking into account the specific dependence of harmonics radiation on the strong wave polarization and when the initial velocity of the particle is parallel to the wave propagation direction. This case of particle–wave parallel propagation is of interest since in this case the interaction length with actual laser beams (or, e.g., wiggler field, which in relation to the relativistic particle is equivalent to a counterpropagating laser field) is maximal, which is especially important for the problem of free electron lasers.

In the case of circular polarization of an incident strong wave ($g = \pm 1$) and $\mathbf{p}_0\mathbf{e}_1 = 0$, $\mathbf{p}_0\mathbf{e}_2 = 0$, carrying out the integration over ω and turning to spherical coordinates in Eq. (1.61) (OZ axis directed along the vector $\overline{\mathbf{p}}$) for the angular distribution of the radiation power for the s-th harmonic we have

$$\frac{dP^{(s)}}{dO} = \frac{e^2 m^2 c^3 \omega_s^2}{2\pi \overline{\mathcal{E}}^2 (1 - \frac{\mathrm{v}}{c}\cos\vartheta)} \xi_0^2 \left[\left(\frac{s^2}{\alpha_s^2} - 1 - \xi_0^{-2} \right) J_s^2(\alpha_s) + J_s'^2(\alpha_s) \right], \quad (1.63)$$

where

$$\omega_s = s\omega_0 \frac{\overline{\mathcal{E}} - c\nu_0 \overline{\mathbf{p}}}{\overline{\mathcal{E}} - c\nu \overline{\mathbf{p}}} = s\omega_0 \frac{1 - \frac{\mathrm{v}}{c}\cos\vartheta_0}{1 - \frac{\mathrm{v}}{c}\cos\vartheta} \quad (1.64)$$

is the radiated frequency and

$$\alpha_s = \frac{smc^2}{\overline{\mathcal{E}}\left(1 - \frac{\overline{\mathrm{v}}}{c}\cos\vartheta\right)}\xi_0 \sin\vartheta \tag{1.65}$$

is the parameter characterizing nonlinear interaction with the strong EM wave. ϑ_0 and ϑ are the incident and scattering angles of the strong and radiated waves with respect to the direction of the particle mean velocity $\overline{\mathbf{v}} = c^2\overline{\mathbf{p}}/\overline{\mathcal{E}}$.

For a weak EM wave: $\xi_0 << 1$ (linear theory) the argument of the Bessel function $\alpha_s << 1$ and as is known for such values of the argument $J_s(\alpha_s) \sim {\alpha_s}^s$ and $P^{(s)} \sim \xi_0^{2s}$. Therefore, in the linear theory the main contribution to the radiation power gives the first harmonic. In this case $J_1^2(\alpha_1) \simeq \alpha_1^2/4$, $J_1'^2(\alpha_1) \simeq 1/4$, $\overline{\mathcal{E}} \simeq \mathcal{E}_0$, $\overline{\mathrm{v}} \simeq \mathrm{v}_0$, and

$$\frac{dP^{(1)}}{dO} = \frac{e^2 m^2 c^3 \omega_1^2}{8\pi\mathcal{E}_0^2(1 - \frac{\mathrm{v}_0}{c}\cos\vartheta)}\xi_0^2\left[2 - \frac{\alpha_1^2}{\xi_0^2}\right]$$

$$= \frac{e^2 m^2 c^3 \omega_1^2}{8\pi\mathcal{E}_0^2(1 - \frac{\mathrm{v}_0}{c}\cos\vartheta)}\xi_0^2\left[2 - \left(\frac{mc^2}{\mathcal{E}_0}\right)^2 \frac{\sin^2\vartheta}{\left(1 - \frac{\mathrm{v}_0}{c}\cos\vartheta\right)^2}\right]. \tag{1.66}$$

Particularly for the particle initially at rest we have the Thomson formula

$$\frac{dP^{(1)}}{dO} = \frac{e^2\omega_0^2}{8\pi c}\xi_0^2\left[1 + \cos^2\vartheta\right],$$

$$P^{(1)} = \frac{e^2\omega_0^2}{4c}\xi_0^2 \int\limits_{-1}^{1}\left[1 + \cos^2\vartheta\right] d\cos\vartheta = \frac{2e^2\omega_0^2}{3c}\xi_0^2. \tag{1.67}$$

For the moderate relativistic intensities $\xi_0 \sim 1$ (moderate nonlinearity) the power of the low harmonics ($s \sim 10$) exceeds the radiation power of the fundamental frequency ω_1. To show the dependence of the radiation power on the harmonics number the relative differential power

$$P_{rel}^{(s)} = \frac{dP^{(s)}}{dO}\Big/\frac{dP^{(1)}}{dO} = \frac{s^2\left[\left(\frac{s^2}{\alpha_s^2} - 1 - \xi_0^{-2}\right)J_s^2(\alpha_s) + J_s'^2(\alpha_s)\right]}{\left(\frac{1}{\alpha_1^2} - 1 - \xi_0^{-2}\right)J_1^2(\alpha_1) + J_1'^2(\alpha_1)} \tag{1.68}$$

is displayed in Fig. 1.3 for the different harmonics. In Fig. 1.4 the relative differential power is plotted as a function of radiation angle for various harmonics.

For the superstrong EM waves of relativistic intensities (strict nonlinearity): $\xi_0 \gg 1$ a relatively simple analytic formula for the radiation power can

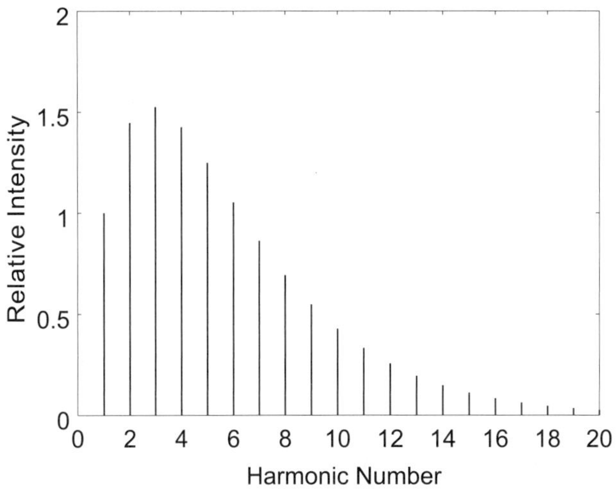

Fig. 1.3. The envelope of the relative differential power of the radiation for the different harmonics is plotted at $\xi_0 = 1$ and $\vartheta\overline{\gamma} = 1$ ($\overline{\gamma} = \overline{\mathcal{E}}/(m^* c^2) = 10$).

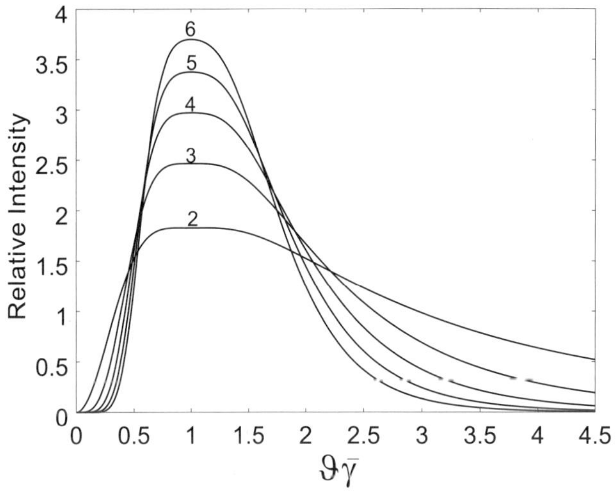

Fig. 1.4. The relative differential power is plotted as a function of radiation angle for various harmonics. The relativistic parameter of intensity is taken to be $\xi_0 = 2$ and $\overline{\gamma} = 10$.

be obtained utilizing the properties of the Bessel function. The argument of the latter in Eq. (1.63) reaches its maximal value

$$\alpha_{s\,\max} = \frac{\xi_0}{\sqrt{1+\xi_0^2}} s$$

at the angle $\cos\vartheta_m = \overline{\text{v}}/c$. Therefore, at $\xi_0 \gg 1$ the harmonics with $s \sim \alpha_s >> 1$ furnish the main contribution to the radiation power. At the angle $\theta = \theta_m$ we have a peak in angular distribution of the radiation power. Besides, in this limit (always $\alpha_s < s$) one can approximate the Bessel function by the Airy one

$$J_s(\alpha_s) \simeq \left(\frac{2}{s}\right)^{1/3} Ai\left(Z\right);\ Z = \left(\frac{s}{2}\right)^{2/3}\left(1 - \frac{\alpha_s^2}{s^2}\right), \tag{1.69}$$

$$J_s' \simeq -\left(\frac{2}{s}\right)^{2/3} Ai'\left(Z\right),$$

and taking into account that

$$\overline{\mathcal{E}} = \frac{m^* c^2}{\sqrt{1 - \frac{\overline{\text{v}}^2}{c^2}}}$$

for the angular distribution of the radiation power we have

$$\frac{dP^{(s)}}{dO} \simeq \frac{e^2\omega_s^2\left(1 - \frac{\overline{\text{v}}^2}{c^2}\right)}{2\pi c(1 - \frac{\overline{\text{v}}}{c}\cos\vartheta)}\left(\frac{2}{s}\right)^{4/3}$$

$$\times\left[\left(\frac{s^2}{\alpha_s^2} - 1 - \xi_0^{-2}\right)\left(\frac{s}{2}\right)^{2/3} Ai^2\left(Z\right) + Ai'^2\left(Z\right)\right]. \tag{1.70}$$

As far as the Airy function exponentially decreasing with increasing of the argument, one can conclude that the cutoff harmonic s_c is determined from the condition $Z_{\min} \sim 1$, where

$$Z_{\min} = \left(\frac{s}{2}\right)^{2/3}\left(1 - \frac{\alpha_{s\max}^2}{s^2}\right) \simeq \left(\frac{s}{2\xi_0^3}\right)^{2/3},$$

which gives $s_c \sim \xi_0^3$.

Consider now the case of linear polarization of the incident strong EM wave. Taking into account the recurrence relation in Eq. (1.62)

$$G_{s-2}(\alpha,\beta) + G_{s+2}(\alpha,\beta) = \frac{s}{\beta}G_s(\alpha,\beta) + \frac{\alpha}{2\beta}\left[G_{s-1}(\alpha,\beta) + G_{s+1}(\alpha,\beta)\right],$$

the differential radiation power in this case can be represented in the form

$$\frac{dP^{(s)}}{dO} = \frac{e^2 m^2 c^3 \omega_s^2}{8\pi\overline{\mathcal{E}}^2 (1 - \frac{\overline{\mathrm{v}}}{c}\cos\vartheta)} \xi_0^2$$

$$\times \left[(G_{s-1} + G_{s+1}) \left(G_{s-1} + G_{s+1} - \frac{\alpha}{2\beta} G_s \right) - \left(2 + \frac{4}{\xi_0^2} + \frac{s}{\beta} \right) G_s^2 \right]. \quad (1.71)$$

The arguments of the generalized Bessel functions when $\mathbf{p}_0\mathbf{e}_1 = 0$ are

$$\alpha_s = \frac{smc^2}{\overline{\mathcal{E}} \left(1 - \frac{\overline{\mathrm{v}}}{c}\cos\vartheta\right)} \xi_0 \left|\nu\mathbf{e}_1\right|,$$

$$\beta_s = \frac{s\xi_0^2}{8 + 4\xi_0^2} \frac{1 - \frac{\overline{\mathrm{v}}^2}{c^2}}{1 - \frac{\overline{\mathrm{v}}}{c}\cos\vartheta_0} \frac{\cos\vartheta_r - 1}{1 - \frac{\overline{\mathrm{v}}}{c}\cos\vartheta}, \quad (1.72)$$

where ϑ_r is the angle between the incident and radiated EM waves.

For the weak EM wave $\xi_0 << 1$ the arguments of the generalized Bessel function $\alpha_s, \beta_s << 1$ and $P^{(s)} \sim \xi_0^{2s}$, therefore the main contribution to the radiation power gives the first harmonic. In this case

$$\frac{dP^{(1)}}{dO} = \frac{e^2 m^2 c^3 \omega_1^2}{8\pi\mathcal{E}_0^2 (1 - \frac{\mathrm{v}_0}{c}\cos\vartheta)} \xi_0^2 \left[1 - \frac{\alpha_1^2}{\xi_0^2} \right]. \quad (1.73)$$

For the particle initially at rest we have the Thomson formula

$$\frac{dP^{(1)}}{dO} = \frac{e^2 \omega_0^2}{8\pi c} \xi_0^2 \left[1 - (\nu\mathbf{e}_1)^2 \right],$$

$$P^{(1)} = \frac{e^2 \omega_0^2}{3c} \xi_0^2. \quad (1.74)$$

In contrast to the circular polarization of the strong wave, for the linear polarization there is no azimuthal symmetry and the asymmetry upon the harmonics parity appears. In particular, in the direction opposite to the strong wave propagation ($\nu\mathbf{e}_1 = 0$ and $\vartheta_r = \pi$) only odd harmonics exist. This is a consequence of the particle dynamics in the strong wave field considered in section 1.2. For this case the generalized Bessel function is reduced to the ordinary Bessel function and we have a relatively simple formula. Thus,

$$G_s(0, \beta, 0) = \sum_{k=-\infty}^{\infty} J_{2k-s}(0) J_k(\beta)$$

$$= \sum_{k=-\infty}^{\infty} \delta_{2k-s,0} J_k(\beta) = \begin{bmatrix} 0, \text{if } s \text{ odd} \\ J_{s/2}(\beta), \text{ if } s \text{ even} \end{bmatrix} \tag{1.75}$$

and for the angular distribution of the radiation power we obtain

$$\left.\frac{dP^{(s)}}{dO}\right|_{\vartheta_r=\pi} = \frac{e^2 m^2 c^3 \omega_s^2 \xi_0^2}{8\pi\overline{\mathcal{E}}^2 (1 - \frac{\mathrm{v}}{c}\cos\vartheta)} \left[J_{\frac{s+1}{2}}\left(\frac{s\xi_0^2}{4+2\xi_0^2}\right) - J_{\frac{s-1}{2}}\left(\frac{s\xi_0^2}{4+2\xi_0^2}\right)\right]^2 . \tag{1.76}$$

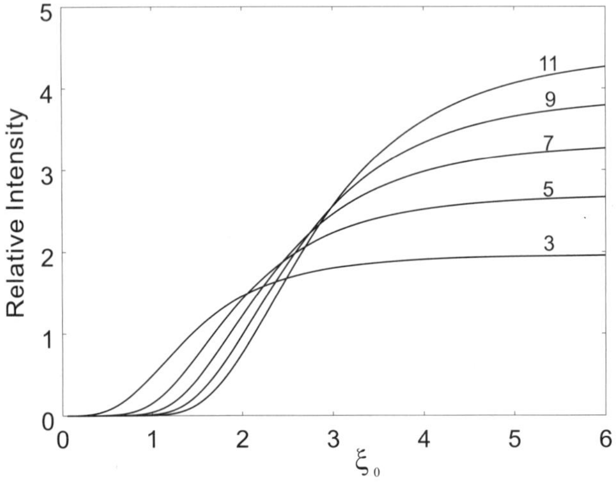

Fig. 1.5. The partial differential power is shown for on axis radiation as a function of ξ_0 for various harmonics ($\overline{\gamma} = 10$).

At $\xi_0 \gg 1$ the argument of the Bessel function tends to the value of the index and as in the case of a wave circular polarization the high harmonics $s >> 1$ give the main contribution to the radiation power and the cutoff harmonic $s_c \sim \xi_0^3$. In Fig. 1.5 the partial differential power is shown for on axis radiation. To show the dependence of the process on the incident wave intensity the relative differential power is plotted as a function of ξ_0 for various harmonics. As we see, with increasing of the wave intensity the power of harmonics well exceeds the power of the fundamental frequency.

1.5 Quantum Description. Volkov Solution of the Dirac Equation

The description of the quantum dynamics of a spinor charged particle (say, electron) in the field of a strong EM wave in vacuum in the scope of relativistic theory requires solution of the Dirac equation, which in the field of arbitrary plane wave allows an exact solution, first obtained by Volkov (1933). This Volkov wave function has the basic role in quantum description of diverse nonlinear electromagnetic processes in superstrong laser fields in vacuum, in particular, major quantum electrodynamic phenomena such as the Compton effect, stimulated bremsstrahlung, and electron–positron pair production, which will be considered in this book. Therefore, this section will be devoted to a description of relativistic wave function of a spinor charged particle in the field of a plane EM wave of arbitrary form and intensity.

The Dirac equation for a spinor particle in a given plane EM wave with arbitrary form of the vector potential $\mathbf{A} = \mathbf{A}(\tau)$ (see Eq. (1.7)) is written as follows:

$$i\hbar\frac{\partial\Psi}{\partial t} = \left[c\alpha\widehat{\mathbf{P}} + mc^2\beta\right]\Psi, \tag{1.77}$$

where

$$\alpha = \begin{pmatrix} \sigma & \mathbf{0} \\ \mathbf{0} & -\sigma \end{pmatrix}, \quad \beta = \begin{pmatrix} 0 & 1 \\ 1 & 0 \end{pmatrix} \tag{1.78}$$

are the Dirac matrices in the spinor representation, $\sigma = (\sigma_x, \sigma_y, \sigma_z)$ are the Pauli matrices

$$\sigma_x = \begin{pmatrix} 0 & 1 \\ 1 & 0 \end{pmatrix}, \quad \sigma_y = \begin{pmatrix} 0 & -i \\ i & 0 \end{pmatrix}, \quad \sigma_z = \begin{pmatrix} 1 & 0 \\ 0 & -1 \end{pmatrix}, \tag{1.79}$$

and

$$\widehat{\mathbf{P}} = \hat{\mathbf{p}} - \frac{e}{c}\mathbf{A}$$

is the operator of the kinetic momentum ($\hat{\mathbf{p}} = -i\hbar\nabla$ is the operator of the generalized momentum).

Looking for the solution of Eq. (1.77) in the form

$$\Psi = \begin{pmatrix} \Psi_1 \\ \Psi_2 \end{pmatrix}, \tag{1.80}$$

for the spinor functions $\Psi_{1,2}$ we obtain the equations

$$i\hbar\frac{\partial\Psi_1}{\partial t} - c\sigma\widehat{\mathbf{P}}\Psi_1 = mc^2\Psi_2,$$

$$i\hbar\frac{\partial\Psi_2}{\partial t} + c\sigma\widehat{\mathbf{P}}\Psi_2 = mc^2\Psi_1. \tag{1.81}$$

Then acting on the first equation by the operator $i\hbar\partial/\partial t + c\sigma\widehat{\mathbf{P}}$ and taking into account the relation

$$(\sigma\mathbf{a})(\sigma\mathbf{b}) = (\mathbf{ab}) + i\sigma[\mathbf{ab}]$$

we obtain the Dirac equation in quadratic form:

$$\left\{\hbar^2\frac{\partial^2}{\partial t^2} - \hbar^2c^2\left(\nu_0\frac{\partial}{\partial\mathbf{r}}\right)^2 + c^2\widehat{\mathbf{P}}_\perp^2 + m^2c^4 - ec\hbar\sigma(\mathbf{H} - i\mathbf{E})\right\}\Psi_1 = 0. \quad (1.82)$$

A similar equation is obtained for Ψ_2:

$$\left\{\hbar^2\frac{\partial^2}{\partial t^2} - \hbar^2c^2\left(\nu_0\frac{\partial}{\partial\mathbf{r}}\right)^2 + c^2\widehat{\mathbf{P}}_\perp^2 + m^2c^4 - ec\hbar\sigma(\mathbf{H} + i\mathbf{E})\right\}\Psi_2 = 0, \quad (1.83)$$

where $\mathbf{E}$ and $\mathbf{H}$ are the electric and magnetic field strengths of the plane EM wave determined by Eq. (1.6). The last terms in these equations $\sigma(\mathbf{H} \mp i\mathbf{E})$ describe the spin interaction (for the scalar particles Eqs. (1.82), (1.83) without these terms are reduced to the Klein–Gordon equation). To solve the problem it is more convenient to pass to the retarding and advanced wave coordinates

$$\tau = t - \nu_0\mathbf{r}/c; \ \eta = t + \nu_0\mathbf{r}/c,$$

then Eq. (1.82) is written as

$$\left\{4\hbar^2\frac{\partial^2}{\partial\tau\partial\eta} + c^2\widehat{\mathbf{P}}_\perp^2 + m^2c^4 - ec\hbar\sigma(\mathbf{H} - i\mathbf{E})\right\}\Psi_1 = 0. \quad (1.84)$$

As the existence of a plane wave does not violate the homogeneity of the space in the plane of the wave polarization $(\mathbf{r}_\perp)$ and the interaction Hamiltonian does not depend on the wave advanced coordinate η, i.e. the variables $\mathbf{r}_\perp$, η are cyclic and the corresponding components of generalized momentum $\mathbf{p}_\perp$ and p_η are conserved. Then the solution of Eq.(1.84) can be represented in the form

$$\Psi_1(\tau, \eta, \mathbf{r}_\perp) = F_1(\tau)\exp\left\{\frac{i}{\hbar}(\mathbf{p}_\perp\mathbf{r}_\perp + p_\eta\eta)\right\}. \quad (1.85)$$

From the initial condition $\mathbf{A}(\tau = -\infty) = 0$ it follows that $\mathbf{p}_\perp$ is the free particle initial transverse momentum and the quantity

$$p_\eta = \frac{1}{2}(c\mathbf{p}\nu_0 - \mathcal{E}), \quad (1.86)$$

where $\mathcal{E}$ and $\mathbf{p}$ are the free particle initial energy and momentum. Note that this quantity coincides with the classical integral of motion (1.10) (with a coefficient).

Substituting Eq.(1.85) into Eq. (1.84) for the function $F_1(\tau)$ yields the equation

$$\left\{\frac{\partial}{\partial \tau}-\frac{ic^2}{4\hbar p_\eta}\left[\left(\mathbf{p}_\perp-\frac{e}{c}\mathbf{A}\right)^2+m^2c^2-\frac{e\hbar}{c}\sigma(\mathbf{H}-i\mathbf{E})\right]\right\}F_1(\tau)=0. \quad (1.87)$$

The solution of Eq. (1.87) can be written in the operator form

$$F_1=\exp\left\{\frac{ic^2}{4\hbar p_\eta}\int_{-\infty}^{\tau}\left[\left(\mathbf{p}_\perp-\frac{e}{c}\mathbf{A}\right)^2+m^2c^2\right]d\tau'\right.$$

$$\left.+\frac{e\,(\sigma\nu_0+1)\,\sigma\mathbf{A}}{4p_\eta}\right\}w_1, \quad (1.88)$$

where w_1 is an arbitrary spinor amplitude.

The operator in the exponent should be understood as a expansion into series

$$e^{\widehat{G}}=1+\widehat{G}+\frac{\widehat{G}^2}{2!}+\cdots.$$

Then it is easy to see that all powers greater than 1 of the operator $(\sigma\nu_0+1)\,\sigma\mathbf{A}$ in Eq. (1.88) are zero because

$$[(\sigma\nu_0+1)\,\sigma\mathbf{A}]^2=\mathbf{A}^2\left(1-\nu_0^2\right)=0.$$

So, the spinor function (1.88) can be written in the form

$$F_1(\tau)=\exp\left\{\frac{ic^2}{4\hbar p_\eta}\int_{-\infty}^{\tau}\left[\left(\mathbf{p}_\perp-\frac{e}{c}\mathbf{A}\right)^2+m^2c^2\right]d\tau'\right\}$$

$$\times\left[1+\frac{e}{4p_\eta}\,(\sigma\nu_0+1)\,\sigma\mathbf{A}\right]w_1. \quad (1.89)$$

In the same way an analogical expression can be written for the spinor function $F_2(\tau)$.

The spinor components of the bispinor wave function of a particle (1.77) will be written as

$$\Psi_1 = \exp\left\{\frac{i}{\hbar} S(\mathbf{r}, t)\right\}\left[1 + \frac{e}{4p_\eta}(\sigma\nu_0 + 1)\,\sigma\mathbf{A}\right] w_1,$$

$$\Psi_2 = \exp\left\{\frac{i}{\hbar} S(\mathbf{r}, t)\right\}\left[1 + \frac{e}{4p_\eta}(\sigma\nu_0 - 1)\,\sigma\mathbf{A}\right] w_2, \tag{1.90}$$

or the ultimate bispinor wave function can be represented via Dirac matrices α

$$\Psi(\mathbf{r}, t) = \exp\left\{\frac{i}{\hbar} S(\mathbf{r}, t)\right\}\left[1 + \frac{e}{4p_\eta}(\alpha\nu_0 + 1)\,\alpha\mathbf{A}\right] w. \tag{1.91}$$

The scalar function $S(\mathbf{r}, t)$ in Eqs. (1.90) and (1.91)

$$S(\mathbf{r}, t) = \frac{c^2}{4p_\eta}\int_{-\infty}^{\tau}\left[\frac{e^2}{c^2}\mathbf{A}^2(\tau') - 2\frac{e}{c}\mathbf{p}\mathbf{A}(\tau')\right] d\tau' + \mathbf{p}\mathbf{r} - \mathcal{E}t \tag{1.92}$$

is the classical action of a charged particle in the plane EM wave field and

$$w = \begin{pmatrix} w_1 \\ w_2 \end{pmatrix}$$

is a constant bispinor, which should be defined from the condition of the particle wave function normalization according to the above stated initial conditions. Namely, we will demand that at $\tau = -\infty$ this wave function should be reduced to the free Dirac equation solution and for a constant bispinor we will set

$$w = \frac{u_\sigma}{\sqrt{2\mathcal{E}}},$$

where u_σ is the bispinor amplitude of a free Dirac particle with polarization σ. It is assumed that

$$\overline{u}u = 2mc^3,$$

where $\overline{u} = u^\dagger\beta$; $u^\dagger$ denotes the transposition and complex conjugation of u (in what follows we will set the volume of the normalization $V = 1$).

In future consideration of the quantum electrodynamic processes it will be reasonable to use the four-dimensional presentation of the Volkov wave function. Therefore, we will represent the wave function (1.91) in the equivalent four-dimensional form. Here and in what follows for the four-component vectors we chose the metric $a \equiv a^\mu = (a_0, \mathbf{a})$ and $ab = a^\mu b_\mu$ for the relativistic scalar product. The vector potential and the phase of the plane EM wave can be written as

$$A = (0, \mathbf{A}); \quad \tau = t - \nu_0 \mathbf{r}/c = \frac{k_\mu x^\mu}{k_0 c},$$

where

$$k = (k_0, \nu_0 k_0)$$

is the four-vector with $k^2 = 0$ and $x = (ct, \mathbf{r})$ is the four-radius vector. Introducing the known $\gamma^\mu = (\gamma_0, \gamma)$ matrices

$$\gamma = \beta\alpha, \ \gamma_0 = \beta$$

and taking into account that

$$p_\eta = -\frac{c}{2k_0} pk; \ p = \left(\frac{\mathcal{E}}{c}, \mathbf{p}\right),$$

$$\frac{e}{4p_\eta} (\alpha\nu_0 + \mathbf{1})\,\alpha\mathbf{A} = \frac{e}{2c(pk)} (\gamma k)(\gamma A),$$

the Volkov wave function may be written as

$$\Psi(x) = \exp\left\{\frac{i}{\hbar} S(x)\right\} \left[1 + \frac{e(\gamma k)(\gamma A)}{2c(pk)}\right] u,$$

$$S(x) = -px - \frac{k_0 c}{2pk} \int_{-\infty}^{\tau} \left[2\frac{e}{c} pA(\tau') - \frac{e^2}{c^2} A^2(\tau')\right] d\tau'. \tag{1.93}$$

Consider the Volkov wave function of a spinor particle in the field of the monochromatic wave (1.48). The latter can be presented in the form

$$\Psi_{\mathbf{p}\sigma} = \left[1 + \frac{e(\gamma k)(\gamma A)}{2c(kp)}\right] \frac{u_\sigma(p)}{\sqrt{2\mathcal{E}}} \exp\left[-\frac{i}{\hbar}\left\{\Pi x - \frac{eA_0}{c(pk)}\right.\right.$$

$$\left.\left.\times(\mathbf{e}_1\mathbf{p}\sin\omega_0\tau - g\mathbf{e}_2\mathbf{p}\cos\omega_0\tau) + \frac{e^2 A_0^2}{8c^2(pk)}(1 - g^2)\sin(2\omega_0\tau)\right]\right\}, \tag{1.94}$$

where $k = (\omega_0/c, \mathbf{k}_0)$ is the four-wave vector and $\Pi = (\Pi_0/c, \mathbf{\Pi})$ is the average four-kinetic momentum or "quasimomentum" of the particle in the periodic field, which is determined via free particle four-momentum $p = (\mathcal{E}/c, \mathbf{p})$ and relativistic invariant parameter of the wave intensity ξ_0 by the equation

$$\Pi = p + k\frac{m^2c^2}{4kp}(1 + g^2)\xi_0^2. \tag{1.95}$$

From this equation it follows that

$$\Pi^2 = m^{*2}c^2;\ m^* = m\left(1 + \frac{1+g^2}{2}\xi_0^2\right)^{1/2}, \tag{1.96}$$

where m^* is the effective mass of the particle in the monochromatic EM wave introduced in Section 1.2 (see Eq. (1.18)). It is seen that quasimomentum $\mathbf{\Pi}=\overline{\mathbf{p}}$ and quasienergy $\Pi_0 = \overline{\mathcal{E}}$ according to Eq. (1.17). The notion of quasimomentum is connected with the space-time translational symmetry - periodicity of the plane wave field as for the electron states in the crystal lattice.

The states (1.94) are normalized by the condition

$$\frac{1}{(2\pi\hbar)^3}\int \Psi^\dagger_{\mathbf{p}'\sigma'}\Psi_{\mathbf{p}\sigma}d\mathbf{r} = \delta(\mathbf{p}-\mathbf{p}')\delta_{\sigma,\sigma'},$$

where $\delta_{\sigma,\sigma'}$ is the Kronecker symbol.

By the analogy of the electron states in the crystal lattice the state of a particle in the monochromatic wave can be characterized by the quasimomentum $\mathbf{\Pi}$ and polarization σ as well:

$$\frac{1}{(2\pi\hbar)^3}\int \Psi^\dagger_{\mathbf{\Pi}'\sigma'}\Psi_{\mathbf{\Pi}\sigma}d\mathbf{r} = \delta(\mathbf{\Pi}-\mathbf{\Pi}')\delta_{\sigma,\sigma'}.$$

In this case the normalization constant should be changed as follows

$$\Psi_{\mathbf{\Pi}\sigma} = \sqrt{\frac{\mathcal{E}}{\Pi_0}}\Psi_{\mathbf{p}\sigma}. \tag{1.97}$$

1.6 Nonlinear Compton Effect

With the help of the Volkov wave function (1.94) one can describe the major quantum process of electron scattering in the field of a strong monochromatic wave — nonlinear Compton effect — as a photon radiation by the electron due to the transitions between the "stationary states" of different quasimomentum $\mathbf{\Pi}$ and polarization σ. The spontaneous radiation of a photon by the electron may be considered by the perturbation theory in the scope of quantum electrodynamics (QED). The first-order Feynman diagram (Fig. 1.6) describes the electron–EM wave scattering process, where the electron lines are described via dynamic wave functions in the strong wave field (1.94) (dressed electron). The probability amplitude of transition from the state with a definite quasimomentum and polarization $\Psi_{\mathbf{\Pi}\sigma}$ to the state $\Psi_{\mathbf{\Pi}'\sigma'}$

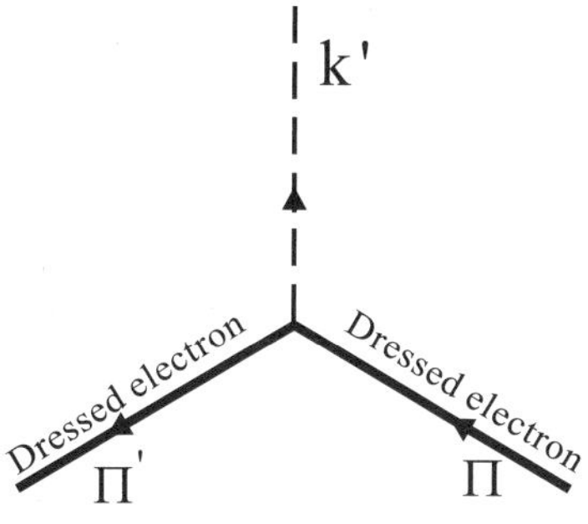

Fig. 1.6. Feynman diagram for nonlinear Compton effect.

with the emission of a photon with the frequency ω' and wave vector $\mathbf{k}'$ is given by

$$S_{if} = -\frac{ie}{\hbar c^2}\int j_{if}(x) A^*_{ph}(x)\, d^4x, \tag{1.98}$$

where

$$A^\mu_{ph}(x) = \sqrt{\frac{2\pi\hbar c^2}{\omega'}}\epsilon^\mu e^{-ik'x} \tag{1.99}$$

is the four-dimensional vector potential of quantized photon field (quantization volume $V = 1$), ϵ^μ is the four-dimensional polarization vector of the photon, and

$$j^\mu_{if} = \overline{\Psi}_{\mathbf{\Pi}'\sigma'}\gamma^\mu\Psi_{\mathbf{\Pi}\sigma}$$

is the four-dimensional transition current ($\overline{\Psi}_{\mathbf{\Pi}'\sigma'} = \Psi^\dagger_{\mathbf{\Pi}'\sigma'}\gamma_0$ and A^* is the complex conjugate of A).

Hence, for the probability amplitude we have

$$S_{if} = -ie\sqrt{\frac{2\pi}{\hbar\omega' c^2}}\int \overline{\Psi}_{\mathbf{\Pi}'\sigma'}\hat{\epsilon}^*\Psi_{\mathbf{\Pi}\sigma}e^{ik'x}d^4x. \tag{1.100}$$

Here and in what follows for arbitrary four-component vector $\widehat{a} = \gamma^\mu a_\mu$. The probability amplitude can be expressed by the generalized Bessel functions $G_s(\alpha, \beta, \varphi)$ introduced in Section 1.3. Thus, taking into account the properties of Dirac γ matrices ($\widehat{k}\widehat{k} = 0$ $\widehat{A}\widehat{k} = -\widehat{k}\widehat{A}$) and Eq. (1.94) one will obtain

$$S_{if} = -i\frac{e}{c}\sqrt{\frac{\pi}{2\hbar\omega'\Pi_0\Pi_0'}}\int \overline{u}_{\sigma'}(p')\left[\widehat{\epsilon^*} + \left(\frac{e\widehat{A}\widehat{k}\widehat{\epsilon^*}}{2c(kp')} + \frac{e\widehat{\epsilon^*}\widehat{k}\widehat{A}}{2c(kp)}\right)\right.$$

$$\left. -\frac{e^2(k\epsilon^*)A^2}{2c^2(kp')(kp)}\widehat{k}\right]u_\sigma(p)e^{i\psi(x)}d^4x. \tag{1.101}$$

Here

$$\psi(x) = \frac{1}{\hbar}\left(\Pi' - \Pi + \hbar k'\right)x + \alpha \sin(kx - \varphi) - \beta \sin 2kx, \tag{1.102}$$

and the parameters α, β, and φ are

$$\alpha = \frac{eA_0}{\hbar c}\left[\left(\frac{\mathbf{e}_1\mathbf{p}}{pk} - \frac{\mathbf{e}_1\mathbf{p}'}{p'k}\right)^2 + g^2\left(\frac{\mathbf{e}_2\mathbf{p}}{pk} - \frac{\mathbf{e}_2\mathbf{p}'}{p'k}\right)^2\right]^{1/2}, \tag{1.103}$$

$$\beta = \frac{e^2A_0^2}{8\hbar c^2}(1 - g^2)\left(\frac{1}{pk} - \frac{1}{p'k}\right), \tag{1.104}$$

$$\tan\varphi = \frac{g\left(\frac{\mathbf{e}_2\mathbf{p}}{pk} - \frac{\mathbf{e}_2\mathbf{p}'}{p'k}\right)}{\left(\frac{\mathbf{e}_1\mathbf{p}}{pk} - \frac{\mathbf{e}_1\mathbf{p}'}{p'k}\right)}. \tag{1.105}$$

After the integration the probability amplitude (1.101) can be represented in the form

$$S_{if} = -i\frac{e}{c}\left(2\pi\hbar\right)^4\sqrt{\frac{\pi}{2\hbar\omega'\Pi_0\Pi_0'}}\,\overline{u}_{\sigma'}(p')\widehat{M}_{if}u_\sigma(p), \tag{1.106}$$

where

$$\widehat{M}_{if} = \left[\widehat{\epsilon^*}Q_0 + \left(\frac{e\widehat{Q}_1\widehat{k}\widehat{\epsilon^*}}{2c(kp')} + \frac{e\widehat{\epsilon^*}\widehat{k}\widehat{Q}_1}{2c(kp)}\right) + \frac{e^2(k\epsilon^*)Q_2}{2c^2(kp')(kp)}\widehat{k}\right] \tag{1.107}$$

with the functions Q_0, Q_1^μ, and Q_2:

$$Q_0 = \sum_{s=-\infty}^{\infty} G_s(\alpha,\beta,\varphi)\delta\left(\Pi' - \Pi + \hbar k' - s\hbar k\right), \tag{1.108}$$

$$Q_1^\mu = (0, \mathbf{Q}_1),$$

$$\mathbf{Q}_1 = \frac{A_0}{2}\sum_{s=-\infty}^{\infty}\left\{\mathbf{e}_1\left(G_{s-1}(\alpha,\beta,\varphi) + G_{s+1}(\alpha,\beta,\varphi)\right)\right.$$

$$\left.+i\mathbf{e}_2 g\left(G_{s-1}(\alpha,\beta,\varphi) - G_{s+1}(\alpha,\beta,\varphi)\right)\right\}\delta\left(\Pi' - \Pi + \hbar k' - s\hbar k\right), \tag{1.109}$$

$$Q_2 = \frac{A_0^2}{2}(1+g^2)Q_0 + \frac{A_0^2}{2}(1-g^2)$$

$$\times \sum_{s=-\infty}^{\infty} \left(G_{s-2}(\alpha,\beta,\varphi) + G_{s+2}(\alpha,\beta,\varphi)\right) \delta\left(\Pi' - \Pi + \hbar k' - s\hbar k\right). \quad (1.110)$$

From the definition of the functions (1.108)–(1.110) follows the useful relation

$$\frac{\mathcal{E}' - \mathcal{E} + \hbar\omega'}{\omega} Q_0 + \frac{e}{c}\left(\frac{p' Q_1}{kp'} - \frac{pQ_1}{kp}\right) + \frac{e^2}{2c^2}\left(\frac{1}{kp'} - \frac{1}{kp}\right) Q_2 = 0 \quad (1.111)$$

We will assume that the Dirac particle is nonpolarized and summation over the final particle polarizations (photon and electron) will be made. Then we need to calculate the sum

$$\frac{1}{2}\sum_{\sigma',\sigma,\epsilon} |S_{if}|^2 = \frac{(2\pi\hbar)^8 \pi e^2}{4\hbar\omega' c^2 \Pi_0 \Pi_0'} \sum_{\sigma',\sigma,\epsilon} \left|\overline{u}_{\sigma'}(p')\widehat{M}_{if} u_\sigma(p)\right|^2$$

$$= \frac{(2\pi\hbar)^8 \pi e^2 c^2}{4\hbar\omega' \Pi_0 \Pi_0'} \sum_{\epsilon} Sp\left[(\widehat{p}' + mc)\widehat{M}_{if}(\widehat{p} + mc)\widehat{\overline{M}}_{if}\right], \quad (1.112)$$

where

$$\widehat{\overline{M}}_{if} = \gamma_0 \widehat{M}_{if}^{\dagger} \gamma_0 .$$

Taking into account that spur of the product of odd number γ matrices is zero we will obtain

$$\frac{1}{2}\sum_{\sigma',\sigma,\epsilon} |S_{if}|^2 = \frac{(2\pi\hbar)^8 \pi e^2 c^2}{4\hbar\omega' \Pi_0 \Pi_0'} \sum_{\epsilon} \left\{ Sp\left[\widehat{p}'\widehat{M}_{if}\widehat{p}\widehat{\overline{M}}_{if}\right] + m^2c^2 Sp\left[\widehat{M}_{if}\widehat{\overline{M}}_{if}\right]\right\}$$

The summation over the photon polarizations is equivalent to the replacements

$$\epsilon_\upsilon^* \epsilon_\mu \to -g_{\upsilon\mu}, \quad \widehat{\epsilon^*}\widehat{a}\widehat{\epsilon} \to 2\widehat{a}, \quad \widehat{\epsilon^*}\widehat{a}\widehat{b}\widehat{c}\widehat{\epsilon} \to 2\widehat{c}\widehat{b}\widehat{a}, \quad (1.113)$$

where $g_{\upsilon\mu}$ is the metric tensor. So,

$$Sp\left[\widehat{M}_{if}\widehat{\overline{M}}_{if}\right] = -16\left|Q_0\right|^2$$

and

$$Sp\left[\widehat{p}'\widehat{M}_{if}\widehat{p}\widehat{\overline{M}}_{if}\right] = 8(p'p)\left|Q_0\right|^2$$

$$+\frac{8e}{c}\left(pk-p'k\right)Re\left(\left(\frac{p'Q_1}{kp'}-\frac{pQ_1}{kp}\right)Q_0^*\right)$$

$$-\frac{4e^2}{c^2}\left[\frac{kp}{kp'}+\frac{kp'}{kp}\right]\left|Q_1\right|^2-\frac{8e^2}{c^2}Re\left(Q_0Q_2^*\right).$$

Then using the relation (1.111) we obtain

$$\frac{1}{2}\sum_{\sigma',\sigma,\epsilon}\left|S_{if}\right|^2=\frac{2\left(2\pi\hbar\right)^8\pi e^2c^2}{\hbar\omega'\Pi_0\Pi_0'}\left[-m^2c^2\left|Q_0\right|^2\right.$$

$$\left.-\frac{e^2}{c^2}\left(1+\frac{\hbar^2\left(kk'\right)^2}{2\left(pk\right)\left(p'k\right)}\right)\left(\left|Q_1\right|^2+Re\left(Q_0Q_2^*\right)\right)\right]. \tag{1.114}$$

For the differential probability per unit time we have

$$dW=\frac{1}{2T}\sum_{\sigma',\sigma,\epsilon}\left|S_{if}\right|^2\frac{d\mathbf{\Pi}'}{\left(2\pi\hbar\right)^3}\frac{d\mathbf{k}'}{\left(2\pi\right)^3}, \tag{1.115}$$

where T is the interaction time. Then taking into account Eqs. (1.108)–(1.110) and the relation

$$\delta\left(\Pi'-\Pi+\hbar k'-s\hbar k\right)\delta\left(\Pi'-\Pi+\hbar k'-s'\hbar k\right)$$

$$=\begin{bmatrix}0, \text{ if } s\neq s', \\ \frac{cT}{\left(2\pi\hbar\right)^4}\delta\left(\Pi'-\Pi+\hbar k'-s\hbar k\right), \text{if } s=s',\end{bmatrix} \tag{1.116}$$

for the differential probability of the nonlinear Compton effect we obtain

$$dW=\sum_{s=1}^{\infty}W^{(s)}\delta\left(\Pi'-\Pi+\hbar k'-s\hbar k\right)d\mathbf{\Pi}'d\mathbf{k}', \tag{1.117}$$

$$W^{(s)}=\frac{e^2m^2c^5}{2\pi\omega'\Pi_0\Pi_0'}\left[-\left|G_s\right|^2+\frac{\xi_0^2}{4}\left(1+\frac{\hbar^2\left(kk'\right)^2}{2\left(pk\right)\left(p'k\right)}\right)\right.$$

$$\times\left(\left(1+g^2\right)\left(\left|G_{s-1}\right|^2+\left|G_{s+1}\right|^2-2\left|G_s\right|^2\right)\right.$$

$$+(1-g^2)Re\left[2G^*_{s-1}G_{s+1}-G^*_s\left(G_{s-2}+G_{s+2}\right)\right]\Bigg)\Bigg]. \tag{1.118}$$

The four-dimensional δ-functions in Eq. (1.117) for differential probability express the conservation laws for quasimomentum and quasienergy of the particle in the nonlinear Compton process. Different s correspond to partial scattering processes with fixed photon numbers and $W^{(s)}$ are the partial probabilities of s-photon absorption by the particle in the strong wave field.

The spectrum of emitted photons is determined from the conservation laws. Taking into account Eqs. (1.95) and (1.96) we will have the following expression for the radiated frequency:

$$\omega' = s\omega\frac{1-\frac{\overline{\mathrm{v}}}{c}\cos\vartheta_0}{1-\frac{\overline{\mathrm{v}}}{c}\cos\vartheta+\frac{s\hbar\omega}{\Pi_0}\left(1-\cos\vartheta_r\right)}, \tag{1.119}$$

where ϑ_0, ϑ are the incident and scattering angles of incident strong wave and radiated photon with respect to the direction of the particle mean velocity $\overline{\mathbf{v}} = c^2\mathbf{\Pi}/\Pi_0$ and ϑ_r is the angle between the incident wave and radiated photon propagation directions. The quantum conservation law of nonlinear Compton effect (1.119) differs from the classical formula (1.64) by the last term in the denominator $\sim s\hbar\omega/\Pi_0$, which is the quantum recoil of emitted photon.

Making the integration over $\mathbf{\Pi}'$ in Eq. (1.117) and multiplying by the photon energy we obtain the radiation power. In the case of circular polarization of an incident strong wave ($g=\pm1$) we have $\left|G_s\right|^2 = J_s^2(\alpha)$ and the radiation power is

$$dP^{(s)}_{\mathbf{k}'} = \frac{\omega'^2 e^2 m^2 c^3}{2\pi\Pi_0\Pi'_0}\left[-J_s^2(\alpha)+\xi_0^2\left(1+\frac{\hbar^2\left(kk'\right)^2}{2\left(pk\right)\left(p'k\right)}\right)\right.$$

$$\left.\times\left[\left(\frac{s^2}{\alpha^2}-1\right)J_s^2(\alpha)+J_s'^2(\alpha)\right]\right]\times\delta\left(\frac{\Pi'_0-\Pi_0}{\hbar}+\omega'-s\omega\right)d\omega' dO,$$

where the Bessel function argument

$$\alpha = \frac{eA_0}{\hbar\omega}\left|\left[\mathbf{k}\left(\frac{\mathbf{p}}{pk}-\frac{\mathbf{p}'}{p'k}\right)\right]\right|. \tag{1.120}$$

Taking into account that

$$\delta\left(\frac{\Pi'_0-\Pi_0}{\hbar}+\omega'-s\omega\right)d\omega' \to \left|\frac{\partial}{\partial\omega'}\left(\frac{\Pi'_0}{\hbar}+\omega'\right)\right|^{-1} = \frac{\Pi'_0\omega'}{c^2\left(\Pi'k'\right)},$$

for the angular distribution of radiation power we obtain

$$\frac{dP^{(s)}}{dO} = \frac{\omega'^3 e^2 m^2 c}{2\pi \Pi_0 (\Pi' k')} \left[-J_s^2(\alpha) + \xi_0^2 \left(1 + \frac{\hbar^2 (kk')^2}{2(pk)(p'k)} \right) \right.$$

$$\left. \times \left(\left(\frac{s^2}{\alpha^2} - 1 \right) J_s^2(\alpha) + J_s'^2(\alpha) \right) \right]. \tag{1.121}$$

This formula differs from the classical one (1.63) only by the terms of quantum recoil, which are of the order of $\hbar kk' / (\Pi' k)$. The maximal value of this parameter is $2s\hbar (\Pi k) / m^{*2} c^2$ and if

$$\frac{2s\hbar (\Pi k)}{m^{*2} c^2} << 1,$$

one can omit the quantum recoil and taking into account that in this case

$$\Pi' k' \simeq \Pi k'; \quad \alpha \simeq \alpha_{classic}; \quad \frac{\hbar^2 (kk')^2}{2(pk)(p'k)} << 1,$$

from Eq. (1.121) we obtain the classical formula for radiation power.

In the limit of weak EM wave when $\xi_0 << 1$ (linear theory) the argument of the Bessel function $\alpha << 1$ and the main contribution to the radiation power gives the first harmonic (as in the classical theory). In this case $J_1^2(\alpha_1) \simeq \alpha_1^2/4$, $J_1'^2(\alpha_1) \simeq 1/4$, $\Pi_0 \simeq \mathcal{E}$, $\Pi_0' \simeq \mathcal{E}'$, and

$$\frac{dP}{dO} = \frac{\omega'^3 e^2 m^2 c}{8\pi \mathcal{E} (p'k')} \left[-\alpha^2 + 2\xi_0^2 \left(1 + \frac{\hbar^2 (kk')^2}{2(pk)(p'k)} \right) \right].$$

Then, using conservation laws, it is easy to see that

$$\left| \left[\mathbf{k} \left(\frac{\mathbf{p}'}{p'k} - \frac{\mathbf{p}}{pk} \right) \right] \right|^2 = 2\hbar \frac{\omega^2}{c^2} \left(\frac{1}{p'k} - \frac{1}{pk} \right) - \omega^2 m^2 \left(\frac{1}{pk} - \frac{1}{p'k} \right)^2,$$

$$\left(1 + \frac{\hbar^2 (kk')^2}{2(pk)(p'k)} \right) = \frac{1}{2} \left[\frac{pk}{p'k} + \frac{p'k}{pk} \right],$$

and for the one-photon Compton effect we obtain

$$\frac{dP}{dO} = \frac{\omega'^3 e^2 m^2 c}{8\pi \mathcal{E} (p'k')} \xi_0^2 \left[\left(\frac{m^2 c^2}{\hbar (p'k)} - \frac{m^2 c^2}{\hbar (pk)} \right)^2 \right.$$

$$-2\left(\frac{m^2c^2}{\hbar\,(p'k)}-\frac{m^2c^2}{\hbar\,(pk)}\right)+\frac{pk}{p'k}+\frac{p'k}{pk}\Bigg]. \tag{1.122}$$

For the differential cross section

$$\frac{d\sigma}{dO}=\frac{1}{\hbar\omega' J}\frac{dP}{dO}$$

one should make the replacement

$$A_0^2 \to \frac{4\pi\hbar c^2}{\omega}, \tag{1.123}$$

corresponding to photon field quantization and

$$J=\frac{c^3 pk}{\omega\mathcal{E}}$$

is the initial flux density (quantization volume $V = 1$). Hence, for the differential cross section of the one-photon Compton effect we obtain

$$\frac{d\sigma}{dO}=\frac{\omega'^2 e^4}{2c^4\,(pk)^2}\left[\left(\frac{m^2c^2}{\hbar\,(p'k)}-\frac{m^2c^2}{\hbar\,(pk)}\right)^2\right.$$

$$\left.-2\left(\frac{m^2c^2}{\hbar\,(p'k)}-\frac{m^2c^2}{\hbar\,(pk)}\right)+\frac{pk}{p'k}+\frac{p'k}{pk}\right]. \tag{1.124}$$

For a particle initially at rest

$$pk = m\omega, \quad pk' = m\omega', \quad \frac{mc^2}{\hbar\omega'}-\frac{mc^2}{\hbar\omega}=1-\cos\vartheta_r,$$

and the differential cross section of the one-photon Compton effect may be written in the known form of Klein and Nishina formula

$$\frac{d\sigma}{dO}=\frac{r_e^2}{2}\left(\frac{\omega'}{\omega}\right)^2\left[\frac{\omega}{\omega'}+\frac{\omega'}{\omega}-\sin^2\vartheta_r\right], \tag{1.125}$$

where $r_e = e^2/mc^2$ is the classical radius of the electron.

1.7 Bremsstrahlung in Superstrong Wave Fields

The other major radiation process with the free electrons in vacuum is bremsstrahlung. In the presence of an external coherent radiation field the bremsstrahlung acquires induced character and stimulated bremsstrahlung (SB) takes place. In the laser fields of relativistic intensities SB becomes essentially multiphoton and the description of nonlinear SB requires relativistic quantum consideration. The latter may be made again via Volkov wave function (1.94), at the electron scattering on a static potential field in the first Born approximation. This process can be described by the first-order Feynman diagram (Fig. 1.7), where the "dressed electron" initial and final states are described by corresponding wave functions (1.94) and the dashed line corresponds to pseudophotons of scattering potential field.

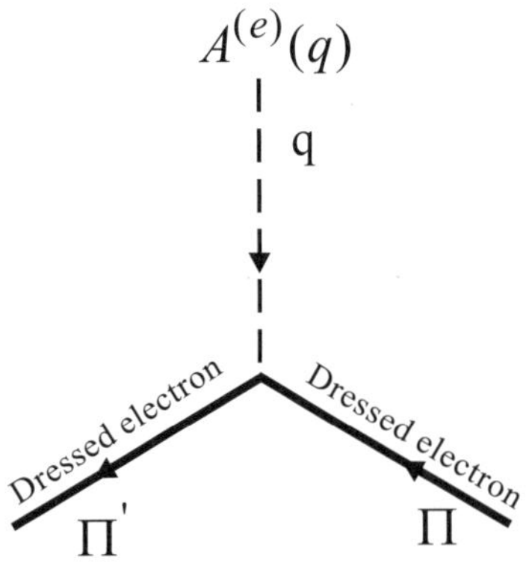

Fig. 1.7. Feynman diagram for bremsstrahlung in superstrong wave field.

For the probability amplitude of the transition $i \to f$ at SB process we have

$$S_{if} = -\frac{ie}{\hbar c^2} \int \overline{\Psi}_{\mathbf{\Pi}'\sigma'} \widehat{A}^{(e)}(x) \Psi_{\mathbf{\Pi}\sigma} d^4x, \qquad (1.126)$$

where $A^{(e)}(x)$ is the four-dimensional vector potential of the scattering field. Upon Fourier transformation

$$A^{(e)}(x) = \frac{1}{(2\pi)^4} \int A^{(e)}(q') e^{-iq'x} d^4q',$$

Eq. (1.126) will have the form

$$S_{if} = -\frac{ie}{\hbar c^2 (2\pi)^4} \int \overline{\Psi}_{\mathbf{\Pi}'\sigma'} \widehat{A}^{(e)}(q') e^{-iq'x} \Psi_{\mathbf{\Pi}\sigma} d^4q' d^4x. \qquad (1.127)$$

The static potential field (for nucleus/ion — as a scattering center — the recoil momentum is neglected) will be described by the scalar potential $\varphi(\mathbf{r})$

$$A^{(e)}(x) = (\varphi(\mathbf{r}), 0)$$

and for the Fourier transform of $A^{(e)}(x)$ we have

$$A^{(e)}(q') = (2\pi\delta(q_0')\varphi(\mathbf{q}'), 0).$$

Then one can conclude that the S-matrix amplitude of this process may be obtained from the S-matrix amplitude of the Compton effect (1.106) by substitutions of the amplitude of vector potential of quantized photon field, as well as four-dimensional polarization and wave vectors of the photon as follows:

$$\sqrt{\frac{2\pi\hbar c^2}{\omega'}} \rightarrow \frac{1}{(2\pi)^3}\delta(q_0')\varphi(\mathbf{q}')\,d^4q',$$

$$\epsilon^* \rightarrow \epsilon_0 = (1,0,0,0), \qquad k' \rightarrow -q'.$$

Hence, making these substitutions in Eq. (1.107) and using δ functions of Eqs. (1.108)–(1.110) for the integration over q', the probability amplitude of SB may be represented in the form

$$S_{if} = -i\pi\frac{e}{Vc\sqrt{\Pi_0\Pi_0'}}\overline{u}_{\sigma'}(p')\widehat{M}_{if}u_\sigma(p) \tag{1.128}$$

with

$$\widehat{M}_{if} = \sum_{s=-\infty}^{\infty}\varphi(\mathbf{q}_s)\left[\widehat{\epsilon}_0 B_s + \left(\frac{e\widehat{B}_{1s}\widehat{k}\widehat{\epsilon}_0}{2c(kp')} + \frac{e\widehat{\epsilon}_0\widehat{k}\widehat{B}_{1s}}{2c(kp)}\right)\right.$$

$$\left. + \frac{e^2(k\epsilon_0)B_{2s}}{2c^2(kp')(kp)}\widehat{k}\right]\delta(\Pi_0' - \Pi_0 - s\hbar\omega), \tag{1.129}$$

where the vector functions $B_{1s}^{\mu} = (0, \mathbf{B}_{1s})$ and scalar functions B_s, B_{2s} are expressed via generalized Bessel functions $G_s(\alpha, \beta, \varphi)$:

$$\mathbf{B}_{1s} = \frac{A_0}{2}\sum_{s=-\infty}^{\infty}\{\mathbf{e}_1(G_{s-1}(\alpha,\beta,\varphi) + G_{s+1}(\alpha,\beta,\varphi))$$

$$+ i\mathbf{e}_2 g(G_{s-1}(\alpha,\beta,\varphi) - G_{s+1}(\alpha,\beta,\varphi))\}, \tag{1.130}$$

$$B_s = G_s(\alpha, \beta, \varphi), \tag{1.131}$$

$$B_{2s} = \frac{A_0^2}{2}(1+g^2)G_0 + \frac{A_0^2}{2}(1-g^2)$$

$$\times \sum_{s=-\infty}^{\infty} \left(G_{s-2}(\alpha, \beta, \varphi) + G_{s+2}(\alpha, \beta, \varphi)\right), \tag{1.132}$$

and

$$\hbar \mathbf{q}_s = \mathbf{\Pi}' - \mathbf{\Pi} - s\hbar \mathbf{k} \tag{1.133}$$

is the recoil momentum. The definitions of arguments α, β, φ are the same as in Eqs. (1.103)–(1.105).

The differential probability of SB process per unit time, summed over the electron final polarization states and averaged over the initial polarization states, is

$$dW = \frac{1}{2T} \sum_{\sigma', \sigma} |S_{if}|^2 \frac{d\mathbf{\Pi}'}{(2\pi\hbar)^3}. \tag{1.134}$$

The calculation of spur will be made in the same way as has been made for the Compton effect using Eq. (1.111) and the following relations:

$$\widehat{\epsilon}_0 = \gamma_0, \quad \widehat{\epsilon}_0 \widehat{b} \widehat{\epsilon}_0^* = \widehat{\overline{b}}, \quad \overline{b} = (b_0, -\mathbf{b}),$$

$$\delta\left(\Pi_0' - \Pi_0 - s\hbar\omega\right) \delta\left(\Pi_0' - \Pi_0 - s'\hbar\omega\right)$$

$$= \begin{bmatrix} 0, \text{ if } s \neq s', \\ \frac{T}{2\pi\hbar} \delta\left(\Pi_0' - \Pi_0 - s\hbar\omega\right), \text{ if } s = s'. \end{bmatrix}$$

Then we obtain

$$\frac{1}{2} \sum_{\sigma', \sigma} |S_{if}|^2 = \frac{2\pi e^2 T}{\hbar \Pi_0' \Pi_0} \sum_s |\varphi(\mathbf{q}_s)|^2 \left\{ \left| \mathcal{E} B_s - \frac{e\left(\mathbf{p}\mathbf{B}_{1s}\right)\omega}{(kp)\, c} + \frac{e^2 \omega B_{2s}}{2c^2 (kp)} \right|^2 \right.$$

$$+ \frac{e^2 \hbar^2 \left[\mathbf{k}\mathbf{q}_s\right]^2}{4(kp')(kp)} \left[|\mathbf{B}_{1s}|^2 - Re\left(B_{2s} B_s^*\right) \right]$$

$$\left. - \frac{\hbar^2 \mathbf{q}_s^2 c^2}{4} |B_s|^2 \right\} \delta\left(\Pi_0' - \Pi_0 - s\hbar\omega\right). \tag{1.135}$$

Dividing the differential probability of the process (1.134) by initial flux density $|\mathbf{\Pi}| c^2/\Pi_0$, and integrating over Π_0' we obtain the differential cross section of multiphoton SB process

$$\frac{d\sigma}{dO} = \sum_{s>-s_m}^{\infty} \frac{d\sigma^{(s)}}{dO}, \tag{1.136}$$

where

$$\frac{d\sigma^{(s)}}{dO} = \frac{e^2 |\varphi(\mathbf{q}_s)|^2 |\mathbf{\Pi}'|}{4\pi^2\hbar^4 c^4 |\mathbf{\Pi}|} \left\{ \left| \mathcal{E} B_s - \frac{e(\mathbf{pB}_{1s})\omega}{(kp)c} + \frac{e^2\omega}{2c^2(kp)} B_{2s} \right|^2 \right.$$

$$\left. - \frac{\hbar^2 \mathbf{q}_s^2 c^2}{4} |B_s|^2 + \frac{e^2\hbar^2 [\mathbf{kq}_s]^2}{4(kp')(kp)} \left[|\mathbf{B}_{1s}|^2 - Re\left(B_{2s} B_s^*\right) \right] \right\} \tag{1.137}$$

is the partial differential cross section, which describes the s-photon SB process. The final quasimomentum of the electron corresponding to s-photon absorption ($s > 0$) or emission ($s < 0$) processes in the strong wave field is

$$\Pi' = \sqrt{\mathbf{\Pi}^2 + \frac{s\hbar\omega}{c^2}(2\Pi_0 + s\hbar\omega)}, \tag{1.138}$$

and s_m is the maximum number of emitted photons:

$$s_m = \frac{\Pi_0 - m^* c^2}{\hbar\omega}. \tag{1.139}$$

For circular polarization of the incident EM wave

$$G_s(\alpha, 0, \varphi) = (-1)^s J_s(\alpha) e^{is\varphi},$$

and taking into account (1.130)–(1.132), for the partial differential cross section of SB we have

$$\frac{d\sigma^{(s)}}{dO} = \frac{e^2 |\varphi(\mathbf{q}_s)|^2 |\mathbf{\Pi}'|}{4\pi^2\hbar^4 c^4 |\mathbf{\Pi}|} \left\{ \left[\left(\Pi_0 + \frac{s\hbar\omega}{(kp)} \frac{\varkappa[\mathbf{kp}]}{\varkappa^2} \right)^2 - \frac{\hbar^2 \mathbf{q}_s^2 c^2}{4} \right] J_s^2(\alpha) \right.$$

$$+ \frac{\hbar^2 e^2 A_0^2}{4(kp')(kp)} [\mathbf{kq}_s]^2 \left[\left(\frac{s^2}{\alpha^2} - 1 \right) J_s^2(\alpha) + J_s'^2(\alpha) \right]$$

$$+\frac{e^2 A_0^2}{(kp)^2}\frac{[\varkappa\,[\mathbf{kp}]]^2}{\varkappa^2}J_s'^2(\alpha)\Bigg\}, \tag{1.140}$$

where

$$\varkappa=\left[\mathbf{k}\left(\frac{\mathbf{p}}{pk}-\frac{\mathbf{p}'}{p'k}\right)\right] \tag{1.141}$$

and the Bessel function argument is

$$\alpha=\frac{eA_0}{\hbar\omega}\left|\varkappa\right|. \tag{1.142}$$

At the absence of incident EM wave ($A_0=0$) from Eq. (1.140) we obtain the Mott formula for elastic scattering of the electron in the Coulomb field, which corresponds to $s=0$ harmonic. Thus, taking into account the Fourier transform of Coulomb potential

$$\varphi(\mathbf{q})=\frac{4\pi Z_a e}{\mathbf{q}^2}, \tag{1.143}$$

where Z_a is the charge number of the nucleus, and Eq. (1.133) for $\mathbf{q}_0$, Eq. (1.140) becomes

$$\frac{d\sigma_{Mott}}{dO}=\frac{4Z_a^2\alpha_0^2}{\hbar^2c^2\mathbf{q}_0^4}\mathcal{E}^2\left[1-\frac{\hbar^2\mathbf{q}_0^2c^2}{4\mathcal{E}^2}\right], \tag{1.144}$$

where $\alpha_0\equiv e^2/(\hbar c)=1/137$ is the fine structure constant.

Concerning the applied approximation for description of multiphoton SB note that the condition of validity of obtained cross sections (1.137) in the first Born approximation by static potential field holds for electron renormalized velocities in the incident wave field. In particular, for Coulomb potential the known condition for the Born approximation turns into conditions

$$\frac{Z_a e^2}{\hbar\overline{\mathrm{v}}}<<1,\quad \frac{Z_a e^2}{\hbar\overline{\mathrm{v}}'}<<1, \tag{1.145}$$

where $\overline{\mathrm{v}}=c^2\left|\mathbf{\Pi}\right|/\mathit{\Pi}_0$, $\overline{\mathrm{v}}'=c^2\left|\mathbf{\Pi}'\right|/\mathit{\Pi}_0'$ are the electron initial and final mean velocities in the EM wave field.

For $\alpha<<1$ the main contribution to the SB cross section produces one-photon emission and absorption processes. In particular, for one-photon stimulated radiation from Eq. (1.140) we have

$$\frac{d\sigma^{(-1)}}{dO}=\frac{e^2\left|\varphi(\mathbf{q}_{-1})\right|^2\left|\mathbf{p}'\right|}{16\pi^2\hbar^4c^4\left|\mathbf{p}\right|}\frac{e^2A_0^2}{\hbar^2\omega^2}\left\{\left[\mathbf{k}\left(\frac{\mathcal{E}'\mathbf{p}}{pk}-\frac{\mathcal{E}\mathbf{p}'}{p'k}\right)\right]^2\right.$$

$$-\frac{\hbar^2 \mathbf{q}_{-1}^2 c^2}{4}\left[\mathbf{k}\left(\frac{\mathbf{p}'}{p'k}-\frac{\mathbf{p}}{pk}\right)\right]^2+\frac{\hbar^4\omega^2\left[\mathbf{k}\mathbf{q}_{-1}\right]^2}{2(kp')(kp)}\Bigg\}. \tag{1.146}$$

From this formula one can obtain the Bethe–Heitler formula for spontaneous bremsstrahlung (one-photon emission) in the Coulomb field. For the latter one needs to make the replacement (1.123) in Eq. (1.146) and multiply the cross section of bremsstrahlung by the density of photon states

$$2\frac{\omega^2}{c^3}d\omega\frac{do}{(2\pi)^3},$$

then we will have the Bethe–Heitler formula

$$d\sigma_{BH}=\frac{\alpha_0^3 Z^2\left|\mathbf{p}'\right|}{\pi^2\hbar^2c^2\omega\left|\mathbf{p}\right|\mathbf{q}_{-1}^4}\left\{\left[\mathbf{k}\left(\frac{\mathcal{E}\mathbf{p}'}{p'k}-\frac{\mathcal{E}'\mathbf{p}}{pk}\right)\right]^2\right.$$

$$\left.-\frac{\hbar^2 \mathbf{q}_{-1}^2 c^2}{4}\left[\mathbf{k}\left(\frac{\mathbf{p}'}{p'k}-\frac{\mathbf{p}}{pk}\right)\right]^2+\frac{\hbar^4\omega^2\left[\mathbf{k}\mathbf{q}_{-1}\right]^2}{2(kp')(kp)}\right\}d\omega do dO. \tag{1.147}$$

For multiphoton SB in the nonrelativistic limit ($\mathrm{v} << c$) one can make dipole approximation for EM wave and omit the terms proportional to $\mathbf{k}^2$ and $\mathbf{q}^2$ in Eq. (1.140). Then we obtain the nonrelativistic factorized cross section of multiphoton SB

$$\frac{d\sigma^{(s)}}{dO}=\frac{d\sigma_R}{dO}J_s^2\left(\frac{eA_0}{\hbar\omega}\left|\left[\frac{\mathbf{k}}{\omega}(\mathbf{v}-\mathbf{v}')\right]\right|\right),$$

where

$$\frac{d\sigma_R}{dO}=\frac{m^2e^2\left|\varphi\left(\mathbf{q}_s\right)\right|^2\left|\mathbf{p}'\right|}{4\pi^2\hbar^4\left|\mathbf{p}\right|} \tag{1.148}$$

is the Reserford cross section.

Comparing the nonrelativistic cross section (1.148) with the relativistic one (1.140) it is easy to see that besides the additional terms, which result from spin–orbital and spin–laser interaction ($\sim q_s^2$), as well as from the intensity effect of strong EM wave ($\sim \xi_0^2$), the relativistic contribution is conditioned by arguments of the Bessel functions. Because of sensitivity of the Bessel function to the relationship of its argument and index the most probable number of emitted or absorbed photons is determined by the condition $|s| \sim |\alpha|$. For this reason the contribution of relativistic effects to the scattering process is already essential for intensities $\xi_0 \sim 0.1$. Hence, the dipole approximation is violated for nonrelativistic parameters of interaction.

Besides, the state of an electron in the field of a strong EM wave and, consequently, the cross section of SB essentially depends on the polarization of the wave. In particular, the cross section for linear polarization of the wave is described by the generalized Bessel function. The cross sections in both cases are complicated and to show some features of multiphoton SB process we present the results of numerical investigation. For the numerical calculations we have chosen the initial electron momentum $\mathbf{p}$ to be colinear with the laser propagation direction. In this case for circular polarization of the wave there is an azimuthal symmetry with respect to propagation direction, which simplifies the calculation of integral quantities. Then we have taken moderate initial electron kinetic energy $\mathcal{E}_k = 2.7$ keV (100 a.u.), neodymium laser ($\hbar\omega \simeq 1.17$ eV), screening Coulomb potential

$$\varphi(\mathbf{q}) = \frac{4\pi Z_a e}{\mathbf{q}^2 + \chi^2},$$

with radius of screening $\chi^{-1} = 4$ a.u. and $Z_a = 1$.

In Fig. 1.8a the envelopes of partial differential cross sections as a function of the number of emitted or absorbed photons for circular polarization of EM wave are shown for the deflection angle $\vartheta \equiv \angle \Pi \Pi' = 10$ mrad. The relativistic parameter of intensity is taken to be $\xi_0 \simeq 0.1$. The dotted and dashed lines correspond to initial electron momentum parallel and antiparallel to the laser propagation direction $\mathbf{k}$, respectively, and the solid line gives the nonrelativistic result. The energy change of a particle is characterized by the absorption/emission (AE) cross section. Partial AE differential cross section will be

$$\frac{d\sigma_{ae}^{(s)}}{dO} = s\left(\frac{d\sigma^{(s)}}{dO} - \frac{d\sigma^{(-s)}}{dO}\right). \tag{1.149}$$

In Fig. 1.8b the envelopes of partial AE differential cross sections for circular polarization of EM wave are shown for the same parameters. It is seen from Fig. 1.8 that the differences between the cases of initial electron momentum parallel or antiparallel to the laser propagation direction $\mathbf{k}$ on the one hand and between nonrelativistic result on the other hand are notable already for $\xi_0 \simeq 0.1$. In particular, the absorption and emission edges and the magnitudes of the peaks are different.

To show the dependence of the SB process on laser intensity in Fig. 1.9a the summed differential cross section

$$\frac{d\sigma}{dO} = \sum_{s>-s_m}^{\infty} \frac{d\sigma^{(s)}}{dO} \tag{1.150}$$

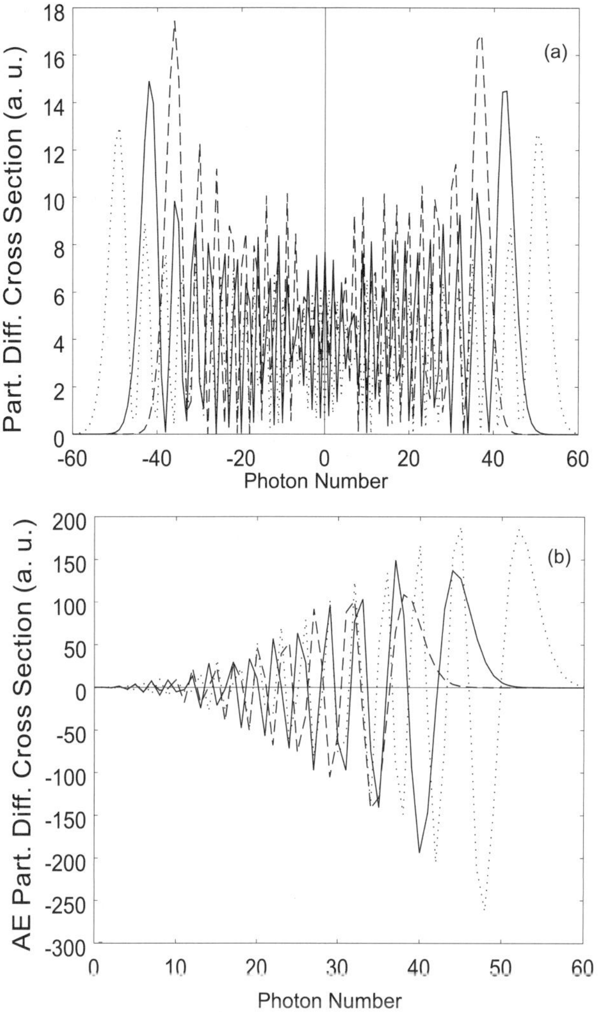

Fig. 1.8. (a) Envelopes of partial differential cross sections in atomic units as a function of the number of emitted or absorbed photons for circular polarization of EM wave for the deflection angle $\vartheta \equiv \angle \Pi\Pi' = 10$ mrad. The relativistic parameter of intensity is $\xi_0 \simeq 0.1$. (b) Envelopes of partial absorption/emission differential cross sections for the same parameters. The dotted and dashed lines correspond to initial electron momentum parallel and antiparallel to the laser propagation direction $\mathbf{k}$, respectively, and the solid line gives the nonrelativistic result.

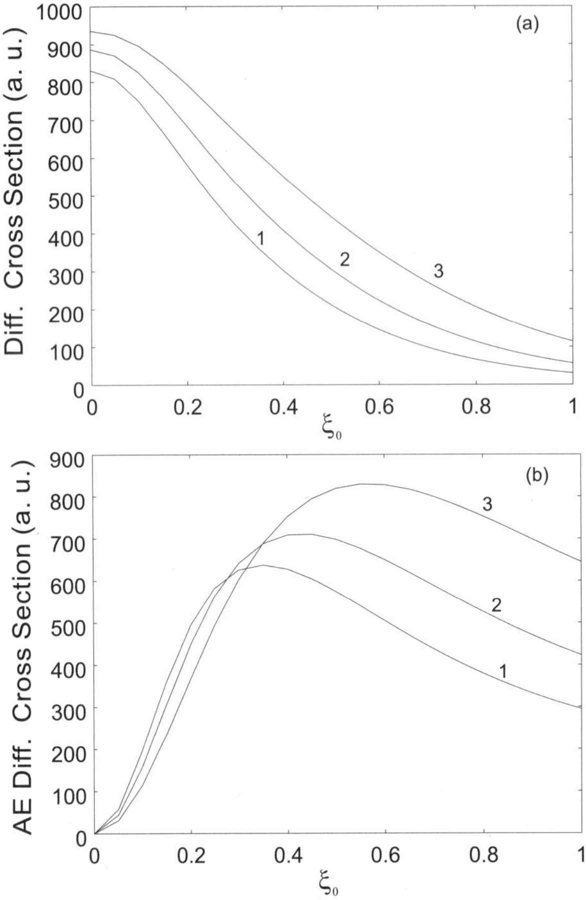

Fig. 1.9. The summed differential cross sections for circular polarization of EM wave are plotted as a function of relativistic parameter of intensity ξ_0 in the range $0 \leq \xi_0 \leq 1$. The initial electron momentum is parallel to the laser propagation direction **k**. (a) SB differential cross section $d\sigma/d\Omega$; (b) absorption/emission differential cross section $d\sigma_{ae}/d\Omega$. Numbers denote different values of deflection angle: 1, $\vartheta = 6$ mrad; 2, $\vartheta = 5$ mrad; 3, $\vartheta = 4$ mrad.

is plotted for various deflection angles as a function of relativistic parameter of intensity ξ_0. The initial electron momentum is parallel to the laser propagation direction **k**. In Fig. 1.9b summed AE differential cross section is shown. We see from Fig. 1.9 that SB as well as AE cross sections decrease with increasing wave intensity. This is a consequence of the SB process being essentially nonlinear in contrast to perturbation theory where s-photon SB cross section $\sim \xi_0^{2s}$.

For the integral quantities such as the total scattering cross section σ and total emission/absorption cross section (σ_T) which characterizes net energy change, one should integrate partial differential cross section of SB process $d\sigma^{(s)}/dO$ over solid angle and perform summation over photon numbers:

$$\sigma = \sum_{s>-s_m}^{\infty} \sigma^{(s)}, \tag{1.151}$$

and total AE cross section (σ_T) will be

$$\sigma_T = \sum_{s>-s_m}^{\infty} s\sigma^{(s)}. \tag{1.152}$$

Note that for these quantities in the optical range of frequencies one can

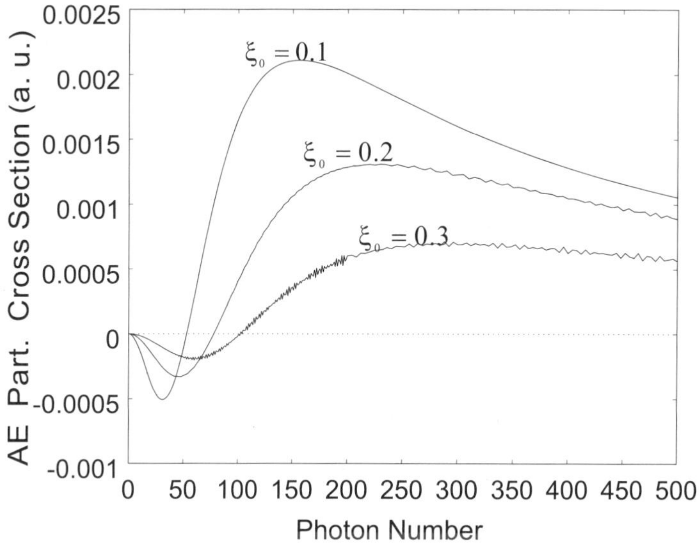

Fig. 1.10. The envelopes of integrated absorption/emission partial cross sections $\sigma_{ae}^{(s)}$ for circular polarization of EM wave as a function of photon number in the range $0 \leq s \leq 500$ for various laser intensities. The initial electron momentum is parallel to the laser propagation direction $\mathbf{k}$. Negative values correspond to net emission, while positive values correspond to net absorption.

neglect the contribution from the spin interaction. The latter is essential for large angle scattering which produces a minor contribution to the total cross sections (for optical frequencies the quantum recoil is negligibly small). For the strong laser fields one should take into account a large number of terms in Eqs. (1.151) and (1.152) since multiphoton absorption/emission processes already play a significant role for moderate laser intensities ($\xi_0 \ll 1$) in contrast, for example, to nonlinear Compton scattering where multiphoton processes become essential for $\xi_0 \sim 1$ and the cutoff number of absorbed photons $\sim \xi_0^3$. This essentially complicates the analysis of total cross sections (1.151) and (1.152). As a first step to exhibit the dependence of SB process on laser intensity, Fig. 1.10 plots the envelopes of integrated AE partial cross sections $\sigma_{ae}^{(s)}$ for various laser intensities as a function of the photon number in the range $0 \leq s \leq 500$. The initial electron momentum is parallel to the laser propagation direction $\mathbf{k}$. Negative values correspond to net emission, while positive values correspond to net absorption. Figure 1.10 reveals that for this initial geometry the absorption process is dominant but with increasing wave intensity the AE cross section decreases.

Bibliography

D.M. Volkov, Z. Phys. **94**, 250 (1935)
A. Vachaspati, Phys. Rev. **128**, 664 (1962)
A. Vachaspati, Phys. Rev. **130**, 2598 (1963)
A.A. Kolomensky, A.N. Lebedev, Zh. Éksp. Teor. Fiz. **44**, 261 (1963)
R.H. Melburn, Phys. Rev. Lett. **10**, 75 (1963)
F.R. Harutyunyan, V.A. Tumanyan, Zh. Éksp. Teor. Fiz. **44**, 2100 (1963)
F.R. Harutyunyan, I.I. Goldman, V.A. Tumanyan, Zh. Éksp. Teor. Fiz. **45**, 312 (1963)
I.I. Goldman, Zh. Éksp. Teor. Fiz. **46**, 1412 (1964)
L.S. Brown, T.W.B. Kibble, Phys. Rev. A **133**, 705 (1964)
T.W.B. Kibble, Phys. Rev. B **138**, 740 (1965)
L.S. Bartell, H.B. Thomson, R.R. Roskos, Phys. Rev. Lett. **14**, 851 (1965)
F.V. Bunkin, M.V. Fedorov, Zh. Éksp. Teor. Fiz. **49**, 4 (1965)
J.J. Sanderson et al., Phys. Lett. **18**, 114 (1965)
G. Toraldo di Francia, Nuovo Cimento **37**, 1553 (1965)
T.W.B. Kibble, Phys. Lett. **20**, 627 (1966)
J.H. Eberly, H.R. Reiss, Phys. Rev. **145**, 1035 (1966)
V.Ya. Davidovski, E.M. Yakushev, Zh. Éksp. Teor. Fiz. **50**, 1101 (1966)
N.D. Sengupta, Phys. Lett. **6**, 642 (1966)
N.J. Philips, J.J. Sanderson, Phys. Lett. **21**, 533 (1966)
J.F. Dawson, Z. Fried, Phys. Rev. Lett. **19**, 467 (1967)
H. Prakash, Phys. Lett. A **24**, 492 (1967)
M.M. Denisov, M.V. Fedorov, Zh. Éksp. Teor. Fiz. **53**, 1340 (1967)
T.W.B. Kibble, Cargése Lect. Phys., 299 (1968)
J.H. Eberly, A. Sleeper, Phys. Rev. **176**, 1570 (1968)
J.H. Eberly, Prog. Opt. **7**, 359 (1969)
Y.W. Chan, Phys. Lett. A **32**, 214 (1970)
A.I. Nikishov, V.I. Ritus, Usp. Fiz. Nauk **100**, 724 (1970)
M.J. Feldman, R.Y. Chiao, Phys. Rev. A **4**, 352 (1971)
H. Brehme, Phys. Rev. C **3**, 837 (1971)

A.I. Nikishov, V.I. Ritus, Ann. Phys. (N.Y.) **69**, 555 (1972)
F.V. Bunkin, A.E. Kazakov, M.V. Fedorov, Usp. Fiz. Nauk **107**, 559 (1972)
R.K. Osborn, Phys. Rev. A **5**, 1660 (1972)
N.M. Kroll, K.M. Watson, Phys. Rev. A **8**, 804 (1973)
J.L. Gersten, M.H. Mittleman, Phys. Rev. A **12**, 1840 (1975)
E.G. Bessonov, V.G. Kurakin, A.V. Serov, Zh. Tekh. Fiz. **46**, 1984 (1976) [in Russian]
A. Weingartshofer et al., Phys. Rev. Lett. **39**, 269 (1977)
V.L. Ritus, Tr. Fiz. Inst. Akad. Nauk SSSR **111**, 141 (1979) [in Russian]
M.H. Mittleman, Phys. Rev. A **19**, 134 (1979)
A. Weingartshofer et al., Phys. Rev. A **19**, 2371 (1979)
G. Ferrante, E. Fiordilino, Il Nuovo Cimento B **56**, 237 (1980)
H.K. Avetissian, H.A. Jivanian, Phys. Lett. A **76**, 5 (1980)
F. De Martini, M. Foresti, J. Appl. Phys. B **28**, 153 (1982)
A. Weingartshofer, J. Phys. B **16**, 1805 (1983)
M. Gavrila, J. Kaminski, Phys. Rev. Lett. **52**, 613 (1984)
H.K. Avetissian et al., Phys. Lett. A **117**, 111 (1986)
H.K. Avetissian et al., J. Phys. B **25**, 3201 (1992)
H.K. Avetissian et al., J. Phys. B **25**, 3217 (1992)
C. Bula et al., Phys. Rev. Lett. **76**, 3116 (1996)
H.K. Avetissian, S.V. Movsissian, Phys. Rev. A **54**, 3036 (1996)
H.K. Avetissian et al., Phys. Rev. A **56**, 4905 (1997)
C. Szymanovsky et al., Phys. Rev. A **56**, 3846 (1997)
C. Szymanovsky, A. Maquet, Optics Express **2**, 262 (1998)
M.V. Fedorov: Atomic and Free Electrons in a Strong Light Field (World Scientific 1998)
H.K. Avetissian et al., Phys. Rev. A **59**, 549 (1999)
T.R. Hovhannisian, A.G. Markossian, G.F. Mkrtchian, Eur. Phys. J. D **20**, 17 (2002)
Y.I. Salamin, C.H. Keitel, Phys. Rev. Lett. **88**, 095005 (2002)
F.V. Hartemann: High-Field Electrodynamics (CRC Press, Boca Raton, FL 2002)
C.A. Brau: Modern Problems in Classical Electrodynamics (Oxford University Press, New York 2004)

2 Interaction of Charged Particles with Strong Electromagnetic Wave in Dielectric Media. Induced Nonlinear Cherenkov Process

What can we expect from particle–strong wave interaction in a medium essentially different from that of a vacuum?

It is well known that in a medium with the refractive index $n(\omega) > 1$ *(dielectric media) the Cherenkov effect takes place — charged particle moving with a velocity* $\mathbf{v} = \mathrm{const}$ *radiates spontaneously transverse EM wave of frequency* ω *at the angle* θ *satisfying the condition of coherency* $\cos\theta = c/\mathrm{v}n(\omega)$. *This means that in the presence of an external plane EM wave of the same frequency* ω *propagating at this angle with respect to the particle motion the spontaneous Cherenkov radiation of the particle will acquire induced character and the inverse process of Cherenkov absorption from the incident wave by the particle is possible as well. This is the general character of arbitrary type spontaneous radiation process in corresponding induced one. However, in contrast to the noncoherent process (e.g., bremsstrahlung), if the spontaneous process is of coherent nature, such as the Cherenkov process, for the satisfaction of the condition of coherency the external wave should be weak enough to not change considerably the particle initial velocity* $\mathbf{v}$ *and violate the mentioned condition of coherency of the spontaneous process. Consequently, this explanation of formation of induced process with the charged particles (induced free–free transitions in quantum terminology) corresponds to the linear theory.*

The behavior of induced Cherenkov process in the strong EM wave field is quite different from the mentioned one. The existence of the threshold value of the particle velocity for the spontaneous Cherenkov radiation $(\mathrm{v} > c/n(\omega))$ *stipulates for the threshold value of the wave intensity essentially changing the character of the dynamics of the particle–wave interaction in a medium and, consequently, the character of electromagnetic processes in dielectriclike media, proceeding in the presence of strong radiation fields. As we will see later, the peculiarities which arise at the nonlinear interaction of charged particles with strong EM waves are the general features of coherent processes like the Cherenkov one.*

To reveal the nonlinear behavior and principal peculiarities of a particle–strong wave interaction in a medium, this chapter will present the nonlinear classical theory of induced Cherenkov process.

2.1 Particle Classical Motion in the Field of Strong Plane EM Wave in a Medium

A plane quasi-monochromatic EM wave in a medium may be described by the vector potential $\mathbf{A}(t,\mathbf{r}) = \mathbf{A}(t - n_0\nu_0\mathbf{r}/c)$, where $n_0 \equiv n(\omega_0)$ is the refractive index of the medium at the carrier frequency of the wave (actually laser radiation). For the electric and magnetic fields we will have respectively

$$\mathbf{E}(t,\mathbf{r}) = \mathbf{E}(t - n_0\nu_0\mathbf{r}/c); \quad \mathbf{H}(t,\mathbf{r}) = \mathbf{H}(t - n_0\nu_0\mathbf{r}/c); \quad \mathbf{H} = n_0\left[\nu_0\mathbf{E}\right]. \tag{2.1}$$

Hereafter we will assume that the frequency ω_0 is far from the main resonance transitions between the atomic levels of the medium to prohibit the wave absorption and nonlinear optical effects in the medium and consequently $n_0 = \sqrt{\varepsilon_0\mu_0} = \text{const}$ will correspond to the linear refractive index of the medium (ε_0 and μ_0 are the dielectric and magnetic permittivities of the medium, respectively).

Without loss of generality we will direct vector ν_0 along the OX axis of a Cartesian coordinate system: $\nu_0 = \{1,0,0\}$ and the relativistic classical equations of motion of a charged particle in the field (2.1) will be written in the form

$$\frac{dp_x}{dt} = n_0\frac{e}{c}\left[\mathrm{v}_y E_y(\tau) + \mathrm{v}_z E_z(\tau)\right], \tag{2.2}$$

$$\frac{dp_y}{dt} = e\left(1 - n_0\frac{\mathrm{v}_x}{c}\right)E_y(\tau); \quad \frac{dp_z}{dt} = e\left(1 - n_0\frac{\mathrm{v}_x}{c}\right)E_z(\tau), \tag{2.3}$$

where $\tau = t - n_0x/c$ is the retarding wave coordinate of the quasi-monochromatic plane EM wave in a medium.

The integration of Eqs. (2.2) and (2.3) is carried out as was done for Eqs. (1.3) and (1.4) and with Eq. (1.9) one can obtain the particle transversal momentum

$$p_y = p_{0y} - \frac{e}{c}A_y(\tau); \quad p_z = p_{0z} - \frac{e}{c}A_z(\tau) \tag{2.4}$$

and integral of motion

$$K \equiv \mathcal{E} - \frac{c}{n_0}p_x = \text{const}, \tag{2.5}$$

which together with the relation $\mathcal{E}^2 = \mathbf{p}^2c^2 + m^2c^4$ determine the energy of the particle in the field of strong quasi-monochromatic plane EM wave in a medium:

$$\mathcal{E} = \frac{\mathcal{E}_0}{n_0^2 - 1}\left\{n_0^2\left(1 - \frac{\mathrm{v}_{0x}}{cn_0}\right) \mp \left[\left(1 - n_0\frac{\mathrm{v}_{0x}}{c}\right)^2\right.\right.$$

$$-\frac{\left(n_0^2-1\right)}{\mathcal{E}_0^2}\left(e^2\mathbf{A}^2(\tau)-2ec\mathbf{p}_0\mathbf{A}(\tau)\right)\Bigg]^{1/2}\Bigg\}. \tag{2.6}$$

Here $\mathbf{p}_0 = \{p_{0x}, p_{0y}, p_{0z}\}$, $\mathcal{E}_0$, and v_{0x} are the particle initial momentum, energy, and longitudinal velocity, respectively, at $\tau = -\infty$ ($\mathbf{A}(\tau)\mid_{\tau=-\infty}=0$ according to unique definition of the vector potential of the wave (1.7)).

Equation (2.6) describes the energy exchange between the charged particle and plane transverse EM wave of arbitrary intensity in a medium in the general case. However, besides the formula of the energy for the description of the particle nonlinear dynamics in this process we will need the formula for the longitudinal velocity of the particle in the field — a major characteristic of the induced Cherenkov process. The latter can be defined from the relation $\mathrm{v}_x = c^2 p_x/\mathcal{E}$ within the expression for the longitudinal momentum of the particle p_x, which is determined by the integral of motion (2.5) and Eq. (2.6). Then for the longitudinal velocity of the particle we will have

$$\mathrm{v}_x = cn_0\frac{1-\mathrm{v}_{0x}/cn_0 \mp \sqrt{D}}{n_0^2\left(1-\mathrm{v}_{0x}/cn_0\right)\mp\sqrt{D}}, \tag{2.7}$$

where

$$D \equiv \left(1-n_0\mathrm{v}_{0x}/c\right)^2-\left(\left(n_0^2-1\right)/\mathcal{E}_0^2\right)\left(e^2\mathbf{A}^2(\tau)-2ec\mathbf{p}_0\mathbf{A}(\tau)\right). \tag{2.8}$$

Further, for the consideration of radiation processes we will need the formulas for transversal velocities of the particle, which can be defined from Eqs. (2.4) and (2.6):

$$\mathrm{v}_{y,z} = \frac{c}{\mathcal{E}_0}\frac{\left(n_0^2-1\right)\left(cp_{0y,z}-eA_{y,z}(\tau)\right)}{n_0^2\left(1-\mathrm{v}_{0x}/cn_0\right)\mp\sqrt{D}}. \tag{2.9}$$

As is seen from Eqs. (2.6)–(2.9) the expressions determining the particle energy or velocity in the wave field are, first, not single-valued and, second, may become imaginary depending on particle and wave parameters. The peculiarity arising in the induced Cherenkov process because of particle–strong wave nonlinear interaction is connected with this fact. Hence, treatment of the particle dynamics in this process should start by clarification of these questions.

2.2 Nonlinear Cherenkov Resonance and Critical Field. Threshold Phenomenon of Particle "Reflection"

To consider the behavior of a particle upon nonlinear interaction with a strong wave in a medium on the basis of Eq. (2.6) we will analyze the case where the initial velocity of the particle is directed along the wave propagation direction for which the picture of the particle nonlinear dynamics is physically more evident. In this case Eq. (2.6) becomes

$$\mathcal{E} = \frac{\mathcal{E}_0}{n_0^2 - 1}\left[n_0^2\left(1 - \frac{\mathrm{v}_0}{cn_0}\right)\right.$$

$$\left.\mp\sqrt{\left(1 - n_0\frac{\mathrm{v}_0}{c}\right)^2 - \left(n_0^2 - 1\right)\left(\frac{mc^2}{\mathcal{E}_0}\right)^2 \xi^2(\tau)}\right], \tag{2.10}$$

where $\xi^2(\tau)$ is the relativistic invariant parameter of a plane EM wave intensity, determined by Eq. (1.19).

As is seen, Eq. (2.10) is twovalence and, at first, we shall provide the unique definition of the particle energy in accordance with the initial condition. In the case of plasma ($n_0 < 1$) or vacuum ($n_0 = 1$) the term under the root is always positive, hence, in these cases one has to take before the root only the upper sign ($-$) to satisfy the initial condition $\mathcal{E}(\tau) = \mathcal{E}_0$ when $\xi(\tau) = 0$. In the case of a vacuum, Eq. (2.10) yields results obtained in Chapter 1 (see Eq. (1.13) or Eqs. (1.24) and (1.36) for the circular and linear polarizations of the wave).

Further investigation is devoted to the case of a medium with refractive index $n_0 > 1$. In this case the nature of the particle motion essentially depends on the initial conditions and the value of the parameter $\xi(\tau)$ as far as the expression under the root in Eq. (2.10) may become negative, while the energy of the particle should be a real quantity and uniquely defined as well. To solve this problem one needs to pass the complex plane, according to which we represent Eq. (2.10) in the form of known inverse Jukowski function (to determine also the sign before the root corresponding to initial condition $\mathcal{E}(\tau)|_{\tau=-\infty} = \mathcal{E}_0$ since at $n_0 > 1$ the quantity $1 - n_0\mathrm{v}_0/c$ under the root may be negative as well):

$$\mathcal{E} = \frac{\mathcal{E}_0}{n_0^2 - 1}\left[n_0^2\left(1 - \frac{\mathrm{v}_0}{cn_0}\right) \mp \left(1 - n_0\frac{\mathrm{v}_0}{c}\right)\sqrt{1 - \frac{\xi^2(\tau)}{\xi_{cr}^2}}\right], \tag{2.11}$$

where

$$\xi_{cr} \equiv \frac{\mathcal{E}_0}{mc^2}\frac{\left|1 - n_0\frac{\mathrm{v}_0}{c}\right|}{\sqrt{n_0^2 - 1}}. \tag{2.12}$$

If $\xi_{max} < \xi_{cr}$ (ξ_{max} is the maximum value of the parameter $\xi(\tau)$) the expression under the root in Eq. (2.11) is always positive and in front of the root one has to take the upper sign ($-$) according to the initial condition.

Then $\mathcal{E} = \mathcal{E}_0$ after the interaction ($\xi(\tau) \to 0$) and the particle energy remains unchanged.

If $\xi_{max} > \xi_{cr}$ the particle is unable to penetrate into the wave, i.e., into the region $\xi > \xi_{cr}$ since at $\xi > \xi_{cr}$ the root in Eq. (2.11) becomes a complex one. This complexity now is bypassed via continuously passing from one Riemann sheet to another, which corresponds to changing the inverse Jukowski function from "$-$" to "$+$" before the root. Hence, the upper sign ($-$) in this case stands up to the value of the wave intensity $\xi(\tau) < \xi_{cr}$, then at $\xi(\tau) = \xi_{cr}$ the root changes its sign from "$-$" to "$+$", providing continuous value for the particle energy in the field. The intensity value $\xi(\tau) = \xi_{cr}$ of the wave is a turn point for the particle motion, so that we call it the critical value.

Thus, when the maximum value of the wave intensity exceeds the critical value a transverse plane EM wave in the medium becomes a potential barrier and the "reflection" of the particle from the wave envelope ($\xi(\tau)$) takes place. If now $\xi(\tau) \to 0$, we obtain after the "reflection" for the particle energy

$$\mathcal{E} = \mathcal{E}_0 \left[1 + 2\frac{1 - n_0 \frac{\mathrm{v}_0}{c}}{n_0^2 - 1} \right]. \tag{2.13}$$

If the initial conditions are such that the wave pulse overtakes the particle ($\mathrm{v}_0 < c/n_0$), then after the "reflection" $\mathcal{E} > \mathcal{E}_0$ and the particle is accelerated. But if the particle overtakes the wave ($\mathrm{v}_0 > c/n_0$), then $\mathcal{E} < \mathcal{E}_0$ and particle deceleration takes place.

This nonlinear threshold phenomenon is bounded on the stimulated Cherenkov process. The coherent nature of the Cherenkov process is related to the existence of the critical intensity of the wave ξ_{cr}. Indeed, from Eq. (2.7) it follows that when $\xi = \xi_{cr}$ the longitudinal velocity of the particle in the field becomes equal to the phase velocity of the wave: $\mathrm{v}_x(\xi) \mid_{\xi=\xi_{cr}} = c/n_0$ irrespective of its initial velocity v_0. The latter is the Cherenkov condition of coherency in a dielectric medium. Fulfillment of the Cherenkov condition in the strong wave field leads to the nonlinear Cherenkov resonance, at which the induced absorption or emission of Cherenkov photons becomes essentially multiphoton. As a result, the particle velocity becomes greater or smaller (depending on initial velocity v_0) than the wave phase velocity and it leaves the wave, i.e., the "reflection" from the wave front occurs. In addition, the energy lost by the particle at the deceleration ($\mathrm{v}_0 > c/n_0$) is coherently transferred to the wave via induced Cherenkov radiation. As is seen from Eq. (2.13), for the initial "Cherenkov velocity" $\mathrm{v}_0 = c/n_0$ the energy of the particle after the "reflection" does not change: $\mathcal{E} = \mathcal{E}_0$, which is in congruence with the critical value of the field: $\xi_{cr} = 0$ at the initial Cherenkov velocity of the particle (see Eq. (2.7)). The latter confirms the nonlinear character of Cherenkov resonance in the strong wave field. In this case the induced Cherenkov effect will occur at $\mathrm{v}_x = \mathrm{v}_0 = \mathrm{const}$, i.e., the wave field should not change the particle initial velocity, which can take place approximately, only in the weak fields — induced Cherenkov effect in the linear theory (in accordance with the initial condition $\xi(\tau)\mid_{\tau=-\infty}= 0$ — the wave is turned on adiabatically — it is evident that in this case the linear induced Cherenkov effect is absent as well).

This threshold phenomenon of the particle "reflection" can be more clearly presented in the frame of reference connected with the wave. In this frame the electric field of the wave vanishes ($\mathbf{E}' \equiv 0$) and there is only the static magnetic field ($|\mathbf{H}'| = |\mathbf{H}| \sqrt{n_0^2 - 1}/n_0$). For not very large particle velocities in this frame the magnetic field will turn the particle back — elastic reflection from the standing wave barrier. In the opposite case the particle slips through the magnetic field. Such behavior of the particle in the intrinsic frame of the wave corresponds to the cases $\xi > \xi_{cr}$ (large velocities close to the Cherenkov one at which ξ_{cr} is small and the condition $\xi > \xi_{cr}$ is achievable) and $\xi < \xi_{cr}$ in the laboratory frame of reference, respectively (see Eq. (2.7)). Note that because of the particle reflection from the standing barrier in the frame of reference of the slowed wave we term the revealed nonlinear phenomenon a "reflection" one.

Hence, the threshold-coherent nature of spontaneous Cherenkov effect over the particle velocity ($\mathrm{v}_{th} = c/n_0$) causes the threshold for the external wave intensity ($\xi_{th} \equiv \xi_{cr}$), which in turn causes the phenomenon of particle "reflection" from the plane EM wave. It is worth emphasizing that the latter may be very small ($\xi_{cr} \to 0$) if the particle initial velocity is close to the wave phase velocity ($\mathrm{v}_0 \to c/n_0$), which means that in this case the linear theory is not applicable even for very weak wave fields ($\xi \to 0$), since the nonlinear phenomenon of particle "reflection" will take place ($\xi > \xi_{cr} \to 0$). Also, it is important that due to this phenomenon the induced process at $\xi > \xi_{cr}$ proceeds strictly in a certain direction — either radiation or absorption (inverse induced process), which has a principal meaning for induced free–free transitions related especially to problems of laser acceleration and free electron lasers.

Let us estimate the particle energy change due to "reflection". Note, at first, that the latter does not depend on interaction length or magnitude of the field (it is necessary only that $\xi > \xi_{cr}$). It is a nonlinear acceleration/deceleration of the shock character, which proceeds in short enough time — smaller than the wave pulse duration. As is seen from Eq. (2.13), for a certain value of the refractive index of the medium the stronger the initial velocity of the particle differs from the Cherenkov one and the closer to 1 ($n_0 - 1 << 1$), the larger is the energy change. As follows from Eq. (2.12) in these cases the strong wave fields are necessary. However, as the medium is to be dielectriclike ($n_0 > 1$) the wave intensity is confined to the threshold ionization of the medium. As is known in nonionized media a wave of intensity $\xi^2 < I/mc^2$, where I is the first ionization energy of the medium atoms (for dielectrics, the width of the forbidden zone), can propagate. In the opposite case a tunnel ionization of the atoms can take place. Consequently, the region of intensities where the "reflection" phenomenon in dielectriclike media can be applied is $\xi^2 < \xi_{max}^2 < I/mc^2$. For typical values $I \sim 10$ eV we have $\xi_{max} \sim 5 \times 10^{-3}$. To such values of the wave critical intensity correspond particle velocities near the Cherenkov one, which is possible in the case of relativistic particles in the gases ($n_0 - 1 << 1$), whereas for nonrelativistic ones, in solids ($n_0 - 1 \sim 1$). However, in the last case the negative effects of multiple scattering and ionization loss of the particle in solids also can influence. Thus, this phenomenon can be realized in the gases for relatively low densities. The optimal values of the refractive index of the gaseous media for this phenomenon are $n_0 - 1 \sim 10^{-3} \div 10^{-5}$ (e.g., for CO_2 and He at standard pressure and temperature $n_0 - 1 \sim 4.48 \times 10^{-4}$ and $\sim 3.47 \times 10^{-5}$, respectively).

As the application of large intensities is restricted with ionization threshold of the medium, we express the particle energy change due to "reflection" through the wave critical intensity. If $n_0 - 1 \equiv \mu_1 << 1$ and $1 - \mathrm{v}_0/c \equiv \mu_2 << 1$ from Eqs. (2.12) and (2.13) we have

$$\xi_{cr} \simeq \frac{|\mu_1 - \mu_2|}{2\sqrt{\mu_1 \mu_2}} \; ; \qquad |\Delta\mathcal{E}| \simeq \xi_{cr} mc^2 \sqrt{\frac{2}{\mu_1}} \, . \tag{2.14}$$

Estimations show that an electron with initial energy $\mathcal{E}_0 \sim 10$ MeV after the "reflection" from a laser pulse with $\xi \sim 5 \times 10^{-4}$ (which corresponds to the neodymium laser radiation strength $E \sim 10^7$ V/cm) in a medium with $n_0 - 1 \sim 10^{-3}$ acquires ($\mathrm{v}_0 < c/n_0$) or loses ($\mathrm{v}_0 > c/n_0$) energy $|\Delta\mathcal{E}| \sim 10$ keV. As the particle deceleration occurs because of stimulated Cherenkov radiation in this case the wave amplification takes place. Hence, as a result of the "reflection" of a beam with electron total number $\sim 5 \times 10^{14}$ an energy of ~ 1 J coherently will be radiated into the wave.

The phenomenon of charged particle "reflection" from a plane EM wave may also be used for the monochromatization of particle beams. The fact that above the critical intensity value the induced Cherenkov process occurs in only one direction — either emission or absorption — and for the initial Cherenkov velocity $\mathrm{v}_{0x} = c/n_0$ the energy of the particle after the "reflection" does not change, in principle enables conversion of the energetic or angular spreads of charged particle beams due to "reflection." The latter requires considering the general case of interaction at the arbitrary direction of particle initial motion with respect to wave propagation. So, without repeating the analysis, which has been made in the case of particle–wave parallel propagation we will present the ultimate results of the "reflection" phenomenon in the general case.

Thus, when the particle initial velocity is directed at an angle (ϑ) to the wave propagation direction the energy of the particle is given by Eq. (2.6), which at the linear polarization of the wave reads

$$\mathcal{E}(\tau) = \frac{\mathcal{E}_0}{n_0^2 - 1}\left\{ n_0^2\left(1 - \frac{\mathrm{v}_0}{cn_0}\cos\vartheta\right) \mp \left[\left(1 - n_0\frac{\mathrm{v}_0}{c}\cos\vartheta\right)^2 - \left(n_0^2 - 1\right)\right.\right.$$

$$\left.\left.\times\left(\frac{mc^2}{\mathcal{E}_0}\right)^2\left[\xi^2(\tau)\cos^2\omega_0\tau - 2\frac{p_0\sin\vartheta}{mc}\xi(\tau)\cos\omega_0\tau\right]\right]^{1/2}\right\} \tag{2.15}$$

(the wave is linearly polarized along the axis OY with vector potential $A_y = A(\tau)\cos\omega_0\tau$ and one can assume $\mathbf{p}_0 = \{p_0\cos\vartheta; p_0\sin\vartheta; 0\}$, as far as the coordinate z is free). As is seen from Eq. (2.15), in this case the "reflection" occurs from certain planes of equal phases but from the front of the wave intensity envelope as in the case $\vartheta = 0$. At the actual values of the parameters for induced Cherenkov process (ultrarelativistic particles in gaseous media with refractive index $n_0 - 1 << 1$ and not very small angles ϑ, as well as the wave intensity being confined to ionization threshold of the medium) the second term under the root is much smaller than the third one, that is, $2p_0|\sin\vartheta|/mc >> \xi_{max}$ and for the critical field in this case we have

$$\xi_{cr}(\vartheta) = \frac{c}{2\mathrm{v}_0}\frac{\mathcal{E}_0}{mc^2}\frac{\left(1 - n_0\frac{\mathrm{v}_0}{c}\cos\vartheta\right)^2}{\left(n_0^2 - 1\right)|\cos\vartheta|}; \qquad \vartheta \neq 0 \tag{2.16}$$

(in the case $\vartheta = 0$, ξ_{cr} is determined by Eq. (2.12)).

If the maximal value of the wave intensity $\xi_{max} > \xi_{cr}(\vartheta)$, then the particle energy after the "reflection" is

$$\mathcal{E}(\vartheta) = \mathcal{E}_0\left[1 + \frac{2\left(1 - n_0\frac{\mathrm{v}_0}{c}\cos\vartheta\right)}{n_0^2 - 1}\right]. \tag{2.17}$$

Let the charged particle beam with an initial energetic (Δ_0) and angular (δ_0) spread interact with a plane transverse EM wave of intensity $\xi_{max} > \xi_{cr}(\vartheta)$ in a gaseous medium. To keep the mean energy $\overline{\mathcal{E}_0}$ of the beam unchanged after the interaction (at the adiabatic turning on and turning off of the wave) the axis of the beam with mean velocity $\overline{\mathrm{v}_0}$ must be pointed at the Cherenkov angle (ϑ_0) to the laser beam, i.e., $n_0(\overline{\mathrm{v}_0}/c)\cos\vartheta_0 = 1$. Under this condition the particles with velocities $\mathrm{v}_0\cos\vartheta < c/n_0$ will acquire an energy and the other particles for which the longitudinal velocities exceed the phase velocity of the wave ($\mathrm{v}_0\cos\vartheta > c/n_0$) will loss an energy according to Eq. (2.17). As a result the energies of the particles $\mathcal{E}(\vartheta)$ will approach close to the mean energy $\overline{\mathcal{E}_0}$ of the beam ($\mathcal{E}(\vartheta) \to \overline{\mathcal{E}_0}$) and the final energetic width of the beam will become less than the initial one. As there is one free parameter (for a specified velocity $\overline{\mathrm{v}_0}$ the parameters ϑ_0 and n_0 are related by Cherenkov condition) it is possible to use it to control the exchange in the energy of the particles after the "reflection" (2.17) and to reach the minimal final energy spread of the beam $\Delta << \Delta_0$ — monochromatization. Depending on the relation between the initial energetic and angular spreads and mean energy of the beam, the opposite process may occur, namely angular narrowing of the beam. Physically it is clear that with the monochromatization the angular divergence of the beam will increase and the opposite — the angular narrowing of the beam — leads to demonochromatization (in accordance with Liouville's theorem). More detailed consideration of this effect with the quantitative results can be found in the bibliography of this chapter.

To illustrate the typical picture of nonlinear interaction of a charged particle with a strong EM wave in a medium we present the graphics of numerical solutions of the Eqs. (2.2) and (2.3) for the laser pulse of finite duration, showing the behavior of particle dynamics below and above critical intensity, with the effect of acceleration. At first we will not take into account the dependence of the slowly varying intensity envelope of a laser beam from the transversal coordinates. Thus, a laser beam may be modeled as

$$E_x = 0, \quad E_z = 0, \quad E_y = \frac{E_0}{\cosh\left(\frac{\tau}{\delta\tau}\right)} \cos\omega_0\tau, \tag{2.18}$$

where $\delta\tau$ characterizes the pulse duration. The particle initial energy is taken to be $\mathcal{E}_0 = 40$ MeV and the initial velocity is directed at the angle $\vartheta = 9\times10^{-3}$ rad to the wave propagation direction ($p_{0z} = 0$). The refractive index of the gaseous medium for this calculation has been chosen to be $n_0 - 1 = 10^{-4}$. Figure 2.1 illustrates the evolution of the particle energy: the energy versus the position x is plotted for a neodymium laser ($\hbar\omega_0 \simeq 1.17$ eV) with electric field strength $E_0 = 3\times10^8$ V/cm and $\delta\tau = 4T$ (T is the wave period). For these parameter values the wave intensity is above the critical point and, as we see from this figure, the particle energy is abruptly changed

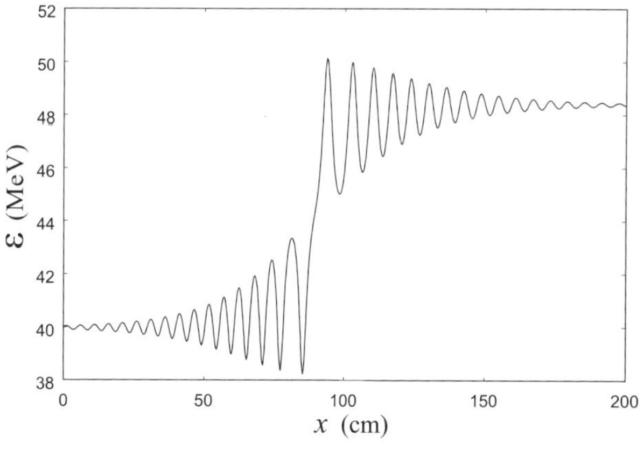

Fig. 2.1. "Reflection" of the particle. The energy versus the position x is plotted when the wave intensity is above the critical point.

corresponding to the "reflection" phenomenon. Figure 2.2a illustrates the evolution of the energies of particles with different initial interaction angles. The initial energies for all particles are $\mathcal{E}_0 \simeq 40$ MeV. Figure 2.2b illustrates the role of initial conditions: the final energy versus the interaction angle is plotted. As follows from Eq. (2.16) the critical intensity, as well as the final energy (2.17), depend on the initial interaction angle and as a consequence we have this picture. Note that the acceleration rate neither depends on the field magnitude (only should be above threshold field) nor on the interaction length.

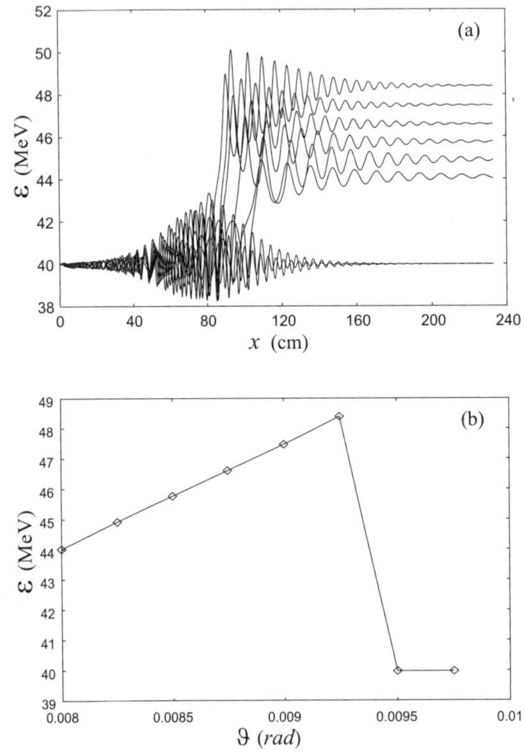

Fig. 2.2. "Reflection" of the particles with different initial interaction angles. Panel (a) displays the evolution of the energies of particles. In (b) the final energy versus the interaction angle is plotted.

To demonstrate the dependence of the considered process on transversal profile of the laser intensity for actual beams in Fig. 2.3 the evolution of the energies of particles with various initial phases (with initial energies $\mathcal{E}_0 \simeq 40$ MeV) is illustrated. The laser beam transversal profile is modeled by the Gaussian function

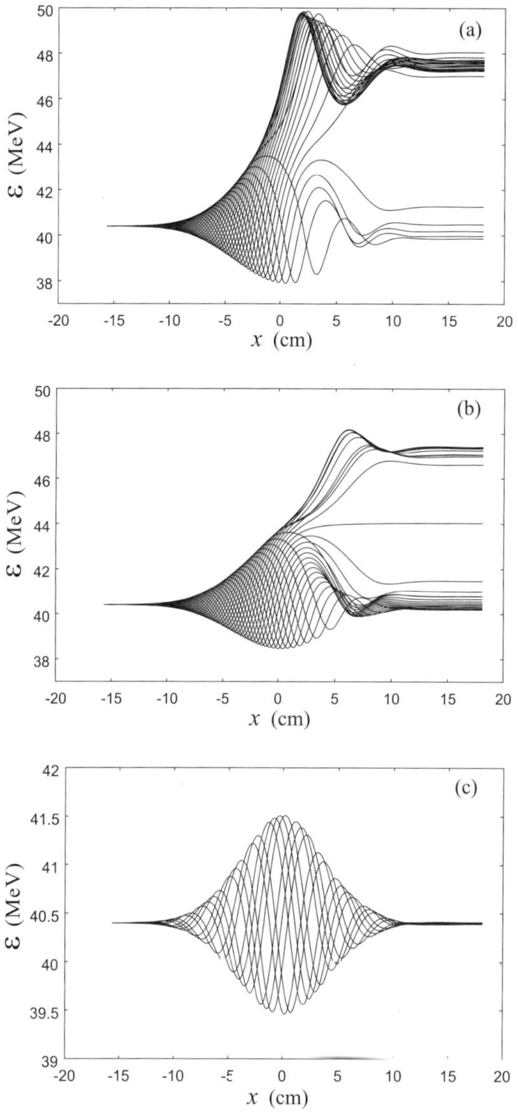

Fig. 2.3. The evolution of the energies of particles with various initial phases are shown for the laser beam with transversal intensity profile for the various entrance cordinates: (a) $z = 0$, (b) $z = d/4$, and (c) $z = d/2$.

$$E_y = E_0 \exp\left(-\frac{4}{d^2}\left(y^2 + z^2\right)\right)\frac{\cos\omega_0\tau}{\cosh\left(\frac{\tau}{\delta\tau}\right)} \tag{2.19}$$

with $d = 10^3\lambda$, $\delta\tau = 50T$. As we see from this figure the acceleration picture is essentially changed depending on the entrance coordinates of the particles. This is the manifestation of the threshold nature of the "reflection" phenomenon.

2.3 Particle Capture by a Plane Electromagnetic Wave in a Medium

If for the intensity exceeding the critical value a plane EM wave becomes a potential barrier for the external particle (with respect to the wave), then for the particle initially situated in the wave it may become a potential well and particle capture by the wave will take place. As the particle state in the wave depends on wave phases we will assume in this case a certain polarization of a monochromatic wave. Let it be linearly polarized with electric field strength along the axis OY :

$$E_y = E_0 \cos\phi; \quad \phi = \omega_0 \left(n_0 \frac{x}{c} - t \right) . \tag{2.20}$$

The solution of equations of motion (2.2) and (2.3) in the field (2.20) may be presented in the form

$$p_x(\phi) = \frac{n_0}{n_0^2 - 1} \frac{\mathcal{E}_0}{c} \left\{ \left(1 - \frac{\mathrm{v}_{0x}}{c n_0} \right) \mp \left[\left(1 - n_0 \frac{\mathrm{v}_{0x}}{c} \right)^2 - \left(n_0^2 - 1 \right) \left(\frac{mc^2}{\mathcal{E}_0} \right)^2 \right. \right.$$

$$\left. \left. \times \xi_0^2 \left(\sin\phi - \sin\phi_0 \right) \left(\sin\phi - \sin\phi_0 - 2 \frac{p_{0y}}{mc\xi_0} \right) \right]^{1/2} \right\} , \tag{2.21}$$

$$p_y(\phi) = p_{0y} - mc\xi_0 (\sin\phi - \sin\phi_0) \ ,$$

$$\mathcal{E}(\phi) = \frac{c}{n_0} p_x(\phi) + \mathcal{E}_0 \left(1 - \frac{\mathrm{v}_{0x}}{c n_0} \right) , \tag{2.22}$$

where $\xi_0 = eE_0/mc\omega_0$ is the intensity parameter of the monochromatic wave (see Eq. (1.25)), $\phi_0 = \omega(n_0 x_0/c - t_0)$ is the initial phase of the particle in the wave. Here without loss of generality it is assumed that the z component of the particle initial momentum $p_{0z} = 0$ as far as the coordinate z is free.

It is seen from Eq. (2.21) that the particle can be in the field region where

$$W(\phi) \equiv (\sin\phi - \sin\phi_0) \left(\sin\phi - \sin\phi_0 - 2 \frac{p_{0y}}{mc\xi_0} \right)$$

$$\leq \left(\frac{\mathcal{E}_0}{mc^2}\right)^2 \frac{\left(1 - n_0 \mathrm{v}_{0x}/c\right)^2}{\left(n_0^2 - 1\right)\xi_0^2} \, . \tag{2.23}$$

If the maximum value of the function $W(\phi)$

$$W_{\max}(\phi) > \left(\frac{\mathcal{E}_0}{mc^2}\right)^2 \frac{\left(1 - n_0 \mathrm{v}_{0x}/c\right)^2}{\left(n_0^2 - 1\right)\xi_0^2} \, , \tag{2.24}$$

then the region (2.23) will be a potential well for the particle and the capture of the latter by the transverse EM wave will take place. The equilibrium phases of the wave (ϕ_s) correspond to the extrema of the function $W(\phi)$:

$$\sin\phi_s = \sin\phi_0 + \frac{p_{0y}}{mc\xi_0} \, ; \qquad \cos\phi_s \neq 0 \, , \tag{2.25}$$

$$\cos\phi_s = 0 \, ; \qquad \sin\phi_s \neq \sin\phi_0 + \frac{p_{0y}}{mc\xi_0} \, . \tag{2.26}$$

The particle moves with a Cherenkov velocity $\mathrm{v}_{xs} = c/n_0$ when it is in the equilibrium phases ϕ_s. Equation (2.22) together with Eqs. (2.25) and (2.26) determine the equilibrated values of the particle transverse momentum p_{ys}. In particular, $p_{ys} = 0$ corresponds to the case (2.25). The motion of the particle in these phases will be stable when

$$|\sin\phi_0 + \frac{p_{0y}}{mc\xi_0}| < 1. \tag{2.27}$$

If the initial velocity of the particle is equal to the Cherenkov one ($\mathrm{v}_{0x} = c/n_0 = \mathrm{v}_{xs}$), then from Eq. (2.24) we have the following condition for the particle capture by the wave:

$$\frac{p_{0y}}{mc\xi_0} < 1 + |\sin\phi_0 + \frac{p_{0y}}{mc\xi_0}|. \tag{2.28}$$

At the fulfillment of Eq. (2.27) the condition of particle capture (2.28) always holds, and therefore the condition of stable motion (2.27) thus determines the capture of the particle in the considered regime. In particular, as is seen from Eqs. (2.25) and (2.27), when $p_{y0} = 0$, then $\phi_s = \phi_0$ and any phase is equilibrated. In this case the phase $\cos\phi_0 = 0$ ($Ey = 0$) is unstable. This is physically clear in the wave frame where the magnetic field of the wave corresponding to this phase is zero: $\mathbf{H}' = 0$, while the stability in the capture regime is due to particle rotation around the vector of the magnetic field (when $p_{ys} = 0$). If the particle initial velocity differs from the Cherenkov value $\mathrm{v}_{0x} = \mathrm{v}_0 = c/n_0 + \Delta\mathrm{v}$, then in the capture regime the particle will undergo stable oscillations close to the equilibrated Cherenkov value. From

Eq. (2.24) one can obtain the following condition for the capture of such particle:

$$|\Delta \mathrm{v}| < \frac{c}{n_0}\frac{mc^2}{\mathcal{E}_0}\xi_0\sqrt{(n^2-1)}\,(1+|\sin\phi_0|) \ . \tag{2.29}$$

The spread tolerances of the unequilibrated particle's initial phase and velocity can be defined from the condition (2.29) ($\Delta \mathrm{v} = (c/n_0\omega_0)|d\phi/dt|$).

Note that the needed value of the field for the particle capture by the wave defined from Eq. (2.29) is the critical value of the field (2.12) for the "reflection" of the external particle ($\phi_0 = 0$).

Consider now the particle capture in equilibrium phases (2.26). With the help of Eqs. (2.22) and (2.23) one can show that the particle motion at the phases $\cos\phi_0 = 0$ will be stable when

$$p_{ys}\sin\phi_s > 0\ ; \quad \phi_s = (2k+1)\pi/2\ ; \quad k = 0; \pm 1; \pm 2; \ldots . \tag{2.30}$$

For the capture of initial Cherenkov particle ($\mathrm{v}_{0x} = c/n_0$) at the phases $\phi_s = (2k+1)\pi/2$ from Eq. (2.24) one can obtain the following condition:

$$W_{\max}(\phi) = 4|\sin\phi_0 + \frac{p_{0y}}{mc\xi_0}| > 0,$$

which always holds. Therefore, the particle capture in this case is determined by condition (2.30). If $p_{y0}\sin\phi_0 > 0$, the phase ϕ_0 is an equilibrated one for any value of the particle transverse momentum ($p_{0y} = p_{ys}$). But if $\mathrm{v}_{0x} = c/n_0 + \Delta\mathrm{v}_x$ the condition for capture is

$$|\Delta\mathrm{v}_x| < \frac{2c}{n_0}\sqrt{n^2-1}\frac{mc^2}{\mathcal{E}_0}\xi_0|\sin\phi_0 + \frac{p_{0y}}{mc\xi_0}|^{1/2}. \tag{2.31}$$

From Eq. (2.31) the critical value of the field can be defined for unequilibrated particle "capture" at the wave phases $\phi_0 = (2k+1)\pi/2$.

If $\cos\phi_0 \neq 0$ from Eq. (2.24) one can obtain that when $p_{0y}/mc\xi_0 > 2$ the Cherenkov particle capture is defined again by condition (2.30).

2.4 Laser Acceleration in Gaseous Media. Cherenkov Accelerator

The phenomenon of charged particle "reflection" and capture by a transverse EM wave can be used for particle acceleration in laser fields. As the application of large intensities in this process is restricted because of the medium ionization the acceleration owing to "reflection" in the medium with refractive index $n_0 = \text{const}$ — single "reflection" — is relatively small. However, if the refractive index decreases along the wave propagation direction in such a way that the condition of particle synchronous motion with the wave $\mathrm{v}_x(x) = c/n_0(x)$ takes place continuously, the phase velocity of the wave will increase all the time and the particle being in front of the wave barrier (at $\xi > \xi_{cr}$) will continuously be "reflected", i.e., continuously accelerated. The law $n_0 = n_0(x)$ must have an adiabatic character not to allow the particle to leave the wave after the single "reflection". Such variation law of the refractive index can be realized in a gaseous medium adiabatically decreasing the pressure.

For particle acceleration one can also use the capture regime. In this case in the medium with $n_0 = \text{const}$ the particle energy does not change on average (particle makes stable oscillations around the equilibrium phases in the wave moving with average velocity $< \mathrm{v}_x >= c/n_0$). However, if one decreases the refractive index along the propagation direction of the wave, so that the particle does not leave the equilibrium phases, then the wave will continuously accelerate the particle. Then, to realize the capture regime (2.25) one needs $p_{0y}/mc\xi_0 < 2$. For not very strong fields this is sufficiently strict confinement on the transverse momentum of the particle. On the other hand, to accelerate the particle significantly large transverse momenta are needed. Therefore, this regime can be used to pass the particles through the matter and, also, to separate the particles by velocities (parameter ξ defines the region of particle velocities captured by the wave (see Eq. (2.29)).

For particle acceleration by laser fields one can use the capture regime (2.26) corresponding to large transverse momenta of the particle $p_{0y}/mc\xi_0 > 2$. So, we will consider the general case of particle capture with arbitrary initial momentum $\mathbf{p}_0$ and laser acceleration in gaseous medium with varying refractive index $n_0(x)$.

We will use the particle equations of motion (2.2) and (2.3) in the field (2.20) where the refractive index $n_0 \to n_0(x)$ and consequently the wave phase is determined as follows:

$$\phi(x,t) = \frac{\omega_0}{c}\int n_0(x)dx - \omega_0 t. \tag{2.32}$$

Then from the equations

$$\frac{d\phi_s}{dt} = 0\,, \qquad \frac{d^2\phi_s}{dt^2} = 0 \tag{2.33}$$

defining wave equilibrium phases we obtain the variation laws for equilibrium velocity of the particle and refractive index of the medium, respectively:

$$\mathrm{v}_{xs}(x) = \frac{c}{n_0(x)}, \tag{2.34}$$

$$\frac{dn_0(x)}{dx} = -\frac{n_0^3(x)}{c^2}\left(\frac{d\mathrm{v}_x}{dt}\right)_s . \tag{2.35}$$

From Eq. (2.2) and the equation for the particle energy variation

$$\frac{d\mathcal{E}}{dt} = e\mathrm{v}_y E_0 \cos\phi(x,t) \tag{2.36}$$

one can obtain the acceleration of the particle in the longitudinal direction

$$\frac{d\mathrm{v}_x}{dt} = \frac{ecn_0(x)}{\mathcal{E}}\left[1 - \frac{\mathrm{v}_x}{cn_0(x)}\right]\mathrm{v}_y E_0 \cos\phi(x,t). \tag{2.37}$$

The equation of motion (2.3) determines in general for an arbitrary $n_0(x)$ the integral of motion (2.5), from which for the equilibrium transverse momentum of the particle we have (again without loss of generality it is assumed that the z component of the particle initial momentum $p_{0z} = 0$ since the coordinate z is free)

$$p_{ys} = p_{0y} - mc\xi_0 (\sin\phi_s - \sin\phi_0) . \tag{2.38}$$

Defining within Eq. (2.38) the equilibrium transverse velocity of the particle $\mathrm{v}_{ys}(x) = c^2 p_{ys}/\mathcal{E}_s(x)$ and substituting together with Eq. (2.34) into Eq. (2.37) for the equilibrium value of the particle longitudinal acceleration we obtain

$$\left(\frac{d\mathrm{v}_x}{dt}\right)_s = c\omega_0\xi_0 \frac{p_{ys}}{mc}\cos\phi_s \left(\frac{mc^2}{\mathcal{E}_s(x)}\right)^2 \frac{n_0^2(x)-1}{n_0(x)} . \tag{2.39}$$

Substituting Eq. (2.39) into Eq. (2.35) we will have the equation which determines the variation law of the medium refractive index:

$$\frac{dn_0(x)}{dx} = -\frac{\omega_0}{c}\xi_0 \frac{p_{ys}}{mc}\cos\phi_s \left(\frac{mc^2}{\mathcal{E}_s(x)}\right)^2 n_0^2(x)\left[n_0^2(x)-1\right] . \tag{2.40}$$

It is seen from this equation that for the particle acceleration in the capture regime via decreasing refractive index of the medium ($dn_0(x)/dx < 0$) one needs $p_{ys}\cos\phi_s > 0$ (equilibrium transverse momentum of the particle must be directed along the vector of the wave electric field). In the opposite case the continuous deceleration of the particle will take place accompanied by induced Cherenkov radiation (regime of continuous amplification of the wave by the particle beam at $dn_0(x)/dx > 0$).

The energy of equilibrium particle acquired on the distance x is defined by

$$\mathcal{E}_s^2(x) = \frac{n_0^2(x)}{n_0^2(x)-1}\left(m^2c^4 + c^2p_{ys}^2\right) . \qquad (2.41)$$

Integrating Eq. (2.40) within Eq. (2.41) the ultimate formula for the variation law of the medium refractive index becomes

$$\frac{1}{2}\left[\frac{n_0(0)}{n_0^2(0)-1} - \frac{n_0(x)}{n_0^2(x)-1}\right] + \frac{1}{4}\ln\left[\frac{n_0(x)+1}{n_0(x)-1}\cdot\frac{n_0(0)-1}{n_0(0)+1}\right]$$

$$= -\frac{mc^2\xi_0\omega_0 p_{ys}\cos\phi_s}{m^2c^4 + c^2p_{ys}^2}x . \qquad (2.42)$$

Equation (2.41) in the general case defines the particle acceleration in the capture regime when the medium refractive index falls along the wave propagation according to law (2.42). It defines the longitudinal dimension of such "Cherenkov accelerator" as well. The transverse dimension of the latter is defined by

$$\mathcal{E}_s(y) = \mathcal{E}_s(0) + mc\omega_0\xi_0(y - y_0)\cos\phi_s. \qquad (2.43)$$

Here $\mathcal{E}_s(0)$ and y_0 are the initial equilibrium values of the energy and transverse coordinate of the particle ($y - y_0$ is the transverse dimension of "Cherenkov accelerator"). As is seen from Eq. (2.43) the particle acceleration takes place if $(y-y_0)\cos\phi_s > 0$, and in the opposite case deceleration occurs ($\mathcal{E}_s(y) < \mathcal{E}_s(0)$) in accordance with what was mentioned above. For relativistic particles, when $n_0(x) \sim 1$ and $n_0(x) - 1 << n_0(0) - 1$, from Eq. (2.42) we have

$$n_0(x) - 1 \simeq \frac{m^2c^4 + c^2p_{ys}^2}{4mc^2\xi_0\omega_0 p_{ys}\cos\phi_s}\frac{1}{x}. \qquad (2.44)$$

As as this formula is valid at the large variation of the medium refractive index $n_0(x)-1$, then according to Eq. (2.41) it corresponds to large acceleration of the particle: $\mathcal{E}_s(x) >> \mathcal{E}_s(0)$. In particular, Eq. (2.41) determines the initial value of the refractive index $n_0(0)$ as a function of the initial value of the equilibrium energy of the particle $\mathcal{E}_s(0)$:

$$n_0(0) - 1 = \frac{\mathcal{E}_s(0) - \sqrt{\mathcal{E}_s^2(0) - c^2p_{ys}^2 - m^2c^4}}{\sqrt{\mathcal{E}_s^2(0) - c^2p_{ys}^2 - m^2c^4}} \qquad (2.45)$$

(since ϕ_s = const, then p_{ys} = const according to Eq. (2.38)). From the comparison of Eqs. (2.44) and (2.45) ($n_0(x)-1 << n_0(0)-1$; $n_0(0) \sim 1$) one can find the longitudinal dimension of acceleration on which the decreasing law of refractive index (2.44) is valid:

$$x >> \frac{\mathcal{E}_s^2(0) - c^2 p_{ys}^2 - m^2 c^4}{2mc^2 \xi_0 \omega_0 p_{ys} \cos\phi_s}. \tag{2.46}$$

The energy of the equilibrium particle acquired on such distances is

$$\mathcal{E}_s(x) \simeq \sqrt{2mc^2 \xi_0 \omega_0 |p_{ys} \cos\phi_s| x}\ ; \qquad \mathcal{E}_s(x) >> \mathcal{E}_s(0). \tag{2.47}$$

The estimations show that, for example, at electric field strengths of laser radiation $E \sim 10^8$ V/cm an electron with initial energy $\mathcal{E}_s(0) \sim 5$ MeV acquires energy $\mathcal{E}_s(x) \sim 50$ MeV already at the distance $x \sim 1$ cm. The transverse dimension of acceleration $y - y_0$ is of the order of a few millimeters and the longitudinal dimension of the system is of the order of the transverse one (a few times larger). At the distance $x \sim 1$ m the particle energy gain is of the order of 1 GeV . Note that because of multiple scattering on the atoms of the medium the particles can leave the regime of stable motion as a result of change of p_{ys}. The analysis shows that the multiple scattering essentially falls in the above-mentioned gaseous media (see Section 2.2) for laser field strengths $E > 10^7$ V/cm.

To illustrate the particle acceleration in the capture regime we will represent the results of numerical solution of Eqs. (2.2) and (2.3) in the field of an actual laser beam with the electric field strength

$$E_y = E_0 \exp\left(-\frac{4}{d^2}\left(y^2 + z^2\right)\right) \frac{\cos\left(\frac{\omega_0}{c}\int n_0(x)dx - \omega_0 t + \varphi_0\right)}{\cosh\left(\frac{\frac{1}{c}\int n_0(x)dx - t + \varphi_0/\omega_0}{\delta\tau}\right)}, \tag{2.48}$$

$$E_x = 0, \quad E_z = 0,$$

where $\delta\tau$ characterizes the pulse duration and φ_0 is the initial phase. Simulations have been made for neodymium laser ($\hbar\omega_0 \simeq 1.17$ eV) with electric field strength $E_0 = 3 \times 10^8$ V/cm and $\delta\tau = 1000T$, $d = 5 \times 10^3 \lambda$. The variation law for the refractive index of the medium is defined in self-consistent manner (see Eqs. (2.35) and (2.37)), that may be approximated by the function

$$n(x) = \frac{n_0 + n_f}{2} + \frac{(n_f - n_0)}{2} \tanh(\kappa x), \tag{2.49}$$

where n_0, n_f are the initial and final values of the refractive index and κ characterizes the decreasing rate.

Figure 2.4 illustrates the evolution of the particle energy in the capture regime. The particle initial energy is taken to be $\mathcal{E}_0 = 50.5$ MeV and the initial velocity is directed at the angle $\vartheta = 9 \times 10^{-3}$ rad to the wave propagation direction ($p_{0z} = 0$). The initial value of the refractive index has been chosen

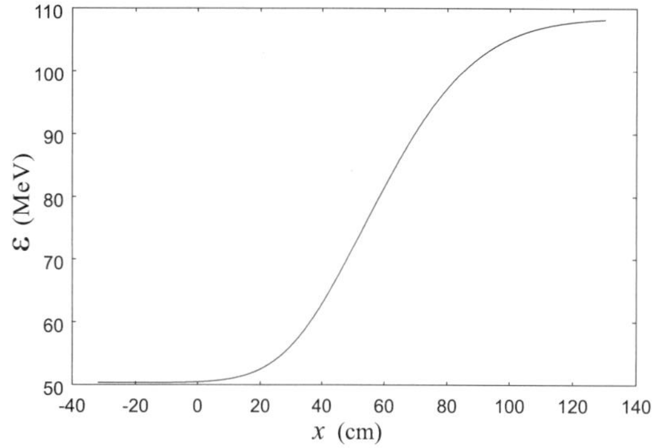

Fig. 2.4. The evolution of the particle energy in the capture regime with variable refractive index.

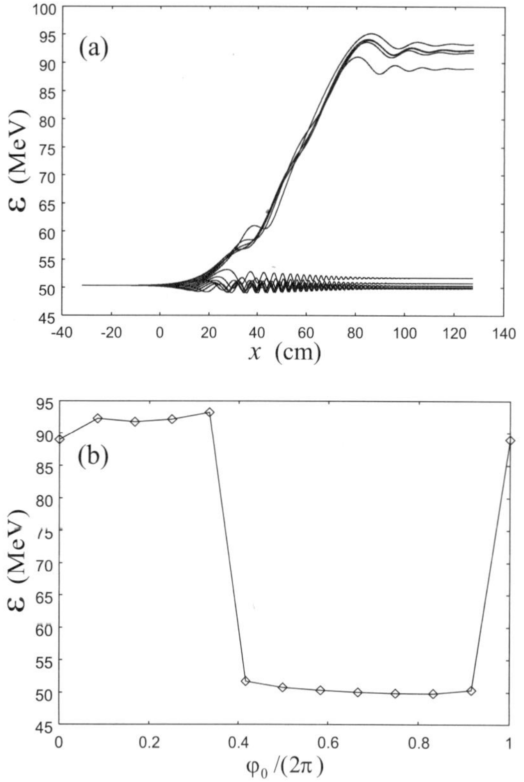

Fig. 2.5. Acceleration of the particles in the capture regime. Panel (a) displays the evolution of the energies of particles with various initial phases. The initial entrance coordinate is $z = 0$. In (b) the final energy versus the initial phase is plotted.

to be $n_0 - 1 \simeq 10^{-4}$. As we see in the capture regime with variable refractive index, one can achieve considerable acceleration.

To show the role of initial conditions in Fig. 2.5a the evolution of the energies of particles with the same initial energies $\mathcal{E}_0 = 50.5$ MeV ($\vartheta = 9 \times 10^{-3}$ rad) and various initial phases is illustrated. The initial entrance coordinate is $z = 0$. Figure 2.5b displays the role of initial conditions: the final energy versus the initial phase is plotted. In Fig. 2.6 the parameters are the same as in Fig. 2.5a except the initial entrance coordinate, which is taken to be $z = 0.25$ mm. As we see from these figures the captured particles are accelerated, while the particles situated in the unstable phases (or if the conditions for capture are not fulfilled) after the interaction remain with the initial energy.

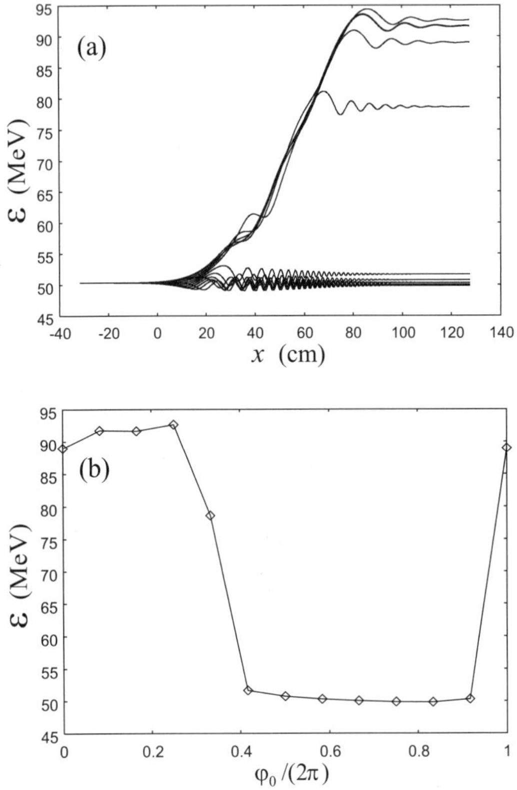

Fig. 2.6. Acceleration of the particles in the capture regime. Panel (a) displays the evolution of the energies of particles with various initial phases. The initial entrance coordinate is $z = 0.25$ mm. In (b) the final energy versus the initial phase is plotted.

2.5 Nonlinear Compton Scattering in a Medium

"Reflection" and capture phenomena are essentially changing the picture of Compton scattering in a medium. The existence of the critical field in a medium with refractive index $n(\omega) > 1$ confines the intensity of external wave on which Compton scattering of a charged particle proceeds. Therefore, one can consider the Compton effect in dielectriclike media only if the wave intensity does not exceed the critical value. On the other hand, as was mentioned above the multiphoton absorption and radiation due to the nonlinear Cherenkov resonance in the field just occurs at wave intensities close to the critical one. Hence, it is important to consider the nonlinear Compton effect in a gaseous medium where the induced Cherenkov radiation will accompany and interfere with the Compton radiation at external wave intensities close to the critical value. At the latter the nonlinear Compton effect (high harmonic radiation) will take place even in very weak wave fields ($\xi \lesssim \xi_{cr} << 1$) in contrast to nonlinear Compton effect in vacuum where for the radiation already of the second harmonic with considerable intensity, superstrong fields ($\xi > 1$) are required, as has been shown in Chapter 1.

The energy radiated by a charged particle in a medium at a frequency ω in the domain $d\omega$ and solid angle dO is given by

$$d\varepsilon_{\mathbf{k}} = \frac{e^2 n\left(\omega\right)}{4\pi^2 c^3}\omega^2 d\omega dO \left| \int_{-\infty}^{+\infty} \left[\nu \mathbf{v}\right] \exp\left[i\mathbf{k}\mathbf{r}(t) - i\omega t\right] dt \right|^2 , \tag{2.50}$$

where $\mathbf{k} = \nu n\left(\omega\right)\omega/c$ is the radiation wave vector in the medium (ν is a unit vector along the radiation direction) and $n\left(\omega\right)$ is the refractive index of the medium at frequency ω.

The particle law of motion $\mathbf{r}(t)$ in the plane monochromatic EM wave of circular polarization is determined by analogy with Eqs. (1.27)–(1.29) and is written as

$$x(t) = \mathrm{v}_x t,$$

$$y(t) = -\xi \frac{c}{\omega_0} \frac{mc^2}{\mathcal{E}\left(1 - n_0 \frac{\mathrm{v}_x}{c}\right)} \cos\omega_0 \left(1 - n_0 \frac{\mathrm{v}_x}{c}\right) t, \tag{2.51}$$

$$z(t) = \xi \frac{c}{\omega_0} \frac{mc^2}{\mathcal{E}\left(1 - n_0 \frac{\mathrm{v}_x}{c}\right)} \sin\omega_0 \left(1 - n_0 \frac{\mathrm{v}_x}{c}\right) t.$$

Here it is assumed that the initial velocity of the particle is directed along the wave propagation ($\mathrm{v}_0 = \mathrm{v}_{0x}$) at which the particle longitudinal velocity v_x and energy $\mathcal{E}$ do not vary in time since it depends only on the wave intensity ξ^2 (see Eqs. (2.7) and (2.10)) and for the circular polarization of the wave $\xi^2 = \text{const}$ (the strong wave intensity effect is responsible for permanent renormalization of these quantities in the field). Then, in the equations for particle energy and velocity (2.7)–(2.10) one should take only the sign minus before the root in accordance with the above discussion.

Substituting Eqs. (2.7), (2.9), and (2.51) into Eq. (2.50) and integrating, the following ultimate formula for the spectral power of the Compton radiation of the s-th harmonic in a medium is obtained:

$$dP_{\mathbf{k}}^{(s)} = \frac{e^2 n(\omega)}{2\pi c} \frac{\omega^2}{\omega_0 \left(1 - n_0 \frac{\mathrm{v}_x}{c}\right)} \left\{ \left[n(\omega) \frac{\mathrm{v}_x}{c} - \cos\theta \right]^2 \frac{J_s^2(\alpha)}{n^2(\omega)\sin^2\theta} \right.$$

$$\left. + \xi^2 \left(\frac{mc^2}{\mathcal{E}} \right)^2 J_s'^2(\alpha) \right\} \delta \left[\omega \frac{1 - n(\omega) \frac{\mathrm{v}_x}{c} \cos\theta}{\omega_0 \left(1 - n_0 \frac{\mathrm{v}_x}{c}\right)} - s \right] d\omega dO, \qquad (2.52)$$

where θ is the angle between the radiation direction and axis OX , and the argument of the Bessel function

$$\alpha = \xi \frac{mc^2}{\mathcal{E}} \frac{\omega n(\omega) \sin\theta}{\omega_0 \left(1 - n_0 \frac{\mathrm{v}_x}{c}\right)}. \qquad (2.53)$$

The δ-function in Eq. (2.52) determines the conservation law of the Compton radiation process in a medium (radiation spectrum)

$$\omega = s\omega_0 \frac{1 - n_0 \frac{\mathrm{v}_x}{c}}{1 - n(\omega) \frac{\mathrm{v}_x}{c} \cos\theta}. \qquad (2.54)$$

First, let us consider the cases of limit intensities of the wave $\xi = 0$ and $\xi = \xi_{cr}$. If in Eq. (2.52) $\xi \to 0$, then the radiation power will differ from zero only for the $s = 0$ harmonic. In that case, the conservation law of Compton process (2.54) becomes the condition of Cherenkov radiation ($\mathrm{v}_x \to \mathrm{v}_{0x} = \mathrm{v}_0$) and Eq. (2.52) after the integration over θ passes to the Tamm–Frank formula

$$dP_\omega^{(0)} = \frac{e^2 \mathrm{v}_0}{c^2} \left(1 - \frac{c^2}{n^2(\omega)\mathrm{v}_0^2} \right) \omega d\omega. \qquad (2.55)$$

In the other limit case of $\xi = \xi_{cr}$, the longitudinal velocity of the particle $\mathrm{v}_x = c/n_0$ and Eq. (2.54) allows the nonzero frequencies of radiation either for infinitely large harmonics ($s = \infty$) or when the condition

$$1 - n(\omega) \frac{\mathrm{v}_x}{c} \cos\theta = 0 \qquad (2.56)$$

is fulfilled. However, it is easy to see that at the satisfaction of condition (2.56) the radiation power becomes zero. Hence, at the value of external wave intensity $\xi = \xi_{cr}$ only the harmonics $s = \infty$ are radiated the power of which differs from zero at the value of the Bessel function argument $\alpha = s$, which gives

$$1 - \frac{\mathbf{k}\mathbf{v}_{cr}}{\omega} = 0 \ ; \qquad \mathbf{k} = \nu n(\omega)\frac{\omega}{c},$$

where

$$\mathbf{v}_{cr} = \left\{ \frac{c}{n_0}, 0, c\sqrt{n_0^2 - 1}\frac{1 - n_0\frac{\mathrm{v}_0}{c}}{n_0^2\left(1 - \frac{\mathrm{v}_0}{cn_0}\right)} \right\}.$$

In that case, Eq. (2.52) again passes to the Tamm–Frank formula (2.55) for a particle moving with the velocity $\mathrm{v}_0 = \mathrm{v}_{cr} > c/n(\omega)$. In this case the radiation of fundamental frequency ω_0 exists as well. So, only in limit cases $\xi = 0$ and $\xi = \xi_{cr}$ does Compton radiation fully turn into Cherenkov radiation and at the values of external wave intensity $0 < \xi < \xi_{cr}$ the radiation of the particle involves superposition of Compton and Cherenkov radiation.

The nonlinear scattering in laser fields of moderate intensities, that is, radiation of high harmonics at $\xi << 1$, is of great interest. In considering this process it is possible even at weak wave fields of intensities $\xi \approx \xi_{cr} << 1$ due to the Cherenkov resonance, i.e., when the radiation is close to the Cherenkov cone with the incident wave. In accordance with Eq. (2.52) significant nonlinearity in the radiation process arises when the argument of the Bessel function $\alpha \sim s$ ($s >> 1$). As is seen from Eqs. (2.53) and (2.54) such large values of α can be reached due to $\mathrm{v}_x \to c/n_0$, i.e., if the intensity of an incident wave is close to the critical value ($\xi \to \xi_{cr}$) and radiation is close to the Cherenkov cone $(1 - n(\omega)(\mathrm{v}_x/c)\cos\theta \to 0)$.

To determine the conditions and quantitative results for high harmonics ($s >> 1$) radiation, one should substitute in Eq. (2.53) the concrete expressions of the particle longitudinal velocity v_x and energy $\mathcal{E}$ in the field. From Eqs. (2.7) and (2.10) we have

$$\alpha = \frac{mc^2}{\mathcal{E}_0}\frac{n(\omega)\omega\sin\theta}{\omega_0\left(1 - n_0\frac{\mathrm{v}_0}{c}\right)\sqrt{1 - \frac{\xi^2}{\xi_{cr}^2}}}\xi. \tag{2.57}$$

In Eq. (2.57), the radiation angle ($\sin\theta$) should be defined from the condition $\theta \simeq \theta_c$, where θ_c is the Cherenkov angle. At fundamental frequency ω_0 the Cherenkov angle $\theta_c << 1$, whereas at other frequencies ω it may not be small depending on the medium dispersion and, consequently, the conditions of nonlinearity will be different. However, the number of harmonics at all frequencies is large enough. The harmonic $s = 0$ at fundamental frequency ω_0 cannot be radiated since $\mathrm{v}_x < c/n_0$. The first harmonic ($s = 1$) at frequency ω_0 is radiated at the angle $\theta = 0$. The negative harmonics ($s = -1, -2, ...$) correspond to anomalous Compton scattering in a medium with refractive index $n(\omega) > 1$. At frequencies $\omega \neq \omega_0$ the harmonic $s = 0$ corresponds to Cherenkov radiation; however, the power of the radiation differs from the Tamm–Frank formula because of the oscillatory character of the particle motion in the wave field (influence of Compton effect).

2.6 Radiation of a Particle in Capture Regime. Cherenkov Amplifier

Consider the radiation of the particle captured by a plane monochromatic wave in a gaseous medium. We will assume that the particle initial velocity is directed along the wave propagation and has a value close to the Cherenkov one:

$$\mathrm{v}_0 = \mathrm{v}_{0x} = \frac{c}{n_0}(1+\mu)\,; \qquad \mu << 1. \tag{2.58}$$

From the equations of motion (2.2) and (2.3) it follows that at $\mu = 0$

$$\mathrm{v}_x = \mathrm{v}_{x0} = \frac{c}{n_0}\,, \qquad \mathrm{v}_y = 0\,, \qquad x = x_0 + \frac{c}{n_0}t, \tag{2.59}$$

where x_0, $y_0 = 0$, $z_0 = 0$ are the initial coordinates of the particle at the moment $t = 0$ in the wave of linear polarization

$$E = E_y = E_0 \cos\left(\omega_0 n_0 \frac{x}{c} - \omega_0 t\right). \tag{2.60}$$

The solution of Eqs. (2.2) and (2.3) at $\mu << 1$ can be represented as

$$\mathrm{v}_x(t) = \frac{c}{n_0}\left(1 + \mu u_x(t)\right)\,, \qquad \mathrm{v}_y(t) = c\mu u_y(t) \tag{2.61}$$

and after the linearization of these equations by parameter μ we have the following set of equations for the functions $u_x(t)$ and $u_y(t)$:

$$\frac{du_x}{dt} = \frac{e\left(n_0^2-1\right)^{3/2}}{n_0^2 mc} E_0 \cos\phi_0 \cdot u_y\,,$$

$$\frac{du_y}{dt} = -\frac{e\left(n_0^2-1\right)^{1/2}}{mc} E_0 \cos\phi_0 \cdot u_x\,. \tag{2.62}$$

Integrating this set of equations at the initial conditions $u_{x0} = 1$ and $u_{y0} = 0$ in accordance with Eq. (2.59), for the particle velocity in the capture regime we obtain

$$\mathrm{v}_x(t) = \frac{c}{n_0}\left(1 + \mu\cos\Omega_0 t\right),$$

$$\mathrm{v}_y(t) = -\frac{c}{\left(n_0^2 - 1\right)^{1/2}}\mu\sin\Omega_0 t, \tag{2.63}$$

$$\Omega_0 = \frac{e\left(n_0^2 - 1\right)E_0|\cos\phi_0|}{n_0 mc}. \tag{2.64}$$

In the derivation of Eqs. (2.63) and (2.64) the following approximation has been made (due to the small parameter μ):

$$\mu\frac{\omega_0}{\Omega_0} << 1, \tag{2.65}$$

which is violated for the wave phase $\cos\phi_0 = 0$. This is connected with the fact that the stability in the capture regime is provided by the action of magnetic field $\mathbf{H}'$ in the frame of reference connected with the wave and $\mathbf{H}' = 0$ in the phase $\cos\phi_0 = 0$, so that this phase is unstable.

As is seen from Eq. (2.63) the particle velocity in the wave oscillates with the frequency Ω_0, which depends on the initial phase ϕ_0. In the particle beam case the various particles being initially in different phases of the wave well will have diverse velocities and space bunching of the particles will occur as a result of which the current density of the beam will be modulated. Equation (2.64) shows that the modulation frequency $\Omega_0 \simeq \omega_0\left(n_0^2 - 1\right)\xi|\cos\phi_0|$ and as even for the strong laser fields $\xi << 1$ (and $n_0^2 - 1 << 1$), then $\Omega_0 << \omega_0$.

To calculate the power of noncoherent radiation by Eq. (2.50) one needs the particle law of motion $\mathbf{r}(t)$ in the capture regime. Defining the latter by integration of Eq. (2.63) with the initial conditions $x(t)\mid_{t=0} = x_0$, $y(t)\mid_{t=0} = 0$

$$x(t) = x_0 + \frac{c}{n_0}t + \mu\frac{c}{n_0\Omega_0}\sin\Omega_0 t\ ,$$

$$y(t) = -\mu\frac{c}{\left(n_0^2 - 1\right)^{1/2}\Omega_0}\left(1 - \cos\Omega_0 t\right) \tag{2.66}$$

and expanding the exponent of Eq. (2.50) into the series over the small parameter μ (taking into account as well that $\mu\omega/\Omega_0 << 1$), after the calculations we will have the following formula for differential power of noncoherent radiation in the capture regime:

$$dP_{\mathbf{k}} = dP_{\mathbf{k}}^{(0)} + dP_{\mathbf{k}}^{(+)} + dP_{\mathbf{k}}^{(-)}\ , \tag{2.67}$$

$$dP_{\mathbf{k}}^{(0)} = \frac{e^2 n(\omega)}{2\pi c n_0^2}\omega^2 \sin^2\theta \cdot \delta\left[\omega\frac{n(\omega)}{n_0}\cos\theta - \omega\right] d\omega dO, \tag{2.68}$$

$$dP_{\mathbf{k}}^{(\pm)} = \mu^2 \frac{e^2 n(\omega)}{8\pi c}\frac{\omega^2}{n_0\left(n_0^2-1\right)}\delta\left[\omega\frac{n(\omega)}{n_0}\cos\theta - \omega \pm \Omega_0\right]$$

$$\times\left\{\left[n_0^2 + \left(\frac{n_0^2}{2}-1\right)\sin^2\theta\right] \pm 2\frac{n(\omega)}{n_0}\left(\frac{n_0^2}{2}-1\right)\frac{\omega}{\Omega_0}\cos\theta\sin^2\theta\right.$$

$$\left. + \frac{n^2(\omega)}{n_0^2}\frac{\omega^2}{\Omega_0^2}\sin^2\theta\left[\frac{n_0^2}{2} + \left(\frac{n_0^2}{2}-1\right)\cos^2\theta\right]\right\} d\omega dO, \tag{2.69}$$

where θ is the angle between the radiation direction and axis OX. The term $dP_{\mathbf{k}}^{(0)}$ corresponds to Cherenkov radiation by the particle moving with the velocity $\mathrm{v} = c/n_0$ in the wave and the terms $dP_{\mathbf{k}}^{(\pm)}$ determine the radiation due to oscillatory motion of the particle. According to the δ-functions in Eqs. (2.68) and (2.69) for the radiation angles we have

$$\cos\theta_0 = \frac{n_0}{n(\omega)}, \qquad \cos\theta_{\pm} = \frac{n_0}{n(\omega)}\left(1 \mp \frac{\Omega_0}{\omega}\right). \tag{2.70}$$

Note that the approximation $\mu\omega/\Omega_0 << 1$ applied in the calculations is necessary only to obtain ultimate analytical formulas (in the general case the particle velocity is expressed by elliptic functions and analytical solution of the problem is complicated).

Integrating Eqs. (2.68) and (2.69) over the solid angle for the spectral distribution of the radiation we obtain

$$dP_{\omega}^{(0)} = \frac{e^2}{c n_0}\left[1 - \frac{n_0^2}{n^2(\omega)}\right]\omega d\omega\ , \tag{2.71}$$

$$dP_{\omega}^{(\pm)} = \mu^2 \frac{e^2}{4c}\frac{1}{n_0\left(n_0^2-1\right)}\left\{n_0^2 + \frac{n_0^2 + n^2(\omega) - 2}{2}\right.$$

$$\left.\times\left[\frac{\omega^2}{\Omega_0^2} - \frac{n_0^2}{n^2(\omega)}\left(1 \mp \frac{\Omega_0}{\omega}\right)^2\right]\right\}\omega d\omega. \tag{2.72}$$

In Eq. (2.72)

$$\omega = \pm\frac{\Omega_0}{1 - \frac{n(\omega)}{n_0}\cos\theta}. \tag{2.73}$$

As Ω_0 depends on initial phase ϕ_0 (see Eq. (2.64)), in the case of a particle beam captured by a wave of linear polarization at a certain angle θ a whole spectrum of frequencies will be radiated, in contrast to common Cherenkov radiation at which only a definite frequency is radiated at that certain angle.

Let us compare the radiation at the fundamental frequency ω_0 with the common Cherenkov radiation at the same frequency (in the absence of the external wave). In this case $dP_{\omega_0}^{(0)} = 0$ and for $dP_{\omega}^{(-)}$ the conservation law for the radiation of frequency ω_0 is violated (see the second expression in Eq. (2.70)). From Eq. (2.72) at $\omega = \omega_0$ we have

$$dP_{\omega_0}^{(+)} = \frac{e^2}{2cn_0}\mu^2\frac{\omega_0}{\Omega_0}\omega_0 d\omega. \tag{2.74}$$

If one substitutes $\mathrm{v} = c(1+\mu)/n_0$ in the Tamm–Frank formula (2.55), then with the linear approximation by parameter μ we will have

$$dP_{\omega_0} = \frac{2e^2}{cn_0}\mu\omega_0 d\omega. \tag{2.75}$$

A comparison of Eqs. (2.74) and (2.75) shows that the radiation of the particle at the fundamental frequency ω_0 in the capture regime is much smaller than the spontaneous Cherenkov radiation (because of condition (2.65)). Such a decrease of radiation is connected with the violation of coherency due to oscillation of particle velocity in the wave field.

The fundamental frequency ω_0 in the capture regime is radiated at the angle $\theta \simeq \sqrt{2\Omega_0/\omega_0}$ (see Eq. (2.73)). The common Cherenkov angle is $\theta_c \simeq \sqrt{\mu/2}$ and as far as $\mu << \Omega_0/\omega_0$ then $\theta >> \theta_c$, i.e., the radiation angle at the frequency of stimulating wave in the capture regime is much larger than the spontaneous Cherenkov angle in the absence of the external wave.

At the other frequencies $\omega \neq \omega_0$ the radiation is mainly determined by $dP_{\omega}^{(0)}$, which practically coincides with the Tamm–Frank formula.

Consider now the case of circular polarization of the incident wave

$$E_y = E_0 \cos\left(\frac{\omega_0 n_0}{c}x - \omega_0 t\right), \qquad E_z = E_0 \sin\left(\frac{\omega_0 n_0}{c}x - \omega_0 t\right). \tag{2.76}$$

Linearizing the equations of motion (2.2) and (2.3) in the field (2.76) under the condition (2.58) for the particle velocity in the capture regime we obtain

$$\mathrm{v}_x = \frac{c}{n_0}\left(1 + \mu\cos\Omega_0' t\right),$$

$$\mathrm{v}_y = -\mu\frac{c}{\left(n_0^2-1\right)^{1/2}}\cos\phi_0 \cdot \sin\Omega_0' t\ , \tag{2.77}$$

$$\mathrm{v}_z = -\mu \frac{c}{\left(n_0^2 - 1\right)^{1/2}} \sin\phi_0 \cdot \sin \Omega_0' t \ ,$$

where the oscillation frequency in the wave well Ω_0' does not depend on the initial phase ϕ_0 in contrast to the case of the linearly polarized wave. If we calculate the radiation power by Eqs. (2.77), then the same formulas (2.67)–(2.73) for the case of wave linear polarization will be obtained. The only difference is that Ω_0' is constant for all particles situated at the difference phases in the wave well, and at the certain angle only one frequency will be radiated in this case.

Equations (2.63) and (2.77) show that the energy of the particle in the field

$$\mathcal{E} = \mathcal{E}_0 + \mu \frac{\mathcal{E}_0}{n_0^2 - 1} \cos \Omega_0 t \ ; \qquad \mathcal{E}_0 = \frac{mc^2 n_0}{\left(n_0^2 - 1\right)^{1/2}} \tag{2.78}$$

oscillates between the values

$$\mathcal{E}_{\min} = \mathcal{E}_0 \left(1 - \frac{\mu}{n_0^2 - 1}\right); \qquad \mathcal{E}_{\max} = \mathcal{E}_0 \left(1 + \frac{\mu}{n_0^2 - 1}\right),$$

consequently the exchange of the energy is

$$\Delta\mathcal{E} = 2\mu \frac{mc^2 n_0}{\left(n_0^2 - 1\right)^{3/2}} . \tag{2.79}$$

According to Eqs. (2.78) the particle captured by the wave periodically acquires and loses such energy $\Delta\mathcal{E}$. Due to the induced Cherenkov effect the energy lost by the particle is coherently radiated into the wave (particularly for this reason the above-considered noncoherent radiation at the frequency of stimulating wave ω_0 is sufficiently suppressed) and the amplification of the initial wave will take place. Hence, the particle capture phenomenon may in principle serve as a FEL mechanism (Cherenkov amplifier). For the latter one needs to solve the self-consistent problem on the basis of the set of Maxwell–Vlasov equations.

Let us now consider the amplitude of the wave field to be a slowly varying function of the space-time coordinates $(x,\, t)$ with respect to the phase. The problem will be investigated first for the circular polarization of the wave

$$E_y(x,t) = E(x,t) \cos\left(\frac{\omega_0 n_0 x}{c} - \omega_0 t\right),$$

$$E_z(x,t) = E(x,t) \sin\left(\frac{\omega_0 n_0 x}{c} - \omega_0 t\right) \tag{2.80}$$

with the boundary conditions

$$E_y(0,t) = E_0 \cos\omega_0 t\ , \qquad E_z(0,t) = -E_0 \sin\omega_0 t. \tag{2.81}$$

Related to particles we will assume that it crosses the boundary of the medium $x = 0$ at the moment $t = t_0$ with the initial velocity (2.58). Linearizing the equations of motion (2.2) and (2.3) in the field (2.80) for a single particle velocity in the field we obtain

$$\mathrm{v}_y = -\frac{c}{\left(n_0^2-1\right)^{1/2}}\mu\cos(\omega_0 t_0)\sin\left[\frac{e\left(n_0^2-1\right)}{mcn_0}\int_{t_0}^{t}E(t',x)dt'\right],$$

$$\mathrm{v}_z = \frac{c}{\left(n_0^2-1\right)^{1/2}}\mu\sin(\omega_0 t_0)\sin\left[\frac{e\left(n_0^2-1\right)}{mcn_0}\int_{t_0}^{t}E(t',x)dt'\right]. \tag{2.82}$$

To define the electric current of the particle stream we assume that the space is continuously filled with the charged particles. Then at the moment t_0 in the point x will be situated only the particles for which $t_0 = t - n_0 x/c$ (with accuracy $\mu\omega_0/\Omega_0 << 1$). Hence, for the electric current of the particle stream we will have

$$j_y(x,t) = -\mu\frac{ec\rho_0}{\left(n_0^2-1\right)^{1/2}}\cos\left(\frac{\omega_0 n_0 x}{c}-\omega_0 t\right)$$

$$\times\sin\left[\frac{e\left(n_0^2-1\right)}{mcn_0}\int_{t-n_0x/c}^{t}E(t',\frac{c}{n_0}(t'-t)+x)dt'\right], \tag{2.83}$$

$$j_z(x,t) = -\mu\frac{ec\rho_0}{\left(n_0^2-1\right)^{1/2}}\sin\left(\frac{\omega_0 n_0 x}{c}-\omega_0 t\right)$$

$$\times\sin\left[\frac{e\left(n_0^2-1\right)}{mcn_0}\int_{t-n_0x/c}^{t}E(t',\frac{c}{n_0}(t'-t)+x)dt'\right],$$

where ρ_0 is the mean density of the particles in the initial stream, which will be assumed constant (since $\mu << 1$ the variation ρ_0 is small and can be neglected).

Because we are investigating the induced radiation, the field of the scalar potential and longitudinal radiation field along the axis OX will not be considered here. Substituting Eqs. (2.83) into the Maxwell equation and taking into account the slow variation of the radiation field amplitude:

$$\left|\frac{\partial E}{\partial t}\right| << \omega_0|E|\ , \qquad \left|\frac{\partial E}{\partial x}\right| << \frac{\omega_0 n_0}{c}|E|\ ,$$

we obtain the equation of the self-consistent field:

$$\frac{\partial E}{\partial x}+\frac{n_0}{c}\frac{\partial E}{\partial t}=\frac{2\pi e\rho_0}{n_0\left(n_0^2-1\right)^{1/2}}\mu$$

$$\times\sin\left[\frac{e\left(n_0^2-1\right)}{mcn_0}\int_{t-n_0x/c}^{t}E(t',\frac{c}{n_0}(t'-t)+x)dt'\right]. \tag{2.84}$$

Equation (2.84) has a simpler form over wave coordinates $\tau=t-n_0x/c$, $\eta=x$. Then, for the field amplitude $E(t,x)=f(\tau,\eta)$ we have

$$\frac{\partial}{\partial\eta}f(\tau,\eta)=\frac{2\pi e\rho_0}{n_0\left(n_0^2-1\right)^{1/2}}\mu\sin\left[\frac{e\left(n_0^2-1\right)}{mc^2}\int_0^{\eta}f(\tau,\eta')d\eta'\right]. \tag{2.85}$$

The simple analytic solution can be received at the incident monochromatic wave: $f(\tau,0)=E_0$. In this case, it follows from Eq. (2.84) that $f(\tau,\eta)$ does not depend on τ, i.e., $f(\tau,\eta)=f(\eta)$, and for the quantity

$$\varphi=\frac{e\left(n_0^2-1\right)}{mc^2}\int_0^{\eta}f(\eta')d\eta' \tag{2.86}$$

we have the nonlinear equation of anharmonic oscillator

$$\varphi''=\frac{2\pi e^2\rho_0\left(n_0^2-1\right)^{1/2}}{mc^2n_0}\mu\sin\varphi, \tag{2.87}$$

the general solution of which is the incomplete elliptic integral of the first kind

$$\frac{1}{2}\left(n_0^2-1\right)\frac{eE_0x}{mc^2}=\int_0^{\varphi/2}\frac{dz}{\sqrt{1+\zeta^2\sin^2 z}},$$

$$\zeta^2=\frac{8\pi\mu}{n_0\left(n_0^2-1\right)^{3/2}}\frac{mc^2\rho_0}{E_0^2}. \tag{2.88}$$

In the linear case when $\varphi<<1$ from Eq. (2.88) we have

$$E(x)=E_0\begin{bmatrix}\cosh\left(\frac{x}{l_c}\right), & \mu>0,\\ \cos\left(\frac{x}{l_c}\right), & \mu<0.\end{bmatrix} \tag{2.89}$$

Hence, for $\mu > 0$, which corresponds to particles' initial velocity $\mathrm{v}_0 > c/n_0$, exponential amplification of the incident wave occurs. For $\mu < 0$, that is, $\mathrm{v}_0 < c/n_0$, the amplification vanishes on average. The quantity in Eq. (2.89)

$$l_c = \left(\frac{mc^2 n_0}{2\pi e^2 \mu \rho_0 \left(n_0^2 - 1\right)^{1/2}} \right)^{1/2} \tag{2.90}$$

is the coherent length of amplification. Equation (2.85) is an analogue of the equation of the quantum amplifier. The role of inverse population in atomic systems here performs detuning of the Cherenkov resonance $\mathrm{v}_0 - c/n_0$ (parameter μ).

Analysis of the obtained formulas shows that the linear regime takes place at the electric field strengths of amplifying radiation

$$E \lesssim e\lambda_0 \rho_0 \left(\frac{mc^2}{\mathcal{E}_0} \right)^3$$

(λ_0 is the wavelength of incident wave) and at the coherent length of amplification

$$l_c \lesssim \frac{mc^2}{e^2 \lambda_0 \rho_0} \left(\frac{\mathcal{E}_0}{mc^2} \right)^2 .$$

In the saturation regime from Eq. (2.85) we have

$$E(x) = E_0 + \mu \frac{2\pi mc^2 \rho_0}{n_0 \left(n_0^2 - 1\right)^{3/2}} \frac{1}{E_0} \left\{ 1 - \cos\left[\left(n_0^2 - 1\right) \frac{eE_0 x}{mc^2} \right] \right\} . \tag{2.91}$$

The wave energy gain found from Eq. (2.91) corresponds to the particle energy exchange in the capture regime (in a unit volume) according to Eq. (2.79):

$$\Delta W = \rho_0 \Delta \mathcal{E} = \frac{2\mu \rho_0 \mathcal{E}_0}{n_0^2 - 1} . \tag{2.92}$$

The saturation regime and Eq. (2.91) is valid when the electric field strengths of amplifying radiation

$$E \gtrsim e\lambda_0 \rho_0 \frac{\mathcal{E}_0}{mc^2} .$$

Consider now the case of linear polarization of incident wave

$$E_y = E(x,t) \cos\left(\frac{\omega_0 n_0 x}{c} - \omega_0 t \right) . \tag{2.93}$$

By analogy with the previous case for the velocity of a single particle in the field (2.93) we obtain

$$\mathrm{v}_x = \frac{c}{n_0}\left(1 + \mu\cos\left[\left[\int_{t_0}^{t} \Omega_0(t', x)dt'\right]\right]\right),$$

$$\mathrm{v}_y = -\frac{c}{\left(n_0^2 - 1\right)^{1/2}}\mu\sin\left[\left|\int_{t_0}^{t} \Omega_0(t', x)dt'\right|\right], \tag{2.94}$$

where the modulation frequency

$$\Omega_0(t, x) = \frac{e\left(n_0^2 - 1\right)}{mcn_0}E(x, t)\cos\omega_0 t_0 \tag{2.95}$$

already depends on initial phase $\phi_0 = \omega_0 t_0$. Therefore, in the particle beam case, all harmonics will be radiated in contrast to circular polarization of the wave. By calculating the electric current of the particle stream and expanding into series over Bessel functions we find that the induced radiation stipulated by the y component of the current (coherent radiation) will include only the odd harmonics and the noncoherent part of the radiation stipulated by the x component of the current (longitudinal field along the axis OX) will include only the even harmonics. As in the previous case we will consider the coherent radiation. Then, substituting y component of the current

$$j_y(x, t) = -\mu\frac{ec\rho_0}{\left(n_0^2 - 1\right)^{1/2}}\sum_{s=-\infty}^{+\infty} i^{s-1}J_s(\alpha)\exp\left[is\omega_0\left(\frac{n_0 x}{c} - t\right)\right],$$

$$s = 2k - 1\ ;\ k = 0, \pm 1, \pm 2, \ldots,$$

$$\alpha(x, t) = \frac{e\left(n_0^2 - 1\right)}{mcn_0}\int_{t - n_0 x/c}^{t} E(t', \frac{c}{n_0}(t' - t) + x)dt' \tag{2.96}$$

into the Maxwell equation for the slowly varying amplitude of the self-consistent field we will have the equation

$$2is\omega_0\left(\frac{n_0}{c}\frac{\partial E_s}{\partial x} + \frac{n_s^2}{c^2}\frac{\partial E_s}{\partial t}\right) + \frac{s^2\omega_0^2}{c^2}\left(n_s^2 - n_0^2\right)E_s$$

$$= i^s\frac{4\pi e\rho_0 s\omega_0}{c\left(n_0^2 - 1\right)^{1/2}}\mu J_s(\alpha), \tag{2.97}$$

where n_s is the medium refractive index at the s-th harmonic of the fundamental frequency ω_0 ($n_s \equiv n(s\omega_0)$).

Consider Eq. (2.97) with regard to the presence and absence of synchronism. In the last case, when $n_s \neq n_0$ taking into account the slow variation of the field amplitude from Eq. (2.97) we obtain

$$E_s = i^s \mu \frac{4\pi e c \rho_0}{\left(n_0^2 - 1\right)^{1/2}} \frac{1}{s\omega_0} \frac{1}{n_s^2 - n_0^2} J_s(\alpha). \tag{2.98}$$

As is seen from this formula in the absence of synchronism, there is a weak dependence of radiation field on harmonics' number.

In the case of synchronism ($n_s = n_0$), Eq. (2.97) becomes

$$\frac{\partial E_s}{\partial x} + \frac{n_0}{c} \frac{\partial E_s}{\partial t} = i^{s-1} \mu \frac{2\pi e \rho_0}{n_0 \left(n_0^2 - 1\right)^{1/2}} J_s(\alpha). \tag{2.99}$$

For the first harmonic (fundamental coherent radiation) the results repeat almost exactly the case of wave circular polarization (Eqs. (2.88)–(2.90)), the only difference being that the coherence length in this case is $\sqrt{2} l_c$.

To determine the radiation on the other harmonics in the case of synchronism consider the problem in the given field. Then, for large x when

$$\frac{e\left(n_0^2 - 1\right) E_0 x}{mc^2} >> 1$$

for the harmonics' amplitudes we have

$$E_s = i^{s-1} \mu \frac{2\pi mc^2 \rho_0}{n_0 \left(n_0^2 - 1\right)^{3/2}} \frac{1}{E_0}. \tag{2.100}$$

Hence, the radiation intensity on the harmonics

$$I_s - \frac{c}{8\pi} |E_s|^2 \simeq e^2 c \frac{(\lambda_0^3 \rho_0)^2}{\lambda_0^4} \left(\frac{\mathcal{E}_0}{mc^2}\right)^2. \tag{2.101}$$

Equation (2.101) as well as Eq. (2.92) and estimation formulas are obtained when $\mu \sim \xi (mc^2/\mathcal{E}_0)^2$, which is defined from the condition of particle capture. As in the linear regime the coherence length increases as energy squared, and the losses of the particles in the medium depend on energy logarithmically, then the energy increase for amplification of weak signals does not give an essential advantage. The optimal energy is $\mathcal{E}_0 \sim mc^2$. Then $l_c \sim (r_0 \lambda_0 \rho_0)^{-1}$, where $r_0 = e^2/mc^2$ is the electron classical radius. The estimations show that for the amplification of optical radiation in the capture regime with $n_0 = \text{const}$, electron beams of large densities are necessary. The situation considerably will be improved if media with varying refractive index $n_0(x)$ are used. Then along the direction of increase of $n_0(x)$ the particles will be continuously decelerated, and the wave continuously amplified (a regime inverse to the one considered in Section 2.4).

Bibliography

R.M. More, Phys. Rev. Lett. **16**, 781 (1966)
V.M. Haroutunian, H.K. Avetissian, Sov. J. Quantum Electron. **2**, 39 (1972)
V.M. Haroutunian, H.K. Avetissian, Zh. Éksp. Teor. Fiz. **62**, 1639 (1972)
A.S. Dementev, A.G. Kulkin, Yu.G. Pavlenko, Zh. Éksp. Teor. Fiz. **62**, 161 (1972)
M.A. Piestrup et al., J. Appl. Phys. **46**, 132 (1975)
H. Dekker, Phys. Lett. A **59**, 369 (1976)
J.E. Walsh, T.C. Marshall, S.P. Schlesinger, Phys. Fluids **20**, 709 (1977)
J.E. Walsh: Stimulated Cerenkov radiation. In: Free Electron Generators of Coherent Radiation, Phys. Quantum Electron. vol 5, ed by S. Jacobs, M. Sargent, M. Scully, R. Spitzer (Addison-Wesley, Reading, MA 1978) p. 357
H.K. Avetissian, Phys. Lett. A **69**, 399 (1978)
J.E. Walsh: Cerenkov and Cerenkov-Raman radiation sources. In Free Electron Generators of Coherent Radiation, Phys. Quantum Electron. vol 7, ed by S. Jacobs, H. Pilloff, M. Sargent, M. Scully, R.Spitzer (Addison-Wesley, Reading, MA 1980) p. 255
K.L. Felch et al., Appl. Phys. Lett. **38**, 601 (1981)
J.A. Edighoffer et al., Phys. Rev. A **23**, 1848 (1981)
J.A. Edighoffer et al., IEEE J. Quantum Electron. **QE-17**, 1507 (1981)
J.E. Walsh, Adv. Electron. and Electron. Phys. **58**, 271 (1982)
J.E. Walsh, J.B. Murphy, IEEE J. Quantum Electron. **18**, 1259 (1982)
W.D. Kimura et al., Appl. Phys. Lett. **40**, 102 (1982)
W.D. Kimura et al., IEEE J. Quantum Electron. **QE-18**, 239 (1982)
W.D. Kimura, J. Appl. Phys. **53**, 5433 (1982)
W. Becker, J.K. McIver, Phys. Rev. A **25**, 956 (1982)
M.A. Piestrup, IEEE J. Quantum Electron. **QE-19**, 1827 (1983)
J.R. Fontana, R.H. Pantell, J. Appl. Phys. **54**, 4285 (1983)
D.Y. Wang et al., IEEE J. Quantum Electron. **QE-19**, 389 (1983)
B. Jhonson, J.E. Walsh, Phys. Rev. A **33**, 3199 (1986)
E.P. Garate et al., Nucl. Instrum. Methods Phys. Res. A **259**, 125 (1987)
F. Ciocci et al., Phys. Rev. Lett. **66**, 699 (1991)
H.K. Avetissian, Usp. Fiz. Nauk (Sov. J.) **167**, 793 (1997)
H.K. Avetissian, K.Z. Hatsagortsian, G.F. Mkrtchian, IEEE J. Quantum Electron. **33**, 897 (1997)

3 Quantum Theory of Induced Multiphoton Cherenkov Process

The existence of critical intensity in the induced Cherenkov process at which nonlinear resonance with a given coherent radiation field takes place leading to threshold phenomena of particle "reflection" and capture, in the quantum description, corresponds to multiphoton absorption/radiation of the particle at free–free transitions. Hence, first it is important to determine the probabilities of induced Cherenkov radiation and absorption below the critical value and close to this one when these probabilities considerably increase. As a result of the multiphoton absorption/radiation the particle quantum state is modulated at the wave harmonics.

Then one should elucidate the role of particle spin in these phenomena since in dielectriclike media the wave periodic electromagnetic field in the intrinsic frame of reference becomes a static magnetic field and spin interaction with such a field should resemble the Zeeman effect.

What other quantum effects may be expected in induced Cherenkov process taking into account that spontaneous Cherenkov effect is of classical nature and has no quantum peculiarity?

The particle "reflection" effect from the wave envelope is also of classical nature, but the quantum state of the reflected particle after the interaction becomes modulated at X-ray frequencies.

The classical phenomenon of particle capture by the wave leads to quantum effect of zone structure of particle states like the particle states in a crystal lattice.

The inelastic diffraction scattering of the particles on the traveling EM wave of intensity below the critical value in induced Cherenkov process takes place like Bragg diffraction (elastic) on a crystal lattice.

The consideration of these quantum problems is the subject of this chapter.

3.1 Quantum Description of Induced Cherenkov Process in Strong Wave Field

The multiphoton interaction of a charged spinor particle with a plane EM wave in induced Cherenkov process should be described in general by the Dirac equation. As will be shown below, the exact solution of the Dirac equation can be obtained only for the particular case when the particle initial velocity is parallel to the wave propagation direction, which is monochromatic and is of circular polarization. In other cases the quantum equations of motions (both nonrelativistic and relativistic) are reduced to ordinary differential equations of the second order of Hill or Mathieu type, the exact solution of which are unknown. In these cases one needs to develop adequate approximations for the quantum description of particle–wave nonlinear interaction.

The Dirac equation for the spinor particle in the given coherent radiation field in a medium is written as

$$i\hbar\frac{\partial\Psi}{\partial t} = \left[c\widehat{\alpha}(\widehat{\mathbf{p}} - \frac{e}{c}\mathbf{A}(t - n_0x/c)) + \widehat{\beta}mc^2\right]\Psi. \tag{3.1}$$

In contrast to the case of interaction in a vacuum where the Dirac equation has been solved in the spinor representation (see Eqs. (1.78) and (1.78)) here it is convenient to solve the problem in the standard representation with the Dirac matrices

$$\widehat{\alpha} = \begin{pmatrix} 0 & \sigma \\ \sigma & 0 \end{pmatrix}; \; \widehat{\beta} = \begin{pmatrix} I & 0 \\ 0 & -I \end{pmatrix}. \tag{3.2}$$

Here $\sigma = (\sigma_x, \sigma_y, \sigma_z)$ are the Pauli matrices (1.79), and I is a two-dimensional unit matrix. In Eq. (3.1) $\mathbf{A} = \mathbf{A}(t - n_0x/c)$ is the vector potential of a linearly polarized plane quasi-monochromatic EM wave propagating in the OX direction in a medium

$$\mathbf{A} = \{0, A_0(\tau)\cos\omega_0\tau, 0\}; \quad \tau = t - n_0x/c. \tag{3.3}$$

As in previous considerations we shall assume that the EM wave is adiabatically switched on at $\tau = -\infty$ and switched off at $\tau = +\infty$.

To solve Eq. (3.1) it is more straightforward to pass to the frame of reference of the rest of the wave (R frame moving with velocity $\mathrm{V} = c/n$). As has been shown in Chapter 2, in the R frame there is only the static magnetic field that will be described according to Eq. (3.3) by the following vector potential:

$$\mathbf{A}_R = \{0, A_0(x')\cos k'x', 0\}, \tag{3.4}$$

where

$$k' = \frac{\omega_0}{c}\sqrt{n_0^2 - 1}. \tag{3.5}$$

The wave function of a particle in the R frame is connected with the wave function in the laboratory frame L by the Lorentz transformation of the bispinors

$$\Psi = \widehat{S}(\vartheta)\Psi_R, \tag{3.6}$$

where the transformation operator

$$\widehat{S}(\vartheta) = ch\frac{\vartheta}{2} + \alpha_x sh\frac{\vartheta}{2}\ ; \qquad th\vartheta = \frac{\mathrm{V}}{c} = \frac{1}{n}. \tag{3.7}$$

For Ψ_R we have the equation

$$i\hbar\frac{\partial \Psi_R}{\partial t'} = \left[c\widehat{\alpha}(\widehat{\mathbf{p}}' - \frac{e}{c}\mathbf{A}_R(\mathbf{x}')) + \widehat{\beta}mc^2\right]\Psi_R. \tag{3.8}$$

Since the interaction Hamiltonian does not depend on the time and transverse (to the direction of the wave propagation) coordinates the eigenvalues of the operators $\widehat{H'}$, $\widehat{p}'_y$, $\widehat{p}'_z$ are conserved: $\mathcal{E}' =$ const, $p'_y =$ const, $p'_z =$ const and the solution of Eq.(3.8) can be represented in the form of a linear combination of free solutions of the Dirac equation with amplitudes $a_i(x')$ depending only on x':

$$\Psi_R(\mathbf{r}',t') = \sum_{i=1}^{4} a_i(x')\Psi_i^{(0)}. \tag{3.9}$$

Here

$$\Psi_{1,2}^{(0)} = \sqrt{\frac{\mathcal{E}' + mc^2}{2\mathcal{E}'}}\begin{bmatrix} \varphi_{1,2} \\ \frac{\sigma_x cp'_x + \sigma_y cp'_y}{\mathcal{E}' + mc^2}\varphi_{1,2}\end{bmatrix}$$

$$\times \exp\left[\frac{i}{\hbar}(p'_x x' + p'_y y' - \mathcal{E}' t)\right],$$

$$\Psi_{3,4}^{(0)} = \sqrt{\frac{\mathcal{E}' + mc^2}{2\mathcal{E}'}}\begin{bmatrix} \varphi_{1,2} \\ \frac{-\sigma_x cp'_x + \sigma_y cp'_y}{\mathcal{E}' + mc^2}\varphi_{1,2}\end{bmatrix}$$

$$\times \exp\left[\frac{i}{\hbar}(-p'_x x' + p'_y y' - \mathcal{E}' t)\right], \tag{3.10}$$

where

$$p'_x = \left(\frac{\mathcal{E}'^2}{c^2} - p'^2_y - m^2c^2\right)^{\frac{1}{2}}, \quad \varphi_1 = \begin{pmatrix}1\\0\end{pmatrix}, \quad \varphi_2 = \begin{pmatrix}0\\1\end{pmatrix}. \tag{3.11}$$

The solution of Eq. (3.8) in the form (3.9) corresponds to the expansion of the wave function in a complete set of the wave functions of a particle with certain energy and transverse momentum p'_y (with longitudinal momenta $\pm(\mathcal{E}'^2/c^2 - p_y'^2 - m^2c^2)^{1/2}$ and spin projections $S_z = \pm 1/2$). The latter are normalized to one particle per unit volume. Since there is symmetry with respect to the direction $\mathbf{A}_R$ (the OY axis), we have taken, without loss of generality, the vector $\mathbf{p}'$ in the XY plane ($p'_z = 0$).

According to Eqs. (3.9) and (3.10) the induced Cherenkov effect in the R frame corresponds to elastic scattering process by which the reflection of the particle from the wave field occurs: $p'_x \to -p'_x$. However, in contrast to classical reflection when the periodic wave field becomes a potential barrier for the particle at the intensity $\xi > \xi_{cr}$, this quantum above-barrier reflection takes place regardless of how weak the wave field is. Hence, the probability of multiphoton absorption/radiation of the incident wave photons by the particle in the L frame, that is, induced Cherenkov effect, will be determined by the probability of particle elastic reflection in the R frame.

Substituting Eq. (3.9) into Eq. (3.8) and then multiplying by the Hermitian conjugate functions and taking into account Eqs. (3.10) and (3.2) we obtain a set of differential equations for the unknown functions $a_i(x')$. The equations for a_1, a_3 and a_2, a_4 are separated and for these amplitudes we have the following set of equations:

$$p'_x \frac{da_1(x')}{dx'} = \frac{iep'_y}{\hbar c} A_y(x') a_1(x')$$

$$- \frac{e\left(p'_x - ip'_y\right)}{\hbar c} A_y(x') \exp\left(-\frac{2i}{\hbar} p'_x x'\right) a_3(x'),$$

$$p'_x \frac{da_3(x')}{dx'} = -\frac{ie}{\hbar c} p'_y A_y(x') a_3(x')$$

$$- \frac{e\left(p'_x + ip'_y\right)}{\hbar c} A_y(x') \exp\left(\frac{2i}{\hbar} p'_x x'\right) a_1(x'). \tag{3.12}$$

A similar set of equations is also obtained for the amplitudes $a_2(x')$ and $a_4(x')$. For simplicity we shall assume that before the interaction there are only particles with specified longitudinal momentum and spin state, i.e.,

$$|a_1(-\infty)|^2 = 1,\ |a_3(+\infty)|^2 = 0,\ |a_2(-\infty)|^2 = 0,\ |a_4(+\infty)|^2 = 0. \tag{3.13}$$

From the condition of conservation of the norm we have

$$|a_1(x')|^2 - |a_3(x')|^2 = \text{const} \tag{3.14}$$

and the probability of reflection is $|a_{3,4}(-\infty)|^2$.

The application of the unitarian transformation

$$a_1(x') = b_1(x') \exp\left(i\frac{ep'_y}{\hbar c p'_x}\int_{-\infty}^{x'} A_y(\eta)d\eta - i\frac{\vartheta'}{2}\right),$$

$$a_3(x') = b_3(x') \exp\left(-i\frac{ep'_y}{\hbar c p'_x}\int_{-\infty}^{x'} A_y(\eta)d\eta + i\frac{\vartheta'}{2}\right) \quad (3.15)$$

simplifies Eq. (3.12). Here ϑ' is the angle between the particle momentum and the direction of the wave propagation in the R frame. The new amplitudes $b_1(x')$ and $b_3(x')$ satisfy the same initial conditions: $|b_1(-\infty)|^2 = 1$, $|b_3(+\infty)|^2 = 0$, according to Eq. (3.13).

From Eqs. (3.12) and (3.15) for $b_1(x')$ and $b_3(x')$ we obtain the following set of equations:

$$\frac{db_1(x')}{dx'} = -f(x')b_3(x'),$$

$$\frac{db_3(x')}{dx'} = -f^*(x')b_3(x'), \quad (3.16)$$

where

$$f(x') = \frac{eA_y(t)p'}{\hbar c p'_x} \exp\left(-\frac{2i}{\hbar}p'_x x' - i\frac{2ep_y}{\hbar c p'_x}\int_{-\infty}^{x'} A_y(\eta)d\eta\right), \quad (3.17)$$

$$p' = \sqrt{p_y'^2 + p_x'^2}.$$

Using the following expansion by the Bessel functions

$$\exp\left(-i\alpha \sin k'x'\right) = \sum_{N=-\infty}^{\infty} J_N\left(\alpha\right) \exp\left(-iNk'x'\right),$$

we can reduce Eq. (3.16) to the form

$$\frac{db_1(x')}{dx'} = -\sum_{N=-\infty}^{\infty} f_N \exp\left[-\frac{i}{\hbar}(2p'_x - N\hbar k')x'\right] b_3(x'),$$

$$\frac{db_3(x')}{dx'} = -\sum_{N=-\infty}^{\infty} f_N \exp\left[\frac{i}{\hbar}(2p'_x - N\hbar k')x'\right] b_1(x'), \quad (3.18)$$

where

$$f_N = \frac{p'}{2p'_y} N k' J_N \left(2\xi \frac{mc}{p'_x} \frac{p'_y}{\hbar k'} \right). \tag{3.19}$$

Because of conservation of particle energy and transverse momentum (in R frame) the real transitions in the field will occur from a p'_x state to the $-p'_x$ one and, consequently, the probabilities of multiphoton scattering will have maximal values for the resonant transitions

$$2p'_x = s\hbar k' \qquad (s = \pm 1, \pm 2 ...). \tag{3.20}$$

The latter expresses the condition of exact resonance between the particle de Broglie wave and the incident "wave lattice". In the L frame the inelastic scattering of the particle on the moving phase lattice takes place and Eq. (3.20) corresponds to the known Cherenkov conservation law

$$\frac{2\mathcal{E}_0(1 - n_0 \frac{\mathrm{v}_0}{c} \cos\vartheta)}{(n_0^2 - 1)} = s\hbar\omega_0, \tag{3.21}$$

where ϑ is the angle between the particle momentum and the wave propagation direction (the Cherenkov angle), and v_0 and $\mathcal{E}_0$ are the particle initial velocity and energy in the L frame.

So, we can utilize the resonant approximation keeping only resonant terms in Eq. (3.18). Generally, in this approximation, at the detuning of resonance $|\delta_s| = \left| 2\frac{p'_x}{\hbar} - sk' \right| << k'$, we have the following set of equations for the particular s-photon transition amplitudes $b_1^{(s)}(x')$ and $b_3^{(s)}(x')$:

$$\frac{db_1^{(s)}(x')}{dx'} = -f_s \exp\left[-i\delta_s x'\right] b_3^{(s)}(x'),$$

$$\frac{db_3^{(s)}(x')}{dx'} = -f_s \exp\left[i\delta_s x'\right] b_1^{(s)}(x'). \tag{3.22}$$

This resonant approximation is valid for the slow varying functions $b_1^{(s)}(x')$ and $b_3^{(s)}(x')$, i.e., by the condition

$$\left| \frac{db_{1,3}^{(s)}(x')}{dx'} \right| << \left| b_{1,3}^{(s)}(x') \right| \cdot k'. \tag{3.23}$$

First we shall solve the case of exact resonance ($\delta_s = 0$). According to the boundary conditions (3.14) we have the following solutions for the amplitudes

$$b_1^{(s)}(x') = \frac{\cosh\left[\int_{x'}^{\infty} f_s d\eta\right]}{\cosh\left[\int_{-\infty}^{\infty} f_s d\eta\right]}, \qquad b_3^{(s)}(x') = \frac{\sinh\left[\int_{x'}^{\infty} f_s d\eta\right]}{\cosh\left[\int_{-\infty}^{\infty} f_s d\eta\right]} \tag{3.24}$$

and for the reflection coefficient

$$R^{(s)} = \left|b_3^{(s)}(-\infty)\right|^2 = \tanh^2\left[f_s \triangle x'\right], \tag{3.25}$$

where $\triangle x'$ is the coherent interaction length. The reflection coefficient in the laboratory frame of reference is the probability of absorption at $\mathrm{v}_0 < c/n_0$ or emission at $\mathrm{v}_0 > c/n_0$. The latter can be obtained expressing the quantities f_s and $\triangle x'$ by the quantities in this frame since the reflection coefficient is Lorentz invariant. So

$$R^{(s)} = \tanh^2\left[F_s \triangle \tau\right], \tag{3.26}$$

where

$$F_s = \left[\frac{(1 - n_0 \frac{\mathrm{v}_0}{c}\cos\vartheta)^2}{n_0^2 - 1} + \frac{\mathrm{v}_0^2}{c^2}\sin^2\vartheta\right]^{1/2}$$

$$\times \frac{s\omega_0 c}{2\mathrm{v}_0 \sin\vartheta} J_s\left(\xi \frac{2m\mathrm{v}_0 c \sin\vartheta}{\hbar\omega_0(1 - n_0 \frac{\mathrm{v}_0}{c}\cos\vartheta)}\right) \tag{3.27}$$

and $\triangle\tau$ for actual cases is the laser pulse duration in the L frame. The condition of applicability of this resonant approximation (3.23) is equivalent to the condition

$$|F_s| << \omega_0, \tag{3.28}$$

which restricts the intensity of the wave as well as the Cherenkov angle. Besides, to satisfy condition (3.28) we must take into account the very sensitivity of the parameter F_s toward the argument of the Bessel function, according to Eq. (3.27). For the wave intensities when $F_s \triangle\tau \gtrsim 1$ the reflection coefficient is of the order of one that can occur for a large number of photons $s >> 1$ for the argument of the Bessel function $\alpha \sim s \gg 1$ in Eq. (3.27) (according to the asymptotic behavior of Bessel function $J_s(\alpha)$ at $\alpha \simeq s \gg 1$).

For the off resonant solution, when $\delta_s \neq 0$, but $f_s^2 > \delta_s^2/4$ from Eq. (3.22) we obtain the following expression for $R^{(s)}$:

$$R^{(s)} = \frac{f_s^2}{\Omega_s^2} \frac{\sinh^2[\Omega_s \triangle x']}{1 + \frac{f_s^2}{\Omega_s^2}\sinh^2[\Omega_s \triangle x']}; \qquad \Omega_s = \sqrt{f_s^2 - \delta_s^2/4}, \tag{3.29}$$

which has the same behavior as in the case of exact resonance. In the opposite case when $f_s^2 \leq \delta_s^2/4$ the reflection coefficient is an oscillating function of interaction length.

3.2 Quantum Description of "Reflection" Phenomenon. Particle Beam Quantum Modulation at X-Ray Frequencies

Though the phenomenon of particle "reflection" from the front of a plane EM wave is of classical nature, which means that quantum effects of tunnel passage and above-barrier reflection should be small enough, nevertheless the quantum consideration of this phenomenon is worthy of note in relation to the appearance of an important coherent quantum effect as a result of classical "reflection" of particles. The influence of spin interaction is not essential here; on the other hand, it is quantitatively small enough in the induced Cherenkov process (for optical frequencies) and may be neglected. The qualitative aspect of spin effects in the induced Cherenkov process will be considered below.

Neglecting the spin interaction, the Dirac equation in quadratic form becomes the Klein–Gordon equation, so we will consider the problem on the basis of the equation

$$-\hbar^2\frac{\partial^2\Psi}{\partial t^2}=\left\{c^2\left[-i\hbar\nabla-\frac{e}{c}\mathbf{A}(t-n_0\frac{x}{c})\right]^2+m^2c^4\right\}\Psi. \tag{3.30}$$

Equation (3.30) over wave coordinates $\tau=t-n_0x/c$ and $\eta=t+n_0x/c$ is written as

$$\hbar^2\left(n_0^2-1\right)\frac{\partial^2\Psi}{\partial\tau^2}-2\hbar^2\left(n_0^2+1\right)\frac{\partial^2\Psi}{\partial\tau\partial\eta}+\hbar^2\left(n_0^2-1\right)\frac{\partial^2\Psi}{\partial\eta^2}$$

$$=c^2\left[-i\hbar\nabla-\frac{e}{c}\mathbf{A}(\tau)\right]^2\Psi+m^2c^4\Psi. \tag{3.31}$$

As the coordinate η is cyclic (as the transverse coordinates $\mathbf{r}_\perp$), then the corresponding component of generalized momentum p_η is conserved

$$p_\eta=\frac{1}{2}\left(\frac{c}{n_0}p_x-\mathcal{E}\right)=\text{const}, \tag{3.32}$$

which coincides (with a coefficient) with the classical integral of motion (2.5).

Hence, the solution of Eq. (3.30) may be sought in the form

$$\Psi(\tau,\eta,\mathbf{r}_\perp)=\Phi(\tau)\exp\left[\frac{i}{\hbar}\mathbf{p}_{\perp0}\mathbf{r}_\perp+\frac{i}{\hbar}p_\eta\eta\right], \tag{3.33}$$

where $\mathbf{p}_{\perp0}$ is the initial transverse momentum of the particle in the plane of wave polarization. Then for $\Phi(\tau)$ we have the equation

$$\hbar^2\left(n_0^2-1\right)\frac{d^2\Phi}{d\tau^2}-2i\hbar p_\eta\left(n_0^2+1\right)\frac{d\Phi}{d\tau}-p_\eta^2\left(n_0^2-1\right)\Phi$$

$$=c^2\left[\mathbf{p}_{\perp 0}-\frac{e}{c}\mathbf{A}(\tau)\right]^2\Phi+m^2c^4\Phi, \tag{3.34}$$

which within the transformation

$$\Phi\left(\tau\right)=U\left(\tau\right)\exp\left(\frac{i}{\hbar}\frac{n_0^2+1}{n_0^2-1}p_\eta\tau\right) \tag{3.35}$$

turns into the one-dimensional Schrödinger equation for the introduced new function $U\left(\tau\right)$

$$\frac{d^2U}{d\tau^2}+\frac{1}{\hbar^2}\frac{1}{\left(n_0^2-1\right)^2}\left\{4n_0^2p_\eta^2-\left(n_0^2-1\right)c^2\left[\mathbf{p}_{\perp 0}-\frac{e}{c}\mathbf{A}(\tau)\right]^2\right.$$

$$\left.-\left(n_0^2-1\right)m^2c^4\right\}U=0. \tag{3.36}$$

The exact solution of Eq. (3.36) can be obtained when the particle initial velocity is parallel to the wave propagation direction ($\mathbf{p}_{\perp 0}=0$) and the latter is monochromatic of circular polarization ($\mathbf{A}^2(\tau)=$const):

$$U\left(\tau\right)=C_1\exp\left[i\tau\frac{\mathcal{E}_0}{\hbar\left(n_0^2-1\right)}\sqrt{\left(1-n_0\frac{\mathrm{v}_0}{c}\right)^2-\left(n_0^2-1\right)\left(\frac{mc^2}{\mathcal{E}_0}\right)^2\xi^2}\right]$$

$$+C_2\exp\left[-i\tau\frac{\mathcal{E}_0}{\hbar\left(n_0^2-1\right)}\sqrt{\left(1-n_0\frac{\mathrm{v}_0}{c}\right)^2-\left(n_0^2-1\right)\left(\frac{mc^2}{\mathcal{E}_0}\right)^2\xi^2}\right], \tag{3.37}$$

One can define constants C_1 and C_2 by introducing an envelope for the monochromatic wave.

Equations (3.33), (3.35), and (3.37) determine the complete wave function of the particle

$$\Psi(\tau,\eta)=\exp\left[-i\frac{\mathcal{E}_0}{2\hbar}\left(1-\frac{\mathrm{v}_0}{cn_0}\right)\left(\eta+\frac{n_0^2+1}{n_0^2-1}\tau\right)\right]$$

$$\times\left\{C_1\exp\left[i\tau\frac{\mathcal{E}_0}{\hbar\left(n_0^2-1\right)}\sqrt{\left(1-n_0\frac{\mathrm{v}_0}{c}\right)^2-\left(n_0^2-1\right)\left(\frac{mc^2}{\mathcal{E}_0}\right)^2\xi^2}\right]\right.$$

$$+C_2 \exp\left[-i\tau\frac{\mathcal{E}_0}{\hbar\left(n_0^2-1\right)}\sqrt{\left(1-n_0\frac{\mathrm{v}_0}{c}\right)^2-\left(n_0^2-1\right)\left(\frac{mc^2}{\mathcal{E}_0}\right)^2\xi^2}\right]\Bigg\}, \qquad (3.38)$$

that is the superposition of two waves — incident and reflected — with the different energy values. If one moves from coordinates τ, η to t, x, these two values of particle energy will coincide with the classical expressions (2.10) that comprise the "reflection" phenomenon, where the sign "+" before the root corresponds to an incident particle, and the sign "−" to a reflected one.

To calculate the probability of reflection from the wave barrier one needs to consider an EM pulse with the envelope of intensity damped asymptotically at infinity. Let it have the form

$$\xi^2(\tau) = \frac{\xi_0^2}{\cosh^2\frac{\tau}{\tau_0}}, \qquad (3.39)$$

where ξ_0^2 is the maximal value of intensity and τ_0 is the half-width of the pulse.

The wave function of the particle at the interaction with the field (3.39) is expressed by the hypergeometric function and for the passage coefficient we obtain

$$D = \frac{\sinh^2\left(\frac{\pi}{2}\widetilde{\Omega}\tau_0\right)}{\sinh^2\left(\frac{\pi}{2}\widetilde{\Omega}\tau_0\right)+\cos^2\left(\frac{\pi}{2}\sqrt{1-\left(\widetilde{\Omega}\tau_0\right)^2\frac{\xi_0^2}{\xi_{cr}^2}}\right)} \quad \text{if} \quad \widetilde{\Omega}\tau_0\frac{\xi_0}{\xi_{cr}} < 1,$$

$$D = \frac{\sinh^2\left(\frac{\pi}{2}\widetilde{\Omega}\tau_0\right)}{\sinh^2\left(\frac{\pi}{2}\widetilde{\Omega}\tau_0\right)+\cosh^2\left(\frac{\pi}{2}\sqrt{\left(\widetilde{\Omega}\tau_0\right)^2\frac{\xi_0^2}{\xi_{cr}^2}-1}\right)} \quad \text{if} \quad \widetilde{\Omega}\tau_0\frac{\xi_0}{\xi_{cr}} > 1.$$

$$(3.40)$$

Here

$$\widetilde{\Omega} = 2\frac{\mathcal{E}_0}{\hbar\left(n_0^2-1\right)}\left|1-n_0\frac{\mathrm{v}_0}{c}\right| \qquad (3.41)$$

is the quantum frequency corresponding to particle classical energy change due to "reflection" (see Eq. (2.13)) and ξ_{cr} is the classical value of critical intensity (2.12).

The major quantity $\widetilde{\Omega}\tau_0$ in Eqs. (3.40) $\widetilde{\Omega}\tau_0 >> 1$ (for actual parameters of electron and laser beams in a medium with refractive index $n_0 - 1 \sim 10^{-4}$ the parameter $\widetilde{\Omega}\tau_0 \sim 10^{15} \div 10^{11}$ for laser pulse duration $\tau_0 \sim 10^{-8} \div 10^{-12}$ s), hence at $\xi_0 > \xi_{cr}$ for the coefficient of reflection we have

$$R = \frac{\exp\left[\pi\widetilde{\Omega}\tau_0\left(\frac{\xi_0}{\xi_{cr}} - 1\right)\right]}{1 + \exp\left[\pi\widetilde{\Omega}\tau_0\left(\frac{\xi_0}{\xi_{cr}} - 1\right)\right]}. \tag{3.42}$$

This equation shows that $R = 1$ with great accuracy (the coefficient of tunnel passage in this case is of the order $\exp\left[\left(-10^{15}\right) \div \left(-10^{11}\right)\right]$). If $\xi_0 < \xi_{cr}$ then the coefficient of reflection $R = 0$ with the same accuracy, i.e. the above barrier reflection is negligibly small in this case. Thus, the quantum effects of tunnel passage and above barrier reflection do not impact on the classical phenomenon of particle "reflection" from the plane EM wave. This is physically clear since the Compton wavelength of a particle (electron) is much smaller than the space size of actual EM pulses. Nevertheless, due to the particle quantum feature as a result of classical reflection the coherent effect of quantum modulation of the free particle probability density and, consequently, electric current density occurs because of superposition of an incident and reflected particle's waves.

Thus, the particle free state after the reflection ($\xi(\tau) = 0$) will be described by the asymptotic expression of Eq. (3.38), that is,

$$\Psi(x,t) = C_1\left\{\exp\left[\frac{i}{\hbar}\left(p_0x - \mathcal{E}_0t\right)\right]\right.$$

$$\left. + \exp\left[\frac{i}{\hbar}\left(p_0 \pm \frac{n_0\hbar\widetilde{\Omega}}{c}\right)x - \frac{i}{\hbar}\left(\mathcal{E}_0 \pm \hbar\widetilde{\Omega}\right)t + i\varphi_0\right]\right\}. \tag{3.43}$$

Here we have taken into account that the coefficient of reflection $R = |C_2|^2/|C_1|^2 = 1$ and the constant phase $\varphi_0 = \arg\left(C_2/C_1\right)$; constant C_1 is determined by the normalization condition. The signs $(\pm)$ in the exponent correspond to cases $\mathrm{v}_0 < c/n_0$ and $\mathrm{v}_0 > c/n_0$, respectively.

The density of electric current of the particle beam defined by Eq. (3.43) is modulated at frequency $\widetilde{\Omega}$

$$\mathbf{J}(x,t) = \mathbf{J}_0\left\{1 + \cos\left[\widetilde{\Omega}\left(t - n_0\frac{x}{c}\right) - \varphi_0\right]\right\}, \tag{3.44}$$

where $\mathbf{J}_0 = \mathrm{const}$ is the electric current density of the initially homogeneous and monochromatic particle beam. The modulation frequency $\widetilde{\Omega}$ in actual cases lies in the X-ray domain as follows from the estimation of particle classical energy change due to "reflection" $\Delta\mathcal{E}$ in Chapter 2 ($\widetilde{\Omega} = \Delta\mathcal{E}/\hbar$).

Note that quantum modulation in contrast to classical modulation is exceptionally the feature of a single particle and so is conserved after the interaction.

3.3 Exact Solution of the Dirac Equation for Induced Cherenkov Process

Consider the nonlinear quantum dynamics of a spinor particle in the field of a plane monochromatic EM wave in a medium. The exact solution of the Dirac equation can be found for the above-considered case when the particle initial velocity is parallel to the wave propagation direction and the latter is of circular polarization:

$$A_y = A_0 \sin \omega_0 \left(t - n_0 \frac{x}{c}\right); \quad A_z = A_0 \cos \omega_0 \left(t - n_0 \frac{x}{c}\right). \tag{3.45}$$

The Dirac equation in quadratic form for the spinor wave function

$$f = \begin{pmatrix} f_1 \\ f_2 \end{pmatrix}$$

in the field (3.45) is written as

$$\left\{\hbar^2 \frac{\partial^2}{\partial t^2} + c^2 \left(\hat{\mathbf{p}} - \frac{e}{c}\mathbf{A}\right)^2 - \hbar ec\sigma \left(\mathbf{H} + i\mathbf{E}\right) + m^2 c^4\right\} f = 0. \tag{3.46}$$

The complete wave function of the particle is determined by the spinor f as follows:

$$\Psi = \frac{1}{mc^2} \left[i\hbar\widehat{\beta} \frac{\partial}{\partial t} - c\widehat{\beta}\widehat{\alpha} \left(\hat{\mathbf{p}} - \frac{e}{c}\mathbf{A}\right) + mc^2\right] \begin{pmatrix} f \\ -f \end{pmatrix}, \tag{3.47}$$

where $\widehat{\alpha}$, $\widehat{\beta}$ are the Dirac matrices in the standard representation (3.2).

Equation (3.46) is a set of two differential equations of the second order for the spinor components f_1 and f_2. Passing from variables x, t to wave coordinates $\tau = t - n_0 x/c$, $\eta = t + n_0 x/c$ and looking for the solution of Eq. (3.46) in the form

$$f = e^{\frac{i}{\hbar} p_\eta \eta} \begin{pmatrix} f_1(\tau) e^{i\omega_0 \tau} \\ f_2(\tau) \end{pmatrix} \tag{3.48}$$

(the quantity $p_\eta = \text{const}$ is given by Eq. (3.32)), then the variables τ, η are separated and we obtain the following set of equations for f_1 and f_2:

$$\frac{d^2 f_1}{d\tau^2} + 2i \left(\omega_0 - \frac{p_\eta}{\hbar} \frac{n_0^2 + 1}{n_0^2 - 1}\right) \frac{df_1}{d\tau} - \left[\omega_0^2 - 2\omega_0 \frac{p_\eta}{\hbar} \frac{n_0^2 + 1}{n_0^2 - 1}\right.$$

$$
+\frac{\left(n_0^2-1\right)p_\eta^2+e^2A_0^2+m^2c^4}{\hbar^2\left(n_0^2-1\right)}\Bigg]f_1=-\frac{iecH_0}{\hbar n_0\left(n_0+1\right)}f_2, \tag{3.49}
$$

$$
\frac{d^2f_2}{d\tau^2}-2i\frac{p_\eta}{\hbar}\frac{n_0^2+1}{n_0^2-1}\frac{df_2}{d\tau}-\frac{\left(n_0^2-1\right)p_\eta^2+e^2A_0^2+m^2c^4}{\hbar^2\left(n_0^2-1\right)}f_2=\frac{iecH_0}{\hbar n_0\left(n_0-1\right)}f_1.
$$

Here H_0 is the amplitude of the wave magnetic field strength: $H_0 = n_0\omega_0A_0/c$.

This set of differential equations of the second order is equivalent to one differential equation of the fourth order the characteristic equation of which may be reduced to a biquadratic algebraic equation. The roots of the latter are

$$
\Omega_{1,2,3,4}=-\frac{\omega_0}{2}+\frac{p_\eta}{\hbar}\frac{n_0^2+1}{n_0^2-1}\pm\frac{\mathcal{E}_0}{\hbar\left(n_0^2-1\right)}
$$

$$
\times\sqrt{\left[1-n_0\frac{\mathrm{v}_0}{c}\pm\frac{\hbar\omega_0}{2\mathcal{E}_0}\left(n_0^2-1\right)\right]^2-\left(n_0^2-1\right)\left(\frac{mc^2}{\mathcal{E}_0}\right)^2\xi^2}, \tag{3.50}
$$

where the signs "±" before the root correspond to an incident and reflected particle analogously to Eq. (3.38). However, due to relativistic quantum effects (spin–field interaction and quantum recoil of photons) two different values of Ω arise as for the incident particle ($\Omega_{1,2}$) as well as for the reflected one ($\Omega_{3,4}$) corresponding to the signs "±" under the root. Consequently, two critical values of intensity appear here corresponding to different initial spin projections along the direction of particle motion:

$$
\xi_{cr1,2}^2=\left(\frac{\mathcal{E}_0}{mc^2}\right)^2\frac{\left[1-n_0\frac{\mathrm{v}_0}{c}\pm\frac{\hbar\omega_0}{2\mathcal{E}_0}\left(n_0^2-1\right)\right]^2}{\left(n_0^2\quad 1\right)}. \tag{3.51}
$$

From Eq. (3.47), within Eqs. (3.48) and (3.50) we obtain the complete wave function of a spinor particle. We present the ultimate equations for spin projections $-1/2$ and $1/2$. If the particle spin before the interaction is directed opposite to axis OX ($\sigma_x = -1$) we have

$$
\Psi_1\left(x,t\right)=C_1\begin{pmatrix}\left(a_1+a_2\right)e^{i\omega_0\left(t-n_0\frac{x}{c}\right)}\\ a_3+1\\ \left(a_1-a_2\right)e^{i\omega_0\left(t-n_0\frac{x}{c}\right)}\\ a_3-1\end{pmatrix}e^{\frac{i}{\hbar}\left(p_1x-\mathcal{E}_1t\right)}, \tag{3.52}
$$

where

$$\mathcal{E}_1 = \mathcal{E}_0 + \frac{\mathcal{E}_0}{n_0^2 - 1}\left(1 - n_0\frac{\mathrm{v}_0}{c} + \frac{\hbar\omega_0}{2\mathcal{E}_0}\left(n_0^2 - 1\right)\right)\left(1 - \sqrt{1 - \frac{\xi_0^2}{\xi_{cr1}^2}}\right), \quad (3.53)$$

and p_1 is determined by $\mathcal{E}_1$ via conserved quantity p_η. The quantities in the bispinor (3.52) are

$$a_1 = a_2\frac{n_0+1}{n_0-1}\frac{\mathcal{E}_0 - cp_0}{mc^2}; \quad a_2 = i\left(n_0 - 1\right)\frac{\mathcal{E}_1 - \mathcal{E}_0}{mc^2\xi_0}; \quad a_3 = \frac{\mathcal{E}_0 - cp_0}{mc^2}$$

and the coefficient of normalization (one particle in the unit volume)

$$C_1 = \frac{1}{\sqrt{2}}\left(1 + |a_1|^2 + |a_2|^2 + |a_3|^2\right)^{-1/2}.$$

In the case of $\sigma_x = +1$ we have

$$\Psi_2\left(x,t\right) = C_2\begin{pmatrix} b_3 + 1 \\ \left(b_1 + b_2\right)e^{-i\omega_0\left(t - n_0\frac{x}{c}\right)} \\ b_3 - 1 \\ \left(b_1 - b_2\right)e^{-i\omega_0\left(t - n_0\frac{x}{c}\right)} \end{pmatrix} e^{\frac{i}{\hbar}\left(p_2 x - \mathcal{E}_2 t\right)}, \quad (3.54)$$

where

$$\mathcal{E}_2 = \mathcal{E}_0 + \frac{\mathcal{E}_0}{n_0^2 - 1}\left(1 - n_0\frac{\mathrm{v}_0}{c} - \frac{\hbar\omega_0}{2\mathcal{E}_0}\left(n_0^2 - 1\right)\right)\left(1 - \sqrt{1 - \frac{\xi_0^2}{\xi_{cr2}^2}}\right). \quad (3.55)$$

The bispinor (3.54) is determined by the quantities

$$b_1 = b_2\frac{n_0-1}{n_0+1}\frac{\mathcal{E}_0 + cp_0}{mc^2}; \quad b_2 = i\left(n_0 + 1\right)\frac{\mathcal{E}_2 - \mathcal{E}_0}{mc^2\xi_0}; \quad b_3 = \frac{\mathcal{E}_0 + cp_0}{mc^2},$$

and the normalization coefficient

$$C_2 = \frac{1}{\sqrt{2}}\left(1 + |b_1|^2 + |b_2|^2 + |b_3|^2\right)^{-1/2}.$$

The wave functions of reflected particles Ψ_3 and Ψ_4 corresponding to spin projections $\sigma_x = +1$ and $\sigma_x = -1$, respectively, are obtained from the expressions Ψ_2 and Ψ_1 by the replacement $\Omega_2 \to \Omega_3$ and $\Omega_1 \to \Omega_4$ and for $\mathcal{E}_{3,4}$ we have

$$\mathcal{E}_{3,4} = \mp\hbar\omega_0 + \mathcal{E}_0 + \frac{\mathcal{E}_0}{n_0^2-1}\left(1 - n_0\frac{\mathrm{v}_0}{c} \pm \frac{\hbar\omega_0}{2\mathcal{E}_0}\left(n_0^2-1\right)\right)$$
$$\times\left(1+\sqrt{1-\frac{\xi_0^2}{\xi_{cr1,2}^2}}\right) \tag{3.56}$$

In particular, from this equation it follows that in Eq. (3.51) ξ_{cr2} corresponds to a particle with the spin directed along the axis OX, while ξ_{cr1} corresponds to the opposite one. The normalization coefficients can be defined by introducing the wave envelope as was stated in Section 3.2.

The expressions of particle wave functions show that the degeneration of particle states over the spin projection that takes place in vacuum (Volkov states) vanishes in a dielectriclike medium. In that case the wave function Ψ_1 corresponds to superposition state with energies $\mathcal{E}_1$ and $\mathcal{E}_1 - \hbar\omega_0$, while Ψ_2 corresponds to energies $\mathcal{E}_2$ and $\mathcal{E}_2 + \hbar\omega_0$. The removal of degeneration of Volkov states is related to the fact that in a medium with refractive index $n_0 > 1$ in the intrinsic frame of reference of the wave there is only a static magnetic field and the spin interaction with such a field results in the splitting of the particle states as by the Zeeman effect. The splitting value ($\Delta\mathcal{E} = |\mathcal{E}_1 - \mathcal{E}_2| = |\mathcal{E}_4 - \mathcal{E}_3|$) is

$$\Delta\mathcal{E} = \frac{\mathcal{E}_0}{n_0^2-1}\left|\left(1 - n_0\frac{\mathrm{v}_0}{c} + \frac{\hbar\omega_0}{2\mathcal{E}_0}\left(n_0^2-1\right)\right)\left(1-\sqrt{1-\frac{\xi_0^2}{\xi_{cr1}^2}}\right)\right.$$
$$\left. - \left(1 - n_0\frac{\mathrm{v}_0}{c} - \frac{\hbar\omega_0}{2\mathcal{E}_0}\left(n_0^2-1\right)\right)\left(1-\sqrt{1-\frac{\xi_0^2}{\xi_{cr2}^2}}\right)\right|. \tag{3.57}$$

As is seen from Eqs. (3.52)–(3.55), in vacuum this splitting vanishes and the wave functions Ψ_1 and Ψ_2 pass into Volkov wave function (1.93).

The spin interaction in a medium within the nonlinear threshold phenomenon of particle "reflection" may lead to particle beam polarization since the critical intensity (3.51) depends on spin projection along the direction of particle motion. Thus, if the condition $\xi_{cr2}^2 < \xi^2 < \xi_{cr1}^2$ holds, then only the particles with certain direction of the spin (along the axis OX) will be reflected. Since the velocities of reflected particles are different from the nonreflected ones, then by separating the particles after the interaction a polarized beam may be obtained.

3.4 Secular Perturbation at Nonlinear Cherenkov Resonance

The multiphoton induced Cherenkov interaction in the capture regime corresponding to transitions between the particle bound states occurs at the nonzero initial angles of particle motion with respect to the wave propagation direction, at which, as mentioned above, the Dirac or Klein–Gordon equations are of Hill or Mathieu type and unable to solve it exactly. However, as was shown in the quantum description of "reflection" phenomenon (free–free transitions), the interaction at the arbitrary initial angle resonantly connects two states of the particle (in the intrinsic frame of reference of the wave the states with longitudinal momenta p_x of the incident particle and $p_x + s\hbar k$ of the scattered particle; s is the number of absorbed or radiated photons with a wave vector $\mathbf{k}$), which makes available the application of resonant approximation to determine the multiphoton probabilities of free–free transitions in induced nonlinear Cherenkov process. Concerning the quantum description of the particle's bound states in the capture regime one must take into account the degeneration of initial states of free particles in the "longitudinal momentum". Therefore, regardless of how weak the field of the wave is, the usual perturbation theory in stimulated Cherenkov process is not applicable because of such degeneration of the states and the interaction near the resonance is needed for description by the secular equation. The latter, in particular, reveals the zone structure of the particle states in the field of a transverse EM wave in a dielectriclike medium. Note that in contrast to the zone structure for the energy of electron states in a crystal lattice, the zone structure in this process holds for the conserved quantity p_η, as the energy could not be quantum characteristic of the state in the nonstationary field of the wave.

First we will solve the Klein–Gordon equation for a scalar particle (3.30) in the given coherent radiation field in a medium (3.45) or the equivalent one-dimensional equation of the Schrödinger type (3.36) in the wave coordinate τ.

Within Eq. (3.30) the state parameter p_η can be expressed by the initial parameters of a free particle:

$$4n_0^2 p_\eta^2 - (n_0^2 - 1)(\mathbf{p}_{\perp 0}^2 c^2 + m^2 c^4) = \mathcal{E}_0^2 \left(1 - n_0 \frac{\mathrm{v}_0}{c} \cos\vartheta\right)^2$$

and for the circular polarization of the wave (3.45), Eq. (3.36) may be represented in the form

$$\frac{d^2 U(\tau)}{d\tau^2} + \frac{\mathcal{E}_0^2}{\hbar^2 (n_0^2 - 1)^2} \left[\left(1 - n \frac{\mathrm{v}_0}{c} \cos\vartheta\right)^2 + 2(n_0^2 - 1) \right.$$

$$\times\left(\frac{mc^2}{\mathcal{E}_0}\right)^2\frac{p_0}{mc}\xi_0\sin\vartheta\cos\omega\tau-(n_0^2-1)\left(\frac{mc^2}{\mathcal{E}_0}\right)^2\xi_0^2\Bigg]U(\tau)=0. \quad (3.58)$$

($\mathcal{E}_0$, p_0, v_0 are the initial values of energy, momentum, and velocity of a free particle, ϑ is the angle between the initial momentum of a particle and the wave vector of the wave; due to the azimuthal symmetry in the direction of the wave propagation OX, without loss of generality, the initial momentum of the particle is chosen in the plane XZ.)

According to Floquet's theorem the solution of Eq.(3.58) is sought in the form

$$U(\tau)=e^{i\frac{p_\tau}{\hbar}\tau}\sum_{s=-\infty}^{\infty}\Phi_s e^{-is\omega_0\tau}, \quad (3.59)$$

where

$$p_\tau^2\equiv\frac{\mathcal{E}_0^2}{(n_0^2-1)^2}\left[\left(1-n_0\frac{\mathrm{v}_0}{c}\cos\vartheta\right)^2-(n_0^2-1)\left(\frac{mc^2}{\mathcal{E}_0}\right)^2\xi_0^2\right] \quad (3.60)$$

is the major quantity in the induced nonlinear Cherenkov process, which is the renormalized (because of intensity effect) generalized momentum of the particle in the laboratory frame conjugate to wave coordinate τ. It connects the "width of initial Cherenkov resonance" $1-n_0\mathrm{v}_0/c$ and wave intensity (ξ_0^2) as the main relation between the physical quantities of this process determining also the condition of nonlinear resonance (see Chapter 2; $\mathrm{v}_x(\xi)\mid_{\xi=\xi_{cr}}=c/n_0$). In the intrinsic frame of reference of the wave p_τ corresponds to longitudinal momentum p_x of the particle on which the degeneration exists.

From Eqs. (3.58) and (3.59) for the coefficients Φ_s we obtain the recurrent equation

$$(s^2\hbar^2\omega_0^2-2s\hbar\omega_0 p_\tau)\Phi_s=\frac{mc^3p_0\xi_0\sin\vartheta}{(n_0^2-1)}\left[\Phi_{s-1}+\Phi_{s+1}\right], \quad (3.61)$$

which can be solved in approximation of the perturbation theory by the wave function:

$$|\Phi_1|<<|\Phi_0|,\quad |\Phi_2|<<|\Phi_1|,\ldots. \quad (3.62)$$

Then from Eq. (3.61) we find the amplitudes of the particle wave function, corresponding to an s photon process. But for condition (3.62) to hold, it is necessary that

$$\left|s^2\hbar^2\omega_0^2-2s\hbar\omega_0 p_\tau\right|>>\left|\frac{mc^3}{(n_0^2-1)}p_0\xi_0\sin\vartheta\right|. \quad (3.63)$$

Regarding those values p_τ for which condition (3.63) does not hold, the usual perturbation theory is already not applicable. In particular, if the expression on the left-hand side of this condition is zero, i.e., at $s = 0$ and $s = \ell$ ($\ell = 1, 2, 3, \dots$), when

$$2\frac{p_\tau}{\hbar} = \ell\omega_0, \tag{3.64}$$

from Eqs. (3.58) and (3.59) it is evident that we already have two states Φ_0 and Φ_ℓ, which are degenerated in the "longitudinal momentum" p_τ, since $p_\tau^2 = (p_\tau - \ell\hbar\omega_0)^2$. Because of this double degeneration in the state parameter p_τ for the definite p_η of the initial unperturbed system it is necessary to use perturbation theory for the degenerated states on the basis of the secular equation.

Thus, under condition (3.64), Eq.(3.58) within perturbation theory should be solved on the basis of the secular equation, according to which we search the solution in the form

$$U(\tau) = e^{i\frac{p_\tau}{\hbar}\tau}\left(\Phi_0 + \Phi_\ell e^{-i\ell\omega_0\tau}\right) = \Phi_0 e^{i\frac{\ell\omega_0}{2}\tau} + \Phi_\ell e^{-i\frac{\ell\omega_0}{2}\tau} \tag{3.65}$$

and the conserved quantity $p_\eta = p_\eta^{(0)} + p_\eta^{(1)}$, where $p_\eta^{(0)}$ is the value corresponding to the Bragg resonance condition (3.64).

In the case of one-photon interaction ($\ell = 1$), substituting Eq. (3.65) in Eq. (3.58), we obtain

$$\Delta_\tau \Phi_0 e^{i\frac{\omega_0}{2}\tau} + \Delta_\tau \Phi_1 e^{-i\frac{\omega_0}{2}\tau} + 2\alpha_1 \Phi_0 e^{i\frac{\omega_0}{2}\tau}\cos\omega_0\tau$$

$$+2\alpha_1 \Phi_1 e^{-i\frac{\omega_0}{2}\tau}\cos\omega_0\tau = 0, \tag{3.66}$$

where Δ_τ is the correction to the value p_τ^2 at the fulfillment of condition (3.64) for $\ell = 1$:

$$\Delta_\tau \equiv \frac{8n_0^2 p_\eta^{(0)}}{(n_0^2 - 1)^2} p_\eta^{(1)}; \qquad \alpha_1 \equiv \frac{mc^3 p_0 \xi_0 \sin\vartheta}{(n_0^2 - 1)}. \tag{3.67}$$

By the standard method from Eq.(3.67) one can obtain the following set of equations for the amplitudes Φ_0 and Φ_1:

$$\begin{cases} \Delta_\tau \Phi_0 + \alpha_1 \Phi_1 = 0, \\ \Delta_\tau \Phi_1 + \alpha_1 \Phi_0 = 0. \end{cases} \tag{3.68}$$

From the compatibility of Eqs. (3.68) we have $\Delta_\tau = \pm\alpha_1$. The signs "+" and "−" relate to $p_\tau^2 > \hbar^2\omega_0^2/4$ and $0 < p_\tau^2 < \hbar^2\omega_0^2/4$, respectively. Thus, at the fulfillment of condition (3.64) we have a jump in the value of p_τ^2, which is equal to $2\alpha_1$, i.e.,

$$\frac{\mathcal{E}_0^2}{(n_0^2-1)^2}\left\{\left(1-n_0\frac{\mathrm{v}_0}{c}\cos\vartheta\right)^2-(n_0^2-1)\left(\frac{mc^2}{\mathcal{E}_0}\right)^2\xi_0^2\right\}\geq\frac{\hbar^2\omega_0^2}{4}+\alpha_1,$$

$$0\leq\frac{\mathcal{E}_0^2}{(n_0^2-1)^2}\left\{\left(1-n_0\frac{\mathrm{v}_0}{c}\cos\vartheta\right)^2-(n_0^2-1)\left(\frac{mc^2}{\mathcal{E}_0}\right)^2\xi_0^2\right\}$$

$$\leq\frac{\hbar^2\omega_0^2}{4}-\alpha_1. \tag{3.69}$$

For $\ell=1$ the matrix element of transition from state $\varPhi_0$ to state $\varPhi_1$ (here we note the state without a phase) is equal to α_1, which is also evident from Eq. (3.68). For large ℓ ($\ell\geq 2$) the matrix element of transition $\varPhi_0\longleftrightarrow\varPhi_\ell$ is equal to zero in the first order of perturbation theory. In this case it makes sense to take into account the transitions to the states with other energies in higher order. For example, for $\ell=2$ it is necessary to consider the transitions $\varPhi_0\to\varPhi_1$ and $\varPhi_0\to\varPhi_2$. For arbitrary ℓ the matrix element of transition is defined by

$$\alpha_\ell=\frac{\alpha_1^\ell}{((\ell-1)!)^2\,(\hbar\omega_0)^{2(\ell-1)}}. \tag{3.70}$$

It should be noted that here it is also necessary to take into account the corrections to the energy eigenvalue of state $\varPhi_0$ in the appropriate order, however, the latter are only of quantitative character, unlike the qualitative corrections (3.70), and will be omitted.

As is seen from Eq. (3.69), the permitted and forbidden zones arise for the particle states in the wave. The widths of permitted zones in the general case of ℓ-photon resonance are defined from the condition

$$\frac{\ell^2\hbar^2\omega_0^2}{4}+\alpha_\ell\leq\frac{\mathcal{E}_0^2}{(n_0^2-1)^2}\left\{\left(1-n_0\frac{\mathrm{v}_0}{c}\cos\vartheta\right)^2-(n_0^2-1)\left(\frac{mc^2}{\mathcal{E}_0}\right)^2\xi_0^2\right\}$$

$$\leq\frac{(\ell+1)^2\hbar^2\omega_0^2}{4}-\alpha_{\ell+1}. \tag{3.71}$$

Such zone structure for the particle states in the wave arises in dielectriclike media because of particle capture by the wave and periodic character of the field — quantum influence of infinite "potential" wells on the particle states similar to zone structure of electron states in a crystal lattice.

To investigate the particle wave functions on the edges of the forbidden zones we turn to the set of equations (3.68). The latter has two solutions and, hence, from Eq. (3.65) we obtain two wave functions corresponding to the top and bottom borders of the forbidden zone (3.71). Thus, for $\Delta_\tau = \alpha_1$ we obtain

$$U_+(\tau) = 2i\Phi_0 \sin \frac{\omega_0}{2}\tau, \tag{3.72}$$

and at $\Delta_\tau = -\alpha_1$

$$U_-(\tau) = 2\Phi_0 \cos \frac{\omega_0}{2}\tau. \tag{3.73}$$

With the help of Eqs. (3.72) and (3.73) the particle wave function is determined by

$$\Psi_\pm(\mathbf{r}, t) = U_\pm(\tau) \exp\left[\frac{i}{\hbar}\mathbf{p}_{\perp 0}\mathbf{r} + \frac{i}{\hbar}p_\eta \eta + \frac{i}{\hbar}\frac{n_0^2+1}{n_0^2-1}p_\eta \tau\right]. \tag{3.74}$$

The condition at which secular perturbation theory is valid taking into account the above-stated degeneration, is $\alpha_1 << \hbar^2\omega_0^2/4$, or

$$\frac{4mc^3 p_0 \xi_0 \sin\vartheta}{\hbar^2\omega_0^2(n_0^2-1)} << 1. \tag{3.75}$$

Thus, it can be concluded that in the induced Cherenkov process there exists zone structure for the quantum parameters p_η, $p_{\perp 0}$ (or quantity p_τ (3.60) corresponding to multiphoton "Bragg resonance" (3.64)) of the particle state in the wave. The permitted zones for this quantity are determined by condition (3.71).

Consider now the case of spinor particles. Proceeding from the Dirac equation, the wave function of a particle can be presented in the form

$$\Psi = \frac{1}{mc^2}\left[i\hbar\widehat{\beta}\frac{\partial}{\partial t} - c\widehat{\beta}\widehat{\alpha}(\widehat{\mathbf{p}} - \frac{e}{c}\mathbf{A}) + mc^2\right]\begin{pmatrix} U_\sigma \\ -U_\sigma \end{pmatrix}$$

$$\times \exp\left[\frac{i}{\hbar}\mathbf{p}_{\perp 0}\mathbf{r} + \frac{i}{\hbar}p_\eta \eta + \frac{i}{\hbar}\frac{n_0^2+1}{n_0^2-1}p_\eta \tau\right], \tag{3.76}$$

where $\widehat{\alpha}$, $\widehat{\beta}$ are the Dirac matrices (3.2) in the standard representation. The spinor function U_σ satisfies the equation

$$\frac{d^2 U_\sigma(\tau)}{d\tau^2} + \frac{1}{\hbar^2(n_0^2-1)^2}\left[4n_0^2 p_\eta^2 - (n_0^2-1)c^2\left(\mathbf{p}_{\perp 0} - \frac{e}{c}\mathbf{A}(\tau)\right)^2\right.$$

$$-(n_0^2 - 1)m^2c^4 + (n_0^2 - 1)\hbar ec\sigma(\mathbf{H} + i\mathbf{E})\Big] U_\sigma(\tau) = 0, \qquad (3.77)$$

where $\mathbf{E} = -\partial\mathbf{A}/c\partial t$ and $\mathbf{H} = \mathrm{rot}\mathbf{A}$ are the electric and magnetic field strengths of the wave. In the case of a linearly polarized wave (Eq.(3.45) at $A_z = 0$), with the help of a unitarian transformation of spinor wave function it is possible to obtain a system of two independent equations of second order for the components of new spinor function from Eq.(3.77). For the other polarizations of the wave, in particular, circular polarization, the components of spinor function are not separated and Eq.(3.77) is equivalent to a differential equation of fourth order (it is related to the absence of a definite field direction, for which the spin projection could have a definite value, as occurs for linear polarization). The above stated spinor transformation, in the case of a linearly polarized wave, is

$$U_\sigma(\tau) = \left(\cosh\frac{\delta}{2} - \sigma_x \sinh\frac{\delta}{2}\right)\begin{pmatrix} V_1(\tau) \\ V_2(\tau) \end{pmatrix}; \quad \tanh\delta = \frac{E}{H} = \frac{1}{n_0}, \qquad (3.78)$$

which represents the transformation of the spinor in four-dimensional space $(\mathbf{r}, t)$ at a rotation by angle δ. The latter has a simple physical interpretation. It corresponds to the Lorentz transformation in a system of reference moving with a velocity $\mathrm{V} = c/n_0$, where the wave electric field $\mathbf{E}' = 0$ and there is only a static magnetic field $\mathbf{H}'$, directed along the axis Z and the spin projection on it has a definite value, since in the chosen representation the matrix σ_z is diagonal.

Thus, after the transformation (3.78), Eq. (3.77) will be transformed into the following independent equations for the spinor components V_1, V_2:

$$\frac{d^2V_1(\tau)}{d\tau^2} + \frac{1}{\hbar^2(n_0^2-1)^2}\left\{4n_0^2p_\eta^2 - (n_0^2-1)c^2\left(\mathbf{p}_{\perp 0} - \frac{e}{c}\mathbf{A}(\tau)\right)^2 - (n_0^2-1)m^2c^4\right\}V_1(\tau) + \frac{ecH}{\hbar n_0\sqrt{n_0^2-1}}V_1(\tau) = 0, \qquad (3.79)$$

$$\frac{d^2V_2(\tau)}{d\tau^2} + \frac{1}{\hbar^2(n_0^2-1)^2}\left\{4n_0^2p_\eta^2 - (n_0^2-1)c^2\left(\mathbf{p}_{\perp 0} - \frac{e}{c}\mathbf{A}(\tau)\right)^2 - (n_0^2-1)m^2c^4\right\}V_2(\tau) - \frac{ecH}{\hbar n_0\sqrt{n_0^2-1}}V_2(\tau) = 0. \qquad (3.80)$$

The solution of Eq. (3.79) (or (3.80)) is sought in the form

$$V_1(\tau) = e^{i\frac{p_\tau}{\hbar}\tau} \sum_{s=-\infty}^{\infty} K_s e^{-is\omega_0\tau}, \tag{3.81}$$

where

$$p_\tau \equiv \frac{\mathcal{E}_0}{(n_0^2 - 1)} \left[\left(1 - n_0 \frac{\mathrm{v}_0}{c} \cos\vartheta\right)^2 - \frac{1}{2}(n_0^2 - 1) \left(\frac{mc^2}{\mathcal{E}_0}\right)^2 \xi_0^2 \right]^{\frac{1}{2}}$$

is the particle "longitudinal momentum" in the wave of linear polarization.

Repeating the procedure as in the case of scalar particles, we obtain the Bragg condition (3.64), at which it is necessary to use the secular perturbation theory for degenerated states. At $\ell = 1$ we obtain the following system of equations for coefficients K_0 and K_1:

$$\begin{cases} \Delta_\tau K_0 + \left(-i\alpha_1 + \frac{1}{2}\mu\right) K_1 = 0, \\ \left(i\alpha_1 + \frac{1}{2}\mu\right) K_0 + \Delta_\tau K_1 = 0, \end{cases} \tag{3.82}$$

where

$$\mu = \frac{\hbar ecH}{n_0\sqrt{n_0^2 - 1}}. \tag{3.83}$$

From Eq. (3.82) for the correction to p_τ^2 we obtain

$$\Delta_\tau = \pm \left(\frac{1}{4}\mu^2 + \alpha_1^2\right)^{\frac{1}{2}}. \tag{3.84}$$

It is easy to see that $K_1 = \mp K_0 e^{i\varphi}$, where $tg\varphi = 2\alpha_1/\mu$. Hence, each spinor component of particle wave function has two values corresponding to the top and bottom borders of the first forbidden zone:

$$V_1^+(\tau) = K_0 \left(e^{i\frac{\omega_0}{2}\tau} - e^{-i\frac{\omega_0}{2}\tau + i\varphi}\right),$$

$$V_1^-(\tau) = K_0 \left(e^{i\frac{\omega_0}{2}\tau} + e^{-i\frac{\omega_0}{2}\tau + i\varphi}\right). \tag{3.85}$$

For $V_2(\tau)$ we have the same expressions as (3.85), where it is only necessary to replace φ by $-\varphi$.

At $\ell = 2$ we have already two channels for the transition from state K_0 to state K_2. The first is the result of the interaction described by a term quadratic in the field ($\sim A^2$), the matrix element of which at $\ell = 2$ is equal to $\left(mc^2\right)^2 \xi_0^2 / 4\hbar^2 (n_0^2 - 1)$, and the second channel proceeds both in the case of scalar particles via transitions $K_0 \to K_1$ and $K_0 \to K_2$, stipulated by the charge interaction $\sim \mathbf{pA}$, as well as for the spin interaction, the matrix elements of which at each transition are equal to $-i\alpha_1$ and $\mu/2$, respectively. Therefore, for two-photon transition

$$\Delta_\tau = \pm\frac{1}{\hbar^2\omega_0^2}\left[\left(\frac{1}{4}\mu^2 - \alpha_1^2 + \frac{\hbar^2\omega_0^2(mc^2)^2\xi_0^2}{4(n_0^2-1)}\right)^2 + \alpha_1^2\mu^2\right]^{\frac{1}{2}} \tag{3.86}$$

and on the borders of the second forbidden zone for the component of spinor function V we obtain (for top and bottom borders accordingly)

$$V_{1,2}^+ = K_0\left(e^{i\omega_0\tau} - e^{i\omega_0\tau\pm i\varphi}\right); \qquad V_{1,2}^- = K_0\left(e^{i\omega_0\tau} + e^{-i\omega_0\tau\pm i\varphi}\right), \tag{3.87}$$

where

$$tg\varphi = \frac{\alpha_1\mu}{\frac{1}{4}\mu^2 - \alpha_1^2 + \frac{\hbar^2\omega_0^2(mc^2)^2\xi_0^2}{4(n_0^2-1)}}. \tag{3.88}$$

The obtained results for spinor particles are valid at the fulfillment of the condition

$$|\Delta_\tau| << \frac{\hbar^2\omega^2}{4}. \tag{3.89}$$

Thus, the quantum picture of induced Cherenkov interaction for charged spinor particles does not differ qualitatively from the case of scalar particles, i.e., the spin interaction results only in quantitative corrections to the quantities describing the process. However, in the absence of charge interaction ($\mathbf{pA} = 0$) in the first order in the field, i.e., for one-photon interaction, the first forbidden zone ($\ell = 1$) does not exist for scalar particles, but exists for spinor particles due to the spin interaction.

3.5 Inelastic Diffraction Scattering on a Traveling Wave

Up to now we have considered the nonlinear phenomena in induced Cherenkov process at the external wave intensities exceeding the critical one — the threshold value of nonlinear Cherenkov resonance in the strong EM radiation field. However, purely quantum effects at the wave intensities under the critical value in induced Cherenkov process exist. Those are the inelastic diffraction scattering of charged particles on a traveling wave in dielectriclike media and quantum modulation of particle beams at the wave fundamental frequency and its harmonics. This and the next section of the present chapter will consider these effects.

Consider first the diffraction of particles on the phase lattice of a slowed traveling wave in a dielectriclike medium. Neglecting the spin interaction, the

Dirac equation in quadratic form is written as the Klein–Gordon equation for the particle in the field of a plane EM wave with vector potential $\mathbf{A}(\tau)$:

$$-\hbar^2\frac{\partial^2\Psi}{\partial t^2}=\left\{-\hbar^2c^2\nabla^2+m^2c^4+2ie\hbar c\mathbf{A}(\tau)\nabla+e^2A^2(\tau)\right\}\Psi. \qquad (3.90)$$

Equation (3.90) will be solved in the eikonal approximation by particle wave function

$$\Psi(\mathbf{r},t)=\sqrt{\frac{N_0}{2\mathcal{E}_0}}f(x,t)\exp\left[\frac{i}{\hbar}\left(\mathbf{p}_0\mathbf{r}-\mathcal{E}_0t\right)\right], \qquad (3.91)$$

according to which $f(x,t)$ is a slowly varying function with respect to free–particle wave function (the latter is normalized on N_0 particles per unit volume):

$$\left|\frac{\partial f}{\partial t}\right|<<\frac{\mathcal{E}_0}{\hbar}\left|f\right|; \qquad \left|\frac{\partial f}{\partial x}\right|<<\frac{p_{0x}}{\hbar}\left|f\right|. \qquad (3.92)$$

Choosing a concrete polarization of the wave (assume a linear one along the axis OY) and taking into account Eq. (3.90) for $f(x,t)$ we will have a differential equation of the first order:

$$\frac{\partial f}{\partial t}+\mathrm{v}_0\cos\vartheta_0\frac{\partial f}{\partial x}$$

$$=\frac{i}{2\hbar\mathcal{E}_0}\left[2ecp_0\sin\vartheta_0\cdot A_0(\tau)\cos\omega_0\tau-e^2A_0^2(\tau)\cos^2\omega_0\tau\right]f(x,t), \qquad (3.93)$$

where $A_0(\tau)$ is a slowly varying amplitude of the vector potential of quasi-monochromatic wave and ϑ_0 is the angle between the particle velocity and wave propagation direction. As $\xi_{\max}<\xi_{cr}<<1$, then for actual values of parameters $p_0\sin\vartheta_0/mc>>\xi_{\max}$ and the last term $\sim A_0^2$ in Eq. (3.93) will be neglected. Changing to characteristic coordinates $\tau'=t-x/\mathrm{v}_0\cos\vartheta_0$ and $\eta'=t$, it will be obvious that at the fulfillment of the induced Cherenkov condition $\mathrm{v}_0\cos\vartheta_0=c/n_0$ the traveling wave in this frame of coordinates becomes a diffraction lattice over the coordinate τ' and for the scattered amplitude of the particle wave function from Eq. (3.93) we have

$$f(\tau')=\exp\left\{\frac{iecp_0\sin\vartheta_0}{\hbar\mathcal{E}_0}\cos\omega_0\tau'\int_{\eta_1}^{\eta_2}A(\eta')d\eta'\right\}, \qquad (3.94)$$

where η_1 and η_2 are the moments of the particle entrance into the wave and exit, respectively. If one returns to coordinates x and t and expands the exponential (3.94) into a series by Bessel functions for the total wave function (3.91) we will have

$$\Psi(\mathbf{r},t) = \sqrt{\frac{N_0}{2\mathcal{E}_0}} \exp\left(\frac{i}{\hbar} y p_0 \sin\vartheta_0\right)$$

$$\times \sum_{s=-\infty}^{+\infty} i^s J_s(\alpha) \exp\left(\frac{i}{\hbar}\left[p_0\cos\vartheta_0 - \frac{s n_0 \hbar\omega_0}{c}\right] x - \frac{i}{\hbar}\left[\mathcal{E}_0 - s\hbar\omega_0\right] t\right), \quad (3.95)$$

where the argument of the Bessel function

$$\alpha = \frac{e\mathrm{v}_0 \sin\vartheta_0}{\hbar\omega_0} \int_{t_1}^{t_2} E(\eta')d\eta', \quad (3.96)$$

and E is the amplitude of the wave electric field strength. The wave function (3.95) describes inelastic diffraction scattering of the particles on the slowed traveling wave in a dielectriclike medium. The particles' energy and momentum after the scattering are

$$\mathcal{E} = \mathcal{E}_0 - s\hbar\omega_0; \quad p_x = p_0\cos\vartheta_0 - \frac{s n_0 \hbar\omega_0}{c}; \quad p_y = \mathrm{const}; \quad s = 0, \pm 1, \ldots. \quad (3.97)$$

The probability of this process

$$W_s = J_s^2\left[\frac{ec^2 p_0 \sin\vartheta_0}{\hbar\omega_0\mathcal{E}_0} \int_{t_1}^{t_2} E(\eta')d\eta'\right]. \quad (3.98)$$

The condition of the applied eikonal approximation (3.92) with Eq. (3.94) is equivalent to the conditions $|p_x - p_{0x}| << p_{0x}$ and $|\mathcal{E} - \mathcal{E}_0| << \mathcal{E}_0$, which with Eq. (3.97) gives: $|s|\, n_0\hbar\omega_0/c << p_0$.

In the case of a monochromatic wave from Eq. (3.98) we have

$$W_s = J_s^2\left(\xi \frac{mc^2}{\hbar} \frac{c p_0 \sin\vartheta_0}{\mathcal{E}_0} t_0\right), \quad (3.99)$$

where $t_0 = t_2 - t_1$ is the duration of the particle motion in the wave.

As is seen from Eq. (3.99) for the actual values of the parameters $\alpha >> 1$, that is, the process is essentially multiphoton. The most probable number of absorbed/emitted Cherenkov photons is

$$\overline{s} \simeq \xi \frac{mc^2}{\hbar} \frac{\mathrm{v}_0}{c} \sin\vartheta_0 \cdot t_0. \quad (3.100)$$

The energetic width of the main diffraction maximums $\Gamma(\overline{s}) \simeq \overline{s}^{1/3}\hbar\omega_0$ and since $\overline{s} >> 1$ then $\Gamma(\overline{s}) << |\mathcal{E} - \mathcal{E}_0|$.

The scattering angles of the s-photon Cherenkov diffraction are determined by Eq. (3.97):

$$\tan\vartheta_s = \frac{sn_0\hbar\omega_0 \sin\vartheta_0}{cp_0 + sn_0\hbar\omega_0\cos\vartheta_0}. \tag{3.101}$$

From Eq. (3.101) it follows that at the inelastic diffraction there is an asymmetry in the angular distribution of the scattered particle: $|\vartheta_{-s}| > \vartheta_s$, i.e., the main diffraction maximums are situated at different angles with respect to the direction of particle initial motion. However, in accordance with the condition $|s|\, n_0\hbar\omega_0/c << p_0$ of the eikonal approximation this asymmetry is negligibly small and for the scattering angles of the main diffraction maximums from Eq. (3.101) we have $\vartheta_{-s} \simeq -\vartheta_s$. Hence, the main diffraction maximums will be situated at the angles

$$\vartheta_{\pm\overline{s}} = \pm\overline{s}\frac{n_0\hbar\omega_0}{cp_0}\sin\vartheta_0 \tag{3.102}$$

with respect to the direction of the particle initial motion.

3.6 Quantum Modulation of Charged Particles

Coherent interaction of charged particles with a plane EM wave of intensity smaller than the critical one in the induced Cherenkov process leads to quantum modulation of the particles' probability density and, consequently, current density after the interaction at the wave fundamental frequency and its harmonics. In contrast to classical modulation of particles' current density proceeding in the free drift region after the interaction and conserving for short distances, the quantum modulation, being quantum feature of a single particle, is conserved after the interaction unlimitedly long. To reveal this quantum coherent effect it is necessary to take into account the quantum character of particle–wave interaction entirely in contrast to the above-developed eikonal approximation for particle wave function. The mathematical point of view requires taking into account in Eq. (3.90) the second-order derivatives of the wave function as well, which have been neglected in the description of the diffraction effect.

To describe the effect of particle quantum modulation with regard to the wave harmonics we will solve Eq. (3.90) by perturbation theory in the field of monochromatic wave ($\mathbf{A}(\tau) = \{0, A_0\cos\omega_0\tau, A_0\sin\omega_0\tau\}$) of intensity $\xi_0 < \xi_{cr} << 1$ at which one can neglect again the constant term $\sim A_0^2$. Then we look for the solution of Eq. (3.90) in the form

$$\Psi(\mathbf{r},t) = \sqrt{\frac{N_0}{2\mathcal{E}_0}}\exp\left[\frac{i}{\hbar}\left(p_{0x}x - \mathcal{E}_0 t\right) + \frac{i}{\hbar}p_{0y}y\right]$$

$$\sum_{s=-\infty}^{+\infty} \Psi_s \exp\left[is\omega_0\left(t - n_0\frac{x}{c}\right)\right], \tag{3.103}$$

where $N_0 = \text{const}$ is the density of initially uniform particle beam. Substituting Eq. (3.103) into Eq. (3.90) we obtain the recurrent equation

$$\left[\left(n_0^2 - 1\right)\hbar^2 s^2 \omega_0^2 + 2\mathcal{E}_0 s\hbar\omega_0 \left(1 - n_0\frac{\mathrm{v}_{0x}}{c}\right)\right]\Psi_s$$

$$= ecp_{0y}A_0\left[\Psi_{s-1} + \Psi_{s+1}\right], \tag{3.104}$$

which will be solved in the approximation of perturbation theory by wave function:

$$|\Psi_{\pm 1}| << |\Psi_0|; \qquad |\Psi_{\pm 2}| << |\Psi_{\pm 1}|, \ldots .$$

Thus, for the amplitude of the particle wave function corresponding to s-photon induced radiation ($s > 0$) we obtain

$$\Psi_s = \frac{1}{s!}\frac{b^s}{(\mu + \triangle_\hbar)(\mu + 2\triangle_\hbar)\cdots(\mu + s\triangle)}, \tag{3.105}$$

and for s-photon absorption

$$\Psi_{-s} = \frac{(-1)^s}{s!}\frac{b^s}{(\mu - \triangle_\hbar)(\mu - 2\triangle_\hbar)\cdots(\mu - s\triangle)}. \tag{3.106}$$

Here the dimensionless parameter of one-photon interaction

$$b = \frac{1}{2}\frac{eA_0}{\hbar\omega_0}\frac{\mathrm{v}_0}{c}\sin\vartheta_0 \tag{3.107}$$

is the small parameter of perturbation theory: $|b| << 1$ and

$$\mu = 1 - n_0\frac{\mathrm{v}_0}{c}\cos\vartheta_0; \qquad \triangle_\hbar = \left(n_0^2 - 1\right)\frac{\hbar\omega_0}{2\mathcal{E}_0} \tag{3.108}$$

are the dimensionless Cherenkov resonance width and quantum recoil parameter, respectively. Hence, for total wave function of the particle after the interaction we have

$$\Psi(\mathbf{r}, t) = \sqrt{\frac{N_0}{2\mathcal{E}_0}}\left\{1 + \sum_{s=1}^{\infty}\frac{b^s}{s!}\left[\frac{e^{is\omega_0(t - n_0 x/c)}}{(\mu + \Delta_\hbar)(\mu + 2\triangle_\hbar)\cdots(\mu + s\Delta_\hbar)}\right.\right.$$

$$+(-1)^s \frac{e^{-is\omega_0(t-n_0x/c)}}{(\mu-\Delta_\hbar)(\mu-2\Delta_\hbar)\cdots(\mu-s\Delta_\hbar)}\Bigg]\Bigg\} e^{\frac{i}{\hbar}(\mathbf{p}_0\mathbf{r}-\mathcal{E}_0t)}. \tag{3.109}$$

The current density of the particles after the interaction corresponding to obtained wave function will be expressed by

$$\mathbf{j}(t,x) = \mathbf{j}_0 \Bigg\{ 1 + 2\sum_{s=1}^{\infty} \frac{b^s}{s!} \Bigg[\frac{1}{(\mu+\Delta_\hbar)\cdots(\mu+s\Delta_\hbar)}$$

$$+ \frac{(-1)^s}{(\mu-\Delta_\hbar)\cdots(\mu-s\Delta_\hbar)}\Bigg] \cos s\omega_0\left(t-n_0x/c\right)$$

$$+2\sum_{s=1}^{\infty}\sum_{s'=1}^{\infty}(-1)^{s'}\frac{b^{s+s'}}{s!s'!}\cos\left[\left(s+s'\right)\omega_0\left(t-n_0x/c\right)\right]$$

$$\times \frac{1}{(\mu+\Delta_\hbar)\cdots(\mu+s\Delta_\hbar)\cdot(\mu-\Delta_\hbar)\cdots(\mu-s'\Delta_\hbar)}\Bigg\}, \tag{3.110}$$

where $\mathbf{j}_0 = \text{const}$ is the current density of initially uniform particle beam. As is seen from Eq. (3.110) as a result of direct and inverse induced Cherenkov effect the current density of initially uniform particle beam is modulated at the wave fundamental frequency and its harmonics. This is a result of coherent superposition of particle states with various energy and momentum due to absorbed and emitted photons in the radiation field that remains after the interaction unlimitedly long (for a monochromatic beam).

We present in explicit form the expression of modulated current density for the first three harmonics

$$\mathbf{j}(t,x) = \mathbf{j}_0 \Bigg[1 - B\cos\omega_0\left(t-n_0x/c\right) + \frac{3}{4}B^2\frac{\mu^2-\Delta_\hbar^2}{\mu^2-4\Delta_\hbar^2}\cos 2\omega_0\left(t-n_0x/c\right)$$

$$-\frac{5}{8}B^3\frac{\left(\mu^2-\Delta_\hbar^2\right)^2}{\left(\mu^2-4\Delta_\hbar^2\right)\left(\mu^2-9\Delta_\hbar^2\right)}\cos 3\omega_0\left(t-n_0x/c\right)+\cdots, \tag{3.111}$$

where the modulation depth at the fundamental frequency of stimulating wave

$$B = \frac{eA}{\mathcal{E}_0}\frac{\frac{\mathrm{v}_0}{c}\left(n_0^2-1\right)\sin\vartheta_0}{\left(1-n_0\frac{\mathrm{v}_0}{c}\cos\vartheta_0\right)^2-\left(n_0^2-1\right)\frac{\hbar^2\omega_0^2}{4\mathcal{E}_0^2}}. \tag{3.112}$$

The denominators in Eqs. (3.110)-(3.112) becomes zero at the fulfillment of exact quantum conservation law for multiphoton Cherenkov process (3.21).

In this case perturbation theory is not applicable and the consideration in the scope of above-developed secular perturbation is required. However, in actual cases because of nonmonochromaticity of particle beams the width of Cherenkov resonance is rather larger than quantum recoil ($\Delta_{\hbar} << \mu$) and one can neglect the latter in Eqs. (3.111)–(3.112). Then the modulation depth at the wave fundamental frequency (3.112) is expressed via critical intensity (2.16):

$$B = \frac{1}{2}\frac{\xi}{\xi_{cr}(\vartheta)} \tag{3.113}$$

and the current density of modulated beam (3.111) will be represented by the parameter of critical field

$$\mathbf{j}(t,x) = \mathbf{j}_0 \left[1 - \frac{1}{2}\frac{\xi}{\xi_{cr}(\vartheta)}\cos\omega_0\left(t - n_0x/c\right) + \frac{3}{16}\left(\frac{\xi}{\xi_{cr}(\vartheta)}\right)^2 \right.$$

$$\left. \times \cos 2\omega_0\left(t - n_0x/c\right) - \frac{5}{64}\left(\frac{\xi}{\xi_{cr}(\vartheta)}\right)^3 \cos 3\omega_0\left(t - n_0x/c\right) + \cdots \right]. \tag{3.114}$$

The equation for particle modulation being expressed in this form shows that the effect of quantum modulation at the stimulating wave harmonics proceeds at intensities smaller than the critical one when the induced Cherenkov interaction of the particles with the periodic wave field (photons) occurs. In the opposite case the interaction proceeds with the potential barrier, i.e., the particle does not "feel" photons (periodic wave field). Note that in the last case the above-considered quantum modulation of the particles due to "reflection" phenomenon (see Section 3.2) occurs at the frequency (actually X-ray) corresponding to particles' energy exchange as a result of the interaction with the moving barrier. It is clear that a modulated particle beam is a coherent source of EM radiation.

Bibliography

G.S. Sahakyan, Dokl. Acad. Nauk Arm. SSR **28**, 630 (1957) [in Russian]
A.A. Sokolov, Yu.M. Loskutov, Zh. Éksp. Teor. Fiz. **32**, 121 (1959)
D.A. Varshalovich, M.I. Dyakonov, Zh. Éksp. Teor. Fiz. **60**, 90 (1971)
V.M. Haroutunian, H.K. Avetissian, Phys. Lett. A **44**, 281 (1973)
S.G. Hovhanissian, H.K. Avetissian, Izv. Acad. Nauk Arm. SSR Ser. Fiz. **8**, 395 (1973) [in Russian]
V.V. Batygin, N.K. Kouz'menko, Zh. Éksp. Teor. Fiz. **68**, 882 (1975)
H.K. Avetissian, Phys. Lett. A **58**, 144 (1976)
H.K. Avetissian, Phys. Lett. A **63**, 7 (1977)

C. Cronstrom, M. Noga, Phys. Lett. A **60**, 137 (1977)
H.K. Avetissian, Phys. Lett. A **63**, 9 (1977)
H.K. Avetissian et al., Phys. Lett. A **244**, 25 (1998)
H.K. Avetissian et al., Phys. Lett. A **246**, 16 (1998)
H.K. Avetissian, A. Kh.Bagdasarian, G.F.Mkrtchian, Zh. Éksp. Teor. Fiz. **86**, 24 (1998)
H.K. Avetissian, G.F. Mkrtchian, Phys. Rev. E **65**, 016506 (2001)
H.K. Avetissian et al.: Quantum Signatures in Laser-Assisted Cherenkov Radiation. 27th International FEL Conference, Stanford University, USA (2005)

4 Cyclotron Resonance at the Particle–Strong Wave Interaction

In this chapter we will consider a charged particle interaction with a strong EM wave in the presence of a uniform magnetic field along the wave propagation direction when the resonant effect of the wave on the particle rotational motion in the static magnetic field is possible.

In vacuum, as a result of the interaction of a charged particle with a monochromatic EM wave and uniform magnetic field the resonance created at the initial moment for the free-particle velocity automatically holds throughout the interaction process due to the equal Doppler shifts of the Larmor and wave frequencies in the field. This phenomenon is known as "Autoresonance". This property of cyclotron resonance in vacuum makes possible the creation of a generator of coherent radioemission by an electron beam, namely, a cyclotron resonance maser (CRM).

From the point of view of quantum theory the relativistic nonequidistant Landau levels of the particle in the wave field become equidistant in the autoresonance due to the quantum recoil at the absorption/emission of photons by the particle. In addition, the dynamic Stark effect of the wave electric field on the transversal bound states of the particle does not violate the equidistance of Landau levels in the autoresonance. Then the inverse process, that is, multiphoton resonant excitation of Landau levels by strong EM wave and, consequently, the particle acceleration in vacuum due to cyclotron resonance, in principle, is possible.

In a medium with arbitrary refractive properties (dielectric or plasma) because of the different Doppler shifts of the Larmor and wave frequencies in the interaction process the autoresonance is violated. However, the threshold (by the wave intensity) phenomenon of electron hysteresis in a medium due to the nonlinear cyclotron resonance in the field of strong monochromatic EM wave takes place. In contrast to autoresonance, the nonlinear cyclotron resonance in a medium proceeds with a large enough resonant width. This so-called phenomenon of electron hysteresis leads to significant acceleration of particles, especially in the plasmalike media where the superstrong laser fields of relativistic intensities can be applied.

The use of dielectriclike (gaseous) media makes it possible to realize cyclotron resonance in the optical domain (with laser radiation) due to an arbitrarily small Doppler shift of a wave frequency close to the Cherenkov cone, in contrast to the vacuum case where the cyclotron resonance for the existing maximal powerful static magnetic fields is possible only in the radio-frequency domain.

4.1 Autoresonance in the Uniform Magnetic Field in Vacuum

Let a charged particle move in the field of a plane EM wave in the presence of a homogeneous static magnetic field directed along the wave propagation direction $\nu_0 = \{1, 0, 0\}$:

$$\mathbf{H}_0 = \nu_0 H_0. \tag{4.1}$$

Relativistic classical equation of motion of the particle in the fields (1.1), (4.1) will be written in the form

$$\frac{d\mathbf{p}}{dt} = e\mathbf{E}(\tau) + \frac{e}{c}[\mathbf{vH}(\tau)] + \frac{eH_0}{c}[\mathbf{v}\nu_0]. \tag{4.2}$$

For the integration of the equation of motion (4.2) the latter should be written in components:

$$\nu_0 \frac{d\mathbf{p}}{dt} = \frac{e}{c}(\mathbf{vE}(\tau)), \tag{4.3}$$

$$\frac{d\mathbf{p}_\perp}{dt} = e\left(1 - \frac{\mathrm{v}\nu_0}{c}\right)\mathbf{E}(\tau) + \frac{eH_0}{c}[\mathbf{v}_\perp \nu_0], \tag{4.4}$$

where $\mathbf{p}_\perp = \{0, p_y, p_z\}$ and $\mathbf{v}_\perp = \{0, \mathrm{v}_y, \mathrm{v}_z\}$ are the transversal momentum and velocity of the particle in the field.

As we see from Eq. (4.3) the existence of the uniform magnetic field (4.1) does not change the equation for the longitudinal momentum of the particle in the field of a plane EM wave (1.3), nor does the equation for particle energy (1.9) change. Hence, in the considered process the integral of motion $\Lambda = \mathcal{E} - c\mathbf{p}\nu_0$ for a charged particle in the field of a plane EM wave in vacuum (1.10) survives.

For integration of the equation for particle transversal momentum (4.4) we pass from the variable t to wave coordinate $\tau = t - \nu_0 \mathbf{r}/c$. Then Eq. (4.4) becomes

$$\frac{d\mathbf{p}_\perp}{d\tau} + \frac{\Omega}{1 - \frac{\mathrm{v}\nu_0}{c}}[\nu_0 \mathbf{p}_\perp] = e\mathbf{E}(\tau), \tag{4.5}$$

where

$$\Omega = \frac{ecH_0}{\mathcal{E}} \tag{4.6}$$

is the Larmor frequency for a relativistic particle in the uniform magnetic field.

From the integral of motion $\Lambda = \mathcal{E} - c\mathbf{p}\nu_0$ follows the conservation of the quantity in Eq. (4.5)

$$\frac{\Omega}{1 - \frac{\mathbf{v}\nu_0}{c}} = \text{const} \equiv \Omega'. \tag{4.7}$$

The set of equations (4.5) for the transversal components of the particle momentum $\{p_y, p_z\}$ is equivalent to the equation

$$\frac{d\widetilde{p}}{d\tau} + i\Omega'\widetilde{p} = e\widetilde{E}(\tau). \tag{4.8}$$

Here we have introduced the complex quantities related to particle momentum and EM field:

$$\widetilde{p}(\tau) = p_y(\tau) + ip_z(\tau), \tag{4.9}$$

$$\widetilde{E}(\tau) = E_y(\tau) + iE_z(\tau). \tag{4.10}$$

The solution of Eq. (4.8) will be

$$\widetilde{p} = \widetilde{p}_0 e^{-i\Omega'(\tau-\tau_0)} + e\int\limits_{\tau_0}^{\tau} \widetilde{E}(\tau') e^{-i\Omega'(\tau-\tau')} d\tau', \tag{4.11}$$

where $\widetilde{p}_0 = p_{0y} + ip_{0z}$ is defined according to initial condition

$$\widetilde{p}\,|_{\tau=\tau_0} = \widetilde{p}_0. \tag{4.12}$$

Separating the real and imagenary parts of the solution (4.11) we obtain the transversal momentum of the particle:

$$p_y = p_{0y}\cos\Omega'(\tau-\tau_0) + p_{0z}\sin\Omega'(\tau-\tau_0)$$

$$+e\int\limits_{\tau_0}^{\tau}\left[E_y(\tau')\cos\Omega'(\tau-\tau') + E_z(\tau')\sin\Omega'(\tau-\tau')\right]d\tau', \tag{4.13}$$

$$p_z = p_{0z}\cos\Omega'(\tau-\tau_0) - p_{0y}\sin\Omega'(\tau-\tau_0)$$

$$+e\int\limits_{\tau_0}^{\tau}\left[E_z(\tau')\cos\Omega'(\tau-\tau') - E_y(\tau')\sin\Omega'(\tau-\tau')\right]d\tau'. \tag{4.14}$$

Now we can define the particle longitudinal momentum (p_x) and energy with the help of Eq. (4.11) utilizing the dispersion law of the particle energy-momentum and the integral of motion Λ. We obtain the following equations in the field of a plane EM wave of arbitrary form and polarization:

$$p_x = p_{0x} + c\frac{|\widetilde{p}|^2 - |\widetilde{p}_0|^2}{2\Lambda}, \tag{4.15}$$

$$\mathcal{E} = \mathcal{E}_0 + c^2\frac{|\widetilde{p}|^2 - |\widetilde{p}_0|^2}{2\Lambda}, \tag{4.16}$$

where p_{0x} and $\mathcal{E}_0$ are the initial longitudinal momentum and energy of the free particle.

Let us consider the case of a monochromatic wave (1.20) of circular polarization (right- or left-hand) and when the initial velocity of the particle is parallel to the wave propagation direction. For the field (1.20), when $g = \pm 1$ we have

$$\widetilde{E}(\tau) = -igE_0 e^{ig\omega_0\tau}. \tag{4.17}$$

Substituting Eq. (4.17) into Eq. (4.11) and assuming an arbitrarily small damping for the amplitude E_0 to switch on adiabatically the wave at $\tau_0 = -\infty$, we obtain

$$\widetilde{p} = \frac{-geE_0}{\Omega' + g\omega_0} e^{ig\omega_0\tau} \tag{4.18}$$

and by the components

$$p_y = \frac{-geE_0}{\Omega' + g\omega_0}\cos\omega_0\tau, \tag{4.19}$$

$$p_z = \frac{-eE_0}{\Omega' + g\omega_0}\sin\omega_0\tau. \tag{4.20}$$

As we see, for the left-hand circular polarization when $g = -1$ in Eqs. (4.19) and (4.20) a resonant effect of the wave on the particle motion is possible when $\Omega' = \omega_0$, or taking into account Eq. (4.7):

$$\frac{\Omega}{1 - \frac{\mathbf{v}\nu_0}{c}} = \omega_0. \tag{4.21}$$

Condition (4.21) performs the equality of the Larmor and Doppler-shifted wave frequency ω':

$$\Omega = \omega'; \quad \omega' = \omega_0 \left(1 - \frac{\mathbf{v}\nu_0}{c}\right). \tag{4.22}$$

The latter means that the particle and the wave electric field rotate in the same direction with the same frequency and as a result coherent energy exchange between the particle and the wave takes place. In addition, the energy exchange does not violate the resonance condition as the ratio of the Doppler-shifted wave frequency to the Larmor frequency of the particle is conserved:

$$\frac{\omega'}{\Omega} = \frac{\omega_0 \Lambda}{ecH_0} = \text{const} \tag{4.23}$$

and the resonance created at the initial moment automatically holds throughout the interaction. This is the phenomenon referred to as "Autoresonance".

According to Eqs. (4.15) and (4.16) the longitudinal momentum and energy of the particle in this case are given by

$$p_x = p_{0x} + \frac{m^2 c^3}{2\Lambda} \frac{\xi_0^2}{\left(1 - \frac{\Omega}{\omega'}\right)^2}, \tag{4.24}$$

$$\mathcal{E} = \mathcal{E}_0 + \frac{m^2 c^4}{2\Lambda} \frac{\xi_0^2}{\left(1 - \frac{\Omega}{\omega'}\right)^2}. \tag{4.25}$$

By analogy of the renormalization of the particle mass in the field of a plane EM wave for these values of energy and momentum (the average transversal momentum $\overline{\mathbf{p}}_\perp = 0$ in accordance with Eqs. (4.19) and (4.20)) one can introduce the "effective mass" of the particle due to the intensity and resonant effects of the strong wave:

$$m^* = m\sqrt{1 + \frac{\xi_0^2}{\left(1 - \frac{\Omega}{\omega'}\right)^2}}. \tag{4.26}$$

The comparison of Eq. (4.26) with the analogous formula (1.18) in the absence of a static magnetic field shows that instead of the parameter of nonlinearity ξ_0^2 in the strong wave field the effective nonlinearity in this process is determined by the resonant parameter $\xi_0^2 / \left(1 - \frac{\Omega}{\omega'}\right)^2 >> \xi_0^2$.

At the exact resonance the solutions (4.19), (4.20) are not applicable. In this case taking into account the resonance condition before the integration in Eq. (4.11) we have

$$p_y = eE_0 \tau \sin \omega_0 \tau, \tag{4.27}$$

$$p_z = eE_0 \tau \cos \omega_0 \tau \tag{4.28}$$

and for the particle longitudinal momentum and energy we obtain

$$p_x = p_{0x} + \frac{e^2 E_0^2 c}{2\Lambda}\tau^2, \tag{4.29}$$

$$\mathcal{E} = \mathcal{E}_0 + \frac{e^2 E_0^2 c^2}{2\Lambda}\tau^2. \tag{4.30}$$

It is seen that at the resonance the energy of the particle monotonically increases.

Then, taking into account Eq.(1.15) for the law of the particle motion in the parametric form $\mathbf{r} = \mathbf{r}(\tau)$ we obtain

$$y(\tau)=\frac{c^2 eE_0}{\Lambda\omega_0^2}\left(\sin\omega_0\tau - \omega_0\tau\cos\omega_0\tau\right), \tag{4.31}$$

$$z(\tau)=\frac{c^2 eE_0}{\Lambda\omega_0^2}\left(\cos\omega_0\tau + \omega_0\tau\sin\omega_0\tau\right), \tag{4.32}$$

$$y^2(\tau) + z^2(\tau) = \left(\frac{c^2 eE_0}{\Lambda\omega_0^2}\right)^2\left(1 + \omega_0^2\tau^2\right), \tag{4.33}$$

$$x(\tau) = \frac{c^2}{\Lambda}p_{0x}\tau + \frac{e^2 E_0^2 c^3 \tau^3}{6\Lambda^2}. \tag{4.34}$$

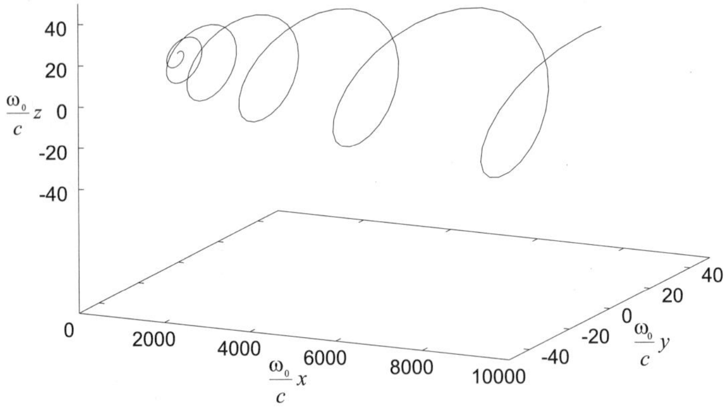

Fig. 4.1. Trajectory of the particle (initially at rest) in the field of circularly polarized EM wave and uniform magnetic field at the cyclotron resonance. The relativistic parameter of intensity is taken to be $\xi_0 = 1$.

Equations (4.31)–(4.34) show that the particle performs helical-like motion (see Fig. 4.1) with increasing radius (in the plane of the wave polarization) and increasing step along the wave propagation direction.

4.2 Exact Solution of the Dirac Equation for Cyclotron Resonance

The quantum description of the dynamics of cyclotron resonance in vacuum in the scope of relativistic theory requires solution of the Dirac equation. The configuration of EM fields when a uniform magnetic field is directed along the axis of propagation of a transverse EM wave is one of those specific cases for which exact solution of the Dirac equation in vacuum has succeeded. The latter has the basic role for quantum description of diverse nonlinear electromagnetic processes in superstrong laser and magnetic fields.

Let a charged particle move in the field of a plane EM wave and uniform magnetic field along the wave propagation direction (OX axis). The vector potential of this configuration of EM fields can be represented in the form

$$\mathbf{A}(\mathbf{r},t) = \mathbf{A}_H(y) + \mathbf{A}_w(\tau), \tag{4.35}$$

where

$$\mathbf{A}_H(y) = (0, 0, yH_0) \tag{4.36}$$

is the vector potential of uniform magnetic field with the strength H_0 (4.1) and

$$\mathbf{A}_w(\tau) = \left\{0, A_y\left(t - \frac{x}{c}\right), A_z\left(t - \frac{x}{c}\right)\right\} \tag{4.37}$$

is the vector potential of a plane transverse EM wave (1.1). The Dirac equation in the field (4.35) is written as

$$i\hbar\frac{\partial\Psi}{\partial t} = \left\{c\alpha\left(-i\hbar\nabla - \frac{e}{c}\mathbf{A}_H(y) - \frac{e}{c}\mathbf{A}_w(\tau)\right) + \beta mc^2\right\}\Psi. \tag{4.38}$$

Here α, β are the Dirac matrices in the standard representation (3.2). As the magnetic field is directed along the X axis for the Pauli matrices we will assume the σ_x to be diagonal:

$$\sigma_x = \begin{pmatrix} 1 & 0 \\ 0 & -1 \end{pmatrix}, \quad \sigma_y = \begin{pmatrix} 0 & 1 \\ 1 & 0 \end{pmatrix}, \quad \sigma_z = \begin{pmatrix} 0 & -i \\ i & 0 \end{pmatrix}. \tag{4.39}$$

Looking for the solution of Eq. (4.38) in the form

$$\Psi = \frac{1}{\sqrt{2}} \begin{pmatrix} \varphi + \chi \\ \varphi - \chi \end{pmatrix} \tag{4.40}$$

and eliminating the spinor φ from the equation for χ and passing to the retarding and advanced wave coordinates

$$\tau = t - \frac{x}{c}; \ \eta = t + \frac{x}{c},$$

we obtain the Dirac equation in the quadratic form for spinor function χ

$$\left\{ 4\hbar^2 \frac{\partial^2}{\partial\tau\partial\eta} + c^2 \left[\widehat{\mathbf{P}}_\perp - \frac{e}{c}\mathbf{A}_w(\tau) \right]^2 + m^2c^4 \right.$$

$$\left. -ec\hbar\sigma(\mathbf{H}_0 + \mathbf{H} + i\mathbf{E}) \right\} \chi = 0, \tag{4.41}$$

where

$$\widehat{\mathbf{P}}_\perp = -i\hbar\nabla_\perp - \frac{e}{c}\mathbf{A}_H(y); \ \nabla_\perp = \left\{ 0, \frac{\partial}{\partial y}, \frac{\partial}{\partial z} \right\}. \tag{4.42}$$

The spinor function φ will be defined via χ as follows:

$$\varphi(\mathbf{r}, t) = \frac{1}{mc^2} \left\{ i\hbar\frac{\partial}{\partial t} - \sigma(ic\hbar\nabla + e\mathbf{A}(\mathbf{r}, t)) \right\} \chi(\mathbf{r}, t). \tag{4.43}$$

The particle quantum motion at $t \to -\infty$ when $\mathbf{A}_w(\tau = -\infty) = 0$ and only the uniform magnetic field exists is separated into the cyclotron (y, z) and the longitudinal (x) degrees of freedom. Since the coordinate z is a cyclic in this issue (also in the presence of a plane EM wave) the cyclotron motion will be described by the set of quantum characteristics of the state $\{l, p_z\}$, where by the number l we indicate the Landau levels and by p_z, the z component of the generalized momentum. Then the longitudinal motion at $t \to -\infty$ will be described by the x component of the particle initial momentum p_x. Concerning the particle transversal initial state we will assume that at $t \to -\infty$ the particle is situated in the $l = s$ Landau level. In addition, there is a fourth quantum number σ which describes the polarization of the particle: $\sigma = \pm\frac{1}{2}$ (spin projections $S_z = \pm\frac{1}{2}$ on the direction of magnetic field $\mathbf{H}_0$). So, the wave function of the particle at $t \to -\infty$ will be given by the equation

$$\psi_{s,\sigma,p_x,p_z}(\mathbf{r}, t) = \psi_{s,\sigma} e^{\frac{i}{\hbar}(p_x x + p_z z - \mathcal{E}_s(p_x)t)}, \tag{4.44}$$

where the bispinors $\psi_{s,\sigma}$, describing the states with the different spin polarizations, are

$$\psi_{s,1/2} = N \begin{pmatrix} (\mathcal{E}_s(p_x) + mc^2)\Phi_s(y) \\ 0 \\ cp_x \Phi_s(y) \\ -i\sqrt{2sc\hbar eH}\Phi_{s-1}(y) \end{pmatrix}, \tag{4.45}$$

$$\psi_{s,-1/2} = N \begin{pmatrix} 0 \\ (\mathcal{E}_s(p_x) + mc^2)\Phi_{s-1}(y) \\ i\sqrt{2sc\hbar eH}\Phi_s(y) \\ -cp_x \Phi_{s-1}(y) \end{pmatrix}, \tag{4.46}$$

and

$$N = \frac{1}{2\pi\hbar\sqrt{2\mathcal{E}_s(p_x)(\mathcal{E}_s(p_x) + mc^2)}} \tag{4.47}$$

is the normalization constant. Here

$$\Phi_s(y) = \sqrt{\frac{a}{2^s s! \sqrt{\pi}}} \exp\left[-\left(ay - \frac{p_z}{\hbar a}\right)^2\right] U_s\left(ay - \frac{p_z}{\hbar a}\right),$$

$$a = \sqrt{\frac{eH_0}{c\hbar}}$$

are the Hermit functions and the dispersion law for the particle energy-momentum is

$$\mathcal{E}_s^2(p_x) = m^2c^4 + p_x^2c^2 + 2ec\hbar H_0 s. \tag{4.48}$$

For the spin projection $\sigma = 1/2$ the quantum numbers for s are: $s = 0, 1, 2, \ldots$, while for $\sigma = -1/2 : s = 1, 2, \ldots$.

Due to the existence of a definite direction of the wave propagation the variable η becomes a cyclic and the conjugate to coordinate η momentum is conserved. This is the known integral of motion (1.10). Hence, the spinor function $\chi(\mathbf{r}, t)$ can be sought in the form

$$\chi(\mathbf{r}, t) = N_f \exp\left\{-\frac{i}{2\hbar}(p_+\tau + \Lambda\eta)\right\} \chi_0(\mathbf{x}_\perp, \tau), \tag{4.49}$$

where

$$p_+ = \mathcal{E}_s(p_x) + cp_x; \quad \mathbf{x}_\perp = \{0, y, z\}. \tag{4.50}$$

Taking into account the dispersion law (4.48) for the spinor function $\chi_0(\mathbf{x}_\perp, \tau)$ we obtain the equation

$$\left\{2i\frac{\hbar\Lambda}{c^2}\frac{\partial}{\partial\tau} - \left[\widehat{\mathbf{P}}_\perp - \frac{e}{c}\mathbf{A}_w(\tau)\right]^2 + 2\frac{e}{c}\hbar H_0 s \right.$$

$$\left. + \frac{e\hbar}{c}\sigma(\mathbf{H}_0 + \mathbf{H} + i\mathbf{E})\right\}\chi_0(\mathbf{x}_\perp, \tau) = 0. \tag{4.51}$$

In Eq. (4.51) the transversal and longitudinal motions are not separated. But after the unitarian transformation for the transformed function the variables are separated. The corresponding unitarian transformation operator is

$$\widehat{U} = e^{\frac{i}{\hbar}\mathbf{K}(\tau)\widehat{\mathbf{P}}_\perp}, \tag{4.52}$$

where the vector function

$$\mathbf{K}(\tau) = \{0, K_y(\tau), K_z(\tau)\} \tag{4.53}$$

will be chosen to separate the cyclotron and longitudinal motions and to satisfy the initial condition. Taking into account that for the Hermitian operator $\widehat{F} = \widehat{F}^\dagger$

$$e^{i\widehat{F}}\widehat{L}e^{-i\widehat{F}} = \widehat{L} + i\left[\widehat{F}, \widehat{L}\right] - \frac{1}{2}\left[\widehat{F}, \left[\widehat{F}, \widehat{L}\right]\right] + \cdots, \tag{4.54}$$

for the transformed operators in Eq. (4.51) we will obtain

$$\widehat{U}\widehat{\mathbf{P}}_\perp\widehat{U}^\dagger = \widehat{\mathbf{P}}_\perp + \frac{e}{c}\left[\mathbf{K}\mathbf{H}_0\right],$$

$$\widehat{U}\frac{\partial}{\partial\tau}\widehat{U}^\dagger = \frac{\partial}{\partial\tau} - \frac{i}{\hbar}\left(\frac{d\mathbf{K}}{d\tau}\widehat{\mathbf{P}}_\perp\right) + i\frac{e}{2c\hbar}\left(\mathbf{H}_0\left[\mathbf{K}\frac{d\mathbf{K}}{d\tau}\right]\right).$$

Let us choose the function $\mathbf{K}(\tau)$ in such a form that the coefficient of the term $\sim \widehat{\mathbf{P}}_\perp$ in the equation for transformed function

$$\chi_0' = \widehat{U}\chi_0(\mathbf{x}_\perp, \tau)$$

becomes zero. Then for the function $\mathbf{K}(\tau)$ we will obtain the classical equation of motion for transverse coordinates describing stimulated cyclotron rotation in the EM wave field (see Eq. (4.5)):

$$\frac{d\mathbf{K}}{d\tau} + \Omega' \left[\nu_0 \mathbf{K}\right] = -\frac{ce}{\Lambda}\mathbf{A}_w\left(\tau\right), \tag{4.55}$$

where Ω' is the Doppler-shifted Larmor frequency (4.7). The solution of Eq. (4.55) can be written with the help of the complex quantities

$$\widetilde{K} = K_y + iK_z; \quad \widetilde{A} = A_y + iA_z \tag{4.56}$$

as follows:

$$\widetilde{K} = -\exp\left\{-i\Omega'\tau\right\}\frac{ec}{\Lambda}\int\limits_{-\infty}^{\tau} \widetilde{A}\left(\tau'\right)\exp\left\{i\Omega'\tau'\right\}d\tau'. \tag{4.57}$$

In Eq. (4.57) we have taken into account the initial condition

$$K_y(-\infty) = K_z(-\infty) = 0.$$

Hence, for the transformed spinor function χ_0' we obtain

$$\left\{2i\frac{\hbar\Lambda}{c^2}\frac{\partial}{\partial\tau} - \widehat{\mathbf{P}}_\perp^2 + 2\frac{e}{c}\hbar H_0 s + \frac{e\Lambda}{c^3}\left(\frac{d\mathbf{K}}{d\tau}\mathbf{A}_w\right)\right.$$

$$\left. + \frac{e\hbar}{c}\sigma(\mathbf{H}_0 + \mathbf{H} + i\mathbf{E})\right\}\chi_0' = 0. \tag{4.58}$$

Looking for the solution of Eq. (4.58) in the form

$$\chi_0' = \begin{pmatrix} \chi_1(\mathbf{x}_\perp, \tau) \\ \chi_2(\mathbf{x}_\perp, \tau) \end{pmatrix}, \tag{4.59}$$

we obtain the set of equations for the functions χ_1 and χ_2:

$$\left\{2i\frac{\hbar\Lambda}{c^2}\frac{\partial}{\partial\tau} - \widehat{\mathbf{P}}_\perp^2 + (2s+1)\frac{e}{c}\hbar H_0 + \frac{e\Lambda}{c^3}\left(\frac{d\mathbf{K}}{d\tau}\mathbf{A}_w\right)\right\}\chi_1 = 0, \tag{4.60}$$

$$\left\{2i\frac{\hbar\Lambda}{c^2}\frac{\partial}{\partial\tau} - \widehat{\mathbf{P}}_\perp^2 + (2s-1)\frac{e}{c}\hbar H_0 + \frac{e\Lambda}{c^3}\left(\frac{d\mathbf{K}}{d\tau}\mathbf{A}_w\right)\right\}\chi_2$$

$$+2i\frac{e\hbar}{c}\left(E_y(\tau) + iE_z(\tau)\right)\chi_1 = 0. \tag{4.61}$$

Now in Eq. (4.60) the variables are separated and the solution can be written as

$$\chi_1\left(\mathbf{x}_\perp,\tau\right)=N_1^{(\sigma)}T_s(\mathbf{x}_\perp)\exp\left[i\frac{e}{2\hbar c}\int\limits_{-\infty}^{\tau}\left(\frac{d\mathbf{K}}{d\tau'}\mathbf{A}_w\left(\tau'\right)\right)d\tau'\right], \tag{4.62}$$

where

$$T_s(\mathbf{x}_\perp)=\Phi_s(y)e^{\frac{i}{\hbar}p_z z}$$

describes the free cyclotron motion of the particle. The solution for the second function χ_2 can be obtained in the same way (adding the particular solution of the nonhomogeneous equation). Hence, for the spinor function χ_0' we obtain

$$\chi_0'=\begin{pmatrix} N_1^{(\sigma)}T_s\left(\mathbf{x}_\perp\right) \\ N_2^{(\sigma)}T_{s-1}\left(\mathbf{x}_\perp\right)-N_1^{(\sigma)}\frac{1}{c}\frac{d\widetilde{K}}{d\tau}T_s\left(\mathbf{x}_\perp\right)\end{pmatrix}$$

$$\times\exp\left[i\frac{e}{2\hbar c}\int\limits_{-\infty}^{\tau}\left(\frac{d\mathbf{K}}{d\tau'}\mathbf{A}_w\left(\tau'\right)\right)d\tau'\right]. \tag{4.63}$$

The coefficients $N_1^{(\sigma)}$, $N_2^{(\sigma)}$ will be chosen to satisfy the initial condition. Thus, for the different initial polarization states ($\sigma=\pm\frac{1}{2}$) we have

$$N_1^{(1/2)}=\frac{\Lambda+mc^2}{2mc^2};\quad N_2^{(1/2)}=\frac{i\sqrt{2sc\hbar eH_0}}{2mc^2}, \tag{4.64}$$

$$N_1^{(-1/2)}=-\frac{i\sqrt{2sc\hbar eH_0}}{2mc^2};\quad N_2^{(-1/2)}=\frac{p_+ + mc^2}{2mc^2}. \tag{4.65}$$

Using inverse transformation $\chi_0=\widehat{U}^\dagger\chi_0'(\mathbf{x}_\perp,\tau)$, with the help of the relation

$$e^{\hat{F}+\hat{L}}=e^{-\frac{1}{2}[\hat{F},\hat{L}]}e^{\hat{F}}e^{\hat{L}} \tag{4.66}$$

we obtain the solution of the initial equation (4.41) (taking into account Eq.(4.49)):

$$\chi\left(\mathbf{r},t\right)=N_f\exp\left[\frac{i}{\hbar}\left(p_x x-\mathcal{E}_s(p_x)t\right)\right.$$

$$+i\frac{e}{2\hbar c}\int\limits_{-\infty}^{\tau}\left(\frac{d\mathbf{K}}{d\tau'}\mathbf{A}_w\left(\tau'\right)\right)d\tau' + i\frac{e}{\hbar c}H_0K_z\left(y-\frac{1}{2}K_y\right)\Bigg]$$

$$\times\begin{pmatrix} N_1^{(\sigma)}T_s\left(\mathbf{x}_\perp-\mathbf{K}\right) \\ \\ N_2^{(\sigma)}T_{s-1}\left(\mathbf{x}_\perp-\mathbf{K}\right)-N_1^{(\sigma)}\frac{1}{c}\frac{d\widetilde{K}}{d\tau}T_s\left(\mathbf{x}_\perp-\mathbf{K}\right)\end{pmatrix}. \quad (4.67)$$

Finally, with the help of Eq. (4.43) the solution of Eq. (4.38) for spinor particle wave function can be written as

$$\Psi_{s,\sigma,p_x,p_z}\left(\mathbf{r},t\right)=N_f\exp\Bigg[\frac{i}{\hbar}\left(p_x x-\mathcal{E}_s(p_x)t\right)$$

$$+i\frac{e}{2\hbar c}\int\limits_{-\infty}^{\tau}\left(\frac{d\mathbf{K}}{d\tau'}\mathbf{A}_w\left(\tau'\right)\right)d\tau' + i\frac{e}{\hbar c}H_0K_z\left(y-\frac{1}{2}K_y\right)\Bigg]$$

$$\times\left\{\begin{pmatrix}\left(N_1^{(\sigma)}\left(p_+ + mc^2\right)+iN_2^{(\sigma)}\sqrt{2sc\hbar eH_0}\right)T_s\left(\mathbf{x}_\perp-\mathbf{K}\right) \\ \left(N_2^{(\sigma)}\left(\Lambda + mc^2\right)-iN_1^{(\sigma)}\sqrt{2sc\hbar eH_0}\right)T_{s-1}\left(\mathbf{x}_\perp-\mathbf{K}\right) \\ \left(N_1^{(\sigma)}\left(p_+ - mc^2\right)+iN_2^{(\sigma)}\sqrt{2sc\hbar eH_0}\right)T_s\left(\mathbf{x}_\perp-\mathbf{K}\right) \\ \left(N_2^{(\sigma)}\left(\Lambda - mc^2\right)-iN_1^{(\sigma)}\sqrt{2sc\hbar eH_0}\right)T_{s-1}\left(\mathbf{x}_\perp-\mathbf{K}\right)\end{pmatrix}\right.$$

$$\left.+\begin{pmatrix}\left(N_2^{(\sigma)}\Lambda-iN_1^{(\sigma)}\sqrt{2sc\hbar eH_0}\right)\frac{1}{c}\frac{d\widetilde{K}^*}{d\tau}T_{s-1}\left(\mathbf{x}_\perp-\mathbf{K}\right) \\ -N_1^{(\sigma)}mc\frac{d\widetilde{K}}{d\tau}T_s\left(\mathbf{x}_\perp-\mathbf{K}\right) \\ \left(N_2^{(\sigma)}\Lambda-iN_1^{(\sigma)}\sqrt{2sc\hbar eH_0}\right)\frac{1}{c}\frac{d\widetilde{K}^*}{d\tau}T_{s-1}\left(\mathbf{x}_\perp-\mathbf{K}\right) \\ N_1^{(\sigma)}mc\frac{d\widetilde{K}}{d\tau}T_s\left(\mathbf{x}_\perp-\mathbf{K}\right)\end{pmatrix}\right\}. \quad (4.68)$$

In particular, for the state with the spin projection $\sigma = 1/2$ from Eqs. (4.68) and (4.64) we have

$$\Psi_{s,1/2,p_x,p_z}\left(\mathbf{r},t\right)=N_f\exp\Bigg[\frac{i}{\hbar}\left(p_x x-\mathcal{E}_s(p_x)t\right)$$

$$+i\frac{e}{2\hbar c}\int\limits_{-\infty}^{\tau}\left(\frac{d\mathbf{K}}{d\tau'}\mathbf{A}_w(\tau')\right)d\tau'+i\frac{e}{\hbar c}H_0K_z\left(y-\frac{1}{2}K_y\right)\Bigg]$$

$$\times\begin{pmatrix}\left(mc^2+\mathcal{E}_s(p_x)\right)T_s(\mathbf{x}_\perp-\mathbf{K})-i\sqrt{\frac{s\hbar eH_0}{2c}}\frac{d\widetilde{K}^*}{d\tau}T_{s-1}(\mathbf{x}_\perp-\mathbf{K})\\ -\frac{\Lambda+mc^2}{2c}\frac{d\widetilde{K}}{d\tau}T_s(\mathbf{x}_\perp-\mathbf{K})\\ cp_xT_s(\mathbf{x}_\perp-\mathbf{K})-i\sqrt{\frac{s\hbar eH_0}{2c}}\frac{d\widetilde{K}^*}{d\tau}T_{s-1}(\mathbf{x}_\perp-\mathbf{K})\\ -i\sqrt{2sc\hbar eH_0}T_{s-1}(\mathbf{x}_\perp-\mathbf{K})+\frac{\Lambda+mc^2}{2c}\frac{d\widetilde{K}}{d\tau}T_s(\mathbf{x}_\perp-\mathbf{K})\end{pmatrix}. \tag{4.69}$$

For a quasi-monochromatic EM wave the states (4.68) can be normalized by the condition

$$\int\Psi^\dagger_{s',\sigma',p'_x,p'_z}\Psi_{s,\sigma,p_x,p_z}d\mathbf{r}=\delta(p'_z-p_z)\,\delta(p'_x-p_x)\,\delta_{\sigma,\sigma'}\delta_{s,s'},$$

where $\delta_{l,l'}$ is the Kronecker symbol. Then for the normalization constant we will have

$$N_f=\frac{1}{2\pi\hbar\sqrt{2\overline{\mathcal{E}}_s(p_x)(\mathcal{E}_s(p_x)+mc^2)}},$$

where

$$\overline{\mathcal{E}}_s(p_x)=\mathcal{E}_s(p_x)+\frac{\Lambda}{2c^2}\overline{\left|\frac{d\widetilde{K}}{d\tau}\right|^2}$$

is the average energy of the particle in the field (4.35).

4.3 Multiphoton Excitation of Landau Levels by Strong EM Wave

On the basis of the obtained wave function consider the possibility of multiphoton excitation of Landau levels by a strong quasi-monochromatic EM wave at the cyclotron resonance in vacuum. We will consider the concrete case of circularly polarized EM wave (1.20) with $g=-1$. For a quasi-monochromatic wave it should be $A_0\Rightarrow A_0(\tau)$, where $A_0(\tau)$ is a slowly varying amplitude with respect to the phase oscillations over the $\omega_0\tau$ and the conditions of adiabatic switching on and switching off will take place automatically.

To determine the probabilities of the multiphoton induced transitions between the Landau levels one must first define the function $\mathbf{K}(\tau)$. After the interaction with the wave ($t \to +\infty$) from Eq. (4.57) at the resonance condition (4.21) we have

$$\widetilde{K} = -\frac{ec\overline{A}_0 T}{\Lambda} e^{-i\omega_0 \tau}, \tag{4.70}$$

where T is the coherent interaction time (for actual laser radiation T is the pulse duration) and $\overline{A}_0$ is the average value of the slowly varied envelope. Substituting Eq. (4.70) into the expression for the wave function (4.68) and expanding the latter in terms of the full basis of the particle eigenstates (4.44)

$$\Psi_{s,\sigma,p_x,p_z}(\mathbf{r},t) = \int dp'_x dp'_z \sum_{s',\sigma'} C_{ss'}^{\sigma\sigma'}(p'_x,p'_z)\psi_{s',\sigma',p'_x,p'_z}(\mathbf{r},t), \tag{4.71}$$

we will find the probabilities of the multiphoton induced transitions between the Landau levels (we expand only by positive energy solutions as in this case the Dirac vacuum is not excited). To calculate the expansion coefficients

$$C_{ss'}^{\sigma\sigma'}(p'_x,p'_z) = \int \psi^{\dagger}_{s',\sigma',p'_x,p'_z}(\mathbf{r},t)\,\Psi_{s,\sigma,p_x,p_z}(\mathbf{r},t)\,d\mathbf{r}, \tag{4.72}$$

we will take into account the result of the following integration

$$\int \exp(-ikx)\Phi_s(a^{-1}x+ab)\Phi_{s'}(a^{-1}x+ab')dx$$

$$= \exp\{i\mu + i(s-s')\lambda\}\, I_{ss'}(\alpha), \tag{4.73}$$

where $I_{ss'}(\alpha)$ is the Lagger function and defined via generalized Lagger polynomials $L_n^l(\alpha)$ as follows:

$$I_{s,s'}(\alpha) = \sqrt{\frac{s'!}{s!}} e^{-\frac{\alpha}{2}} \alpha^{\frac{s-s'}{2}} L_{s'}^{s-s'}(\alpha),$$

$$L_n^l(\alpha) = \frac{1}{n!} e^{\alpha} \alpha^{-l} \frac{d^n}{d\alpha^n}\left(e^{-\alpha}\alpha^{n+l}\right). \tag{4.74}$$

The characteristic parameters μ, λ, and α are determined by the expressions

$$\mu = \frac{ka^2(b+b')}{2}; \quad \lambda = \tan^{-1}\frac{k}{b'-b}; \quad \alpha = a^2\frac{k^2+(b-b')^2}{2}. \tag{4.75}$$

Taking into account Eqs.(4.68), (4.70), (4.72), and (4.73) we get the following expansion coefficients:

$$C_{ss'}^{\sigma\sigma'}(p'_x, p'_z) = w_{ss'}^{\sigma\sigma'}(p'_x, p'_z) \exp\left\{ \frac{i}{\hbar}(\mathcal{E}_{s'}(p'_x) - \mathcal{E}_s(p_x) - \hbar\omega_0(s'-s))t \right\}$$

$$\times \delta(p'_z - p_z)\delta(p'_x - p_x - \hbar\frac{\omega_0}{c}(s'-s)). \tag{4.76}$$

Here the Dirac δ-functions express the momentum conservation law. The transition amplitudes $w_{ss'}^{\sigma\sigma'}(p'_x, p'_z)$ for the spin projection of the particle $\sigma = 1/2$ are defined as follows:

$$w_{ss'}^{1/2,1/2}(p'_x, p'_z) = N_f N' (2\pi\hbar)^2 \left[\left\{ c^2 p_x p'_x + \left(\mathcal{E}_s(p_x) + mc^2\right) \right. \right.$$

$$\left. \times \left(\mathcal{E}_{s'}(p'_x) + mc^2\right) \right\} I_{s,s'}(\alpha) - Q\left(p'_+ + mc^2\right)\sqrt{2sc\hbar eH_0} I_{s-1,s'}(\alpha)$$

$$\left. +2c\hbar eH_0\sqrt{ss'} I_{s-1,s'-1}(\alpha) - Q\left(\Lambda + mc^2\right)\sqrt{2s'c\hbar eH_0} I_{s,s'-1}(\alpha) \right], \tag{4.77}$$

and the transition amplitudes with the spin flip $1/2 \to -1/2$ are

$$w_{ss'}^{1/2,-1/2}(p'_x, p'_z) = -iN_f N' (2\pi\hbar)^2 \left[Q\left(p'_+ + mc^2\right) \right.$$

$$\times \left(\Lambda + mc^2\right) I_{s,s'-1}(\alpha) - cp'_x\sqrt{2sc\hbar eH_0} I_{s-1,s'-1}(\alpha)$$

$$\left. +\sqrt{2s'c\hbar eH_0}\, cp_x I_{s,s'}(\alpha) - 2c\hbar eH_0 Q\sqrt{s's} I_{s-1,s'}(\alpha) \right]. \tag{4.78}$$

The analogous formula is obtained for $\sigma = -1/2$:

$$w_{ss'}^{-1/2,-1/2}(p'_x, p'_z) = N_f N' (2\pi\hbar)^2 \left[\left\{ c^2 p_x p'_x + \left(\mathcal{E}_s(p_x) + mc^2\right) \right. \right.$$

$$\left. \times \left(\mathcal{E}_{s'}(p'_x) + mc^2\right) \right\} I_{s-1,s'-1}(\alpha) - Q\left(p'_+ + mc^2\right)\sqrt{2sc\hbar eH_0} I_{s,s'-1}(\alpha)$$

$$\left. +2c\hbar eH_0\sqrt{ss'} I_{s,s'}(\alpha) - Q\sqrt{2s'c\hbar eH_0}\left(\Lambda + mc^2\right) I_{s-1,s'}(\alpha) \right], \tag{4.79}$$

and the transition amplitudes with the spin flip $-1/2 \to 1/2$ are

$$w_{ss'}^{-1/2,1/2}(p'_x, p'_z) = -iN_f N' (2\pi\hbar)^2 \left[Q\left(p'_+ + mc^2\right) \right.$$

$$\times\left(\Lambda+mc^2\right)I_{s-1,s'}(\alpha)-cp'_x\sqrt{2sc\hbar eH_0}I_{s,s'}(\alpha)$$

$$+\sqrt{2s'c\hbar eH_0}cp_xI_{s-1,s'-1}(\alpha)-2Qc\hbar eH_0\sqrt{ss'}I_{s,s'-1}(\alpha)\Big]. \tag{4.80}$$

Here the parameter

$$Q\equiv\frac{\omega_0e\overline{A}_0T}{2\Lambda} \tag{4.81}$$

and the argument of the Lagger function is

$$\alpha\equiv\frac{ceH_0}{2\hbar}\left(\frac{e\overline{A}_0T}{\Lambda}\right)^2. \tag{4.82}$$

According to Eq. (4.76) the transition of the particle from an initial state $\{s,\sigma,p_x,p_z\}$ to a state $\{s',\sigma',p'_x,p'_z\}$ is accompanied by the emission or absorption of $s-s'$ number of photons. Consequently, substituting Eq. (4.76) into Eq. (4.71) and integrating over the momentum we can rewrite the particle wave function as

$$\Psi_{s,\sigma,p_x,p_z}(\mathbf{r},t)=\sum_{s'=0}^{\infty}w_{ss'}^{\sigma,1/2}\exp\left[\frac{i}{\hbar}\delta S_{ss'}(\mathbf{r},t)\right]\psi_{s',1/2}$$

$$+\sum_{s'=1}^{\infty}w_{ss'}^{\sigma,-1/2}\exp\left[\frac{i}{\hbar}\delta S_{ss'}(\mathbf{r},t)\right]\psi_{s',-1/2}, \tag{4.83}$$

where

$$\delta S_{ss'}(\mathbf{r},t)=p_zz+(p_x+\frac{\hbar\omega_0}{c}(s'-s))x-(\mathcal{E}_s(p_x)+\hbar\omega_0(s'-s))t. \tag{4.84}$$

Using Eqs. (4.77)–(4.80) and the momentum conservation law, and taking into consideration the recurrent relations for the Lagger function

$$I_{s,s'-1}(\alpha)=\sqrt{\frac{\alpha}{s'}}\left(\frac{s-s'-\alpha}{2\alpha}I_{s,s'}(\alpha)-I'_{s,s'}(\alpha)\right),$$

$$I_{s-1,s'}(\alpha)=\sqrt{\frac{\alpha}{s}}\left(\frac{s-s'+\alpha}{2\alpha}I_{s,s'}(\alpha)+I'_{s,s'}(\alpha)\right),$$

$$I_{s-1,s'-1}(\alpha)=\frac{\alpha}{\sqrt{ss'}}\left(\frac{s+s'-\alpha}{2\alpha}I_{s,s'}(\alpha)-I'_{s,s'}(\alpha)\right),$$

the transition amplitudes $w_{ss'}^{\sigma\sigma'}(p'_x, p'_z)$ can be written in the compact form

$$w_{ss'}^{1/2,1/2} = N_{ss'} \left\{ I_{s,s'}(\alpha) + \frac{\sqrt{\zeta s'}\hbar\omega_0}{\mathcal{E}_{s'}(p'_x) + mc^2} I_{s,s'-1}(\alpha) \right\}, \tag{4.85}$$

$$w_{ss'}^{1/2,-1/2} = -iN_{ss'} \frac{(\Lambda + mc^2)}{\mathcal{E}_{s'}(p'_x) + mc^2} \sqrt{\frac{\hbar\omega_0}{2\Lambda}} \alpha I_{s,s'-1}(\alpha) \tag{4.86}$$

and

$$w_{ss'}^{-1/2,-1/2} = N_{ss'} \left\{ I_{s-1,s'-1}(\alpha) + \frac{\sqrt{\zeta s'}\hbar\omega_0}{\mathcal{E}_{s'}(p'_x) + mc^2} I_{s-1,s'}(\alpha) \right\}, \tag{4.87}$$

$$w_{ss'}^{-1/2,1/2} = -iN_{ss'} \frac{(\Lambda + mc^2)}{\mathcal{E}_{s'}(p'_x) + mc^2} \sqrt{\frac{\hbar\omega_0}{2\Lambda}} \alpha I_{s-1,s'}(\alpha), \tag{4.88}$$

where

$$N_{ss'} \equiv \sqrt{\frac{\mathcal{E}_{s'}(p'_x)\left(\mathcal{E}_{s'}(p'_x) + mc^2\right)}{\overline{\mathcal{E}}_s(p_x)(\mathcal{E}_s(p_x) + mc^2)}}. \tag{4.89}$$

Now let us consider the concrete case of initial spin polarization $\sigma = 1/2$. The probability of the induced transition $s \to s'$ between the Landau levels is ultimately defined by Eqs. (4.85) and (4.86):

$$W_{ss'} = \left| w_{ss'}^{1/2,1/2} \right|^2 + \left| w_{ss'}^{1/2,-1/2} \right|^2$$

$$= \frac{\mathcal{E}_{s'}(p'_x)}{\overline{\mathcal{E}}_s(p_x)} \left[I_{s,s'}^2(\alpha) + \frac{s\hbar\omega_0}{\mathcal{E}_s(p_x) + mc^2} \left(I_{s-1,s'-1}^2(\alpha) - I_{s,s'}^2(\alpha) \right) \right]. \tag{4.90}$$

For the particle initially situated in the ground state the Lagger function

$$I_{0,s'}^2(\alpha) = \frac{\alpha^{s'}}{s'!} e^{-\alpha},$$

and consequently for the probability of the induced transition $0 \to s'$ we have

$$W_{0s'} = \frac{\mathcal{E}_0(p_x) + \hbar\omega_0 s'}{\mathcal{E}_0(p_x) + \hbar\omega_0 \alpha} \frac{\alpha^{s'}}{s'!} e^{-\alpha}. \tag{4.91}$$

If $\hbar\omega_0 << \mathcal{E}_0(p_x)$ this is the well-known Poisson distribution:

$$W_{0s'}(\alpha) = \frac{\alpha^{s'}}{s'!}e^{-\alpha},$$

at which the mean value of s' is $\overline{s'} = \alpha$ and there is a maximum at $\alpha = s'$. The latter shows that the most probable transitions are

$$\hbar\omega_0 s' = \triangle\mathcal{E}_{cl} = \overline{\mathcal{E}}_0(p_x) - \mathcal{E}_0(p_x), \tag{4.92}$$

i.e., the energy change corresponds to classical dynamics. This is a consequence of the fact that the Poisson distribution describes the coherent state of harmonic oscillator which can be created from the ground state $s = 0$ (a special case of coherent state). In the coherent state the probability distribution in space retains its shape, and its center follows the trajectory of a classical particle in a harmonic well (in the considered case the static magnetic field is equivalent to a harmonic well).

Let us now estimate the average number of emitted (absorbed) photons by the electron at the cyclotron resonance for the high excited Landau levels ($s >> 1$) and for the strong EM wave. In this case the most probable number of photons in the strong EM wave field corresponds to the quasiclassical limit ($|s - s'| >> 1$) when multiphoton processes dominate and the nature of the interaction process is very close to the classical one. In this case the argument of the Lagger function can be represented as

$$\alpha \equiv \frac{1}{4s}\left(\frac{ec\overline{A}_0 p_\perp T}{\hbar\Lambda}\right)^2, \tag{4.93}$$

where $p_\perp \simeq \sqrt{2e\hbar H_0 s/c}$ is the particle mean transverse momentum. The Lagger function is maximal at $\alpha \to \alpha_0 = \left(\sqrt{s'} - \sqrt{s}\right)^2$, exponentially falling beyond α_0. Hence, for the transition $s \to s'$ and when $|s - s'| << s$ we have

$$\alpha_0 \simeq \frac{(s' - s)^2}{4s}. \tag{4.94}$$

The energy change of the particle according to classical perturbation theory (when $e\overline{A}_0\omega_0 T/c << p_\perp$) is

$$\triangle\mathcal{E}_{cl} = \frac{ecp_\perp \overline{A}_0\omega_0 T}{\Lambda}. \tag{4.95}$$

The comparison of this expression with Eqs. (4.93) and (4.94) shows that the most probable transitions are

$$|s - s'| \simeq \frac{\Delta \mathcal{E}_{cl}}{\hbar \omega_0}, \tag{4.96}$$

in accordance with the correspondence principle.

4.4 Cyclotron Resonance in a Medium. Nonlinear Threshold Phenomenon of "Electron Hysteresis"

Consider now the dynamics of cyclotron resonance in the field of a strong EM wave in a medium. In this case the problem can be solved analytically only for the circular polarization of monochromatic wave and if the initial velocity of the particle is directed along the axis of the wave propagation. The particle equations of motion in components in this process are written as

$$\frac{dp_x}{dt} = n_0 \frac{e}{c} \left[\mathrm{v}_y E_y(\tau) + \mathrm{v}_z E_z(\tau) \right], \tag{4.97}$$

$$\frac{dp_y}{dt} = e \left(1 - n_0 \frac{\mathrm{v}_x}{c} \right) E_y(\tau) + e \frac{\mathrm{v}_z}{c} H_0, \tag{4.98}$$

$$\frac{dp_z}{dt} = e \left(1 - n_0 \frac{\mathrm{v}_x}{c} \right) E_z(\tau) - e \frac{\mathrm{v}_y}{c} H_0. \tag{4.99}$$

As far as the equation for the particle longitudinal momentum (4.97) is not changed in the presence of a uniform magnetic field with respect to Eq. (2.2) in the field of a plane EM wave in a medium, and the equation for the particle energy change in the field (1.9) remains unchanged, then we have the same integral of motion (2.5) in this process. Hence, with the help of the latter one can represent the particle longitudinal velocity

$$\mathrm{v}_x = c n_0 \frac{\left(1 - \frac{\mathrm{v}_0}{c n_0}\right) - \left(1 - n_0 \frac{\mathrm{v}_0}{c}\right) \left[1 \mp \frac{\mathbf{p}_\perp^2(\tau)}{(mc\zeta)^2}\right]^{1/2}}{n_0^2 \left(1 - \frac{\mathrm{v}_0}{c n_0}\right) - \left(1 - n_0 \frac{\mathrm{v}_0}{c}\right) \left[1 \mp \frac{\mathbf{p}_\perp^2(\tau)}{(mc\zeta)^2}\right]^{1/2}} \tag{4.100}$$

and energy

$$\mathcal{E} = \frac{\mathcal{E}_0}{n_0^2 - 1} \left\{ n_0^2 \left(1 - \frac{\mathrm{v}_0}{c n_0}\right) - \left(1 - n_0 \frac{\mathrm{v}_0}{c}\right) \left[1 \mp \frac{\mathbf{p}_\perp^2(\tau)}{(mc\zeta)^2}\right]^{1/2} \right\} \tag{4.101}$$

via the transversal momentum $\mathbf{p}_\perp(\tau) = \{0, p_y(\tau), p_z(\tau)\}$ in the field. Here the parameter ζ is

$$\zeta \equiv \frac{\mathcal{E}_0}{mc^2}\frac{\left|1 - n_0\frac{\mathrm{v}_0}{c}\right|}{\sqrt{|n_0^2 - 1|}}. \tag{4.102}$$

Note that ζ is the critical value of the wave intensity (2.12) (at $n_0 > 1$) for the particle "reflection" phenomenon in the absence of a static magnetic field ($H_0 = 0$). The sign "$-$" under the roots in Eqs. (4.100), (4.101) corresponds to the case of the interaction in dielectriclike media with $n_0 > 1$ and the sign "+", plasmalike media with $n_0 < 1$. Note that in contrast to the case $H_0 = 0$ (induced Cherenkov process) in Eqs. (4.100), (4.101) before the root, only the sign "$-$" is taken (in accordance with the initial conditions $\mathrm{v}_x = \mathrm{v}_0$ and $\mathcal{E} = \mathcal{E}_0$ of the free particle) since, as will be shown below, in this case the expression under the root is always positive and consequently the root cannot change its sign. Formally, Eqs. (4.100) and (4.101) have the same form as the analogous equations (2.7) and (2.10) if $\mathbf{p}_\perp^2(\tau)/m^2c^2 \to \xi^2(\tau)$. However, there is a principal difference between these equations because of the above-mentioned fact. In particular, in the presence of a static magnetic field the particle "reflection" and capture phenomena vanish — the particle longitudinal velocity cannot reach the phase velocity of the wave (threshold value for nonlinear Cherenkov resonance in the wave field) due to the particle transversal rotation in the uniform magnetic field.

Now the considered problem reduces to definition of the particle transversal momentum $\mathbf{p}_\perp(\tau)$. To integrate Eqs. (4.98) and (4.99) it is convenient to pass from the variable t to wave coordinate $\tau = t - n_0 x/c$. Then taking into account Eqs. (4.100) and (4.101) for the particle transversal momentum we will have the equations

$$\frac{dp_y}{d\tau} = eE_y(\tau) + \frac{ecH_0}{\mathcal{E}_0\left(1 - n_0\frac{\mathrm{v}_0}{c}\right)\left[1 \mp \frac{\mathbf{p}_\perp^2(\tau)}{(mc\zeta)^2}\right]^{1/2}} p_z(\tau),$$

$$\frac{dp_z}{d\tau} = eE_z(\tau) - \frac{ecH_0}{\mathcal{E}_0\left(1 - n_0\frac{\mathrm{v}_0}{c}\right)\left[1 \mp \frac{\mathbf{p}_\perp^2(\tau)}{(mc\zeta)^2}\right]^{1/2}} p_y(\tau). \tag{4.103}$$

From the set of Eqs. (4.103) one can obtain the equation for the complex quantity

$$Z(\tau) = \frac{p_y(\tau) + ip_z(\tau)}{mc} \tag{4.104}$$

related to the dimensionless parameter of the particle transversal momentum. It is written as

$$\frac{dZ(\tau)}{d\tau} = \frac{eE(\tau)}{mc} - i\frac{\Omega_0}{\left(1 - n_0\frac{\mathrm{v}_0}{c}\right)\left[1 \mp \frac{|Z(\tau)|^2}{\zeta^2}\right]^{1/2}} Z(\tau), \tag{4.105}$$

where

$$E(\tau) = E_y(\tau) + iE_z(\tau)$$

and

$$\Omega_0 = \frac{ecH_0}{\mathcal{E}_0}$$

is the Larmor frequency for the initial velocity of the particle.

For an arbitrary plane EM wave Eq. (4.105) is a nonlinear equation the exact solution of which cannot be found. However, for the monochromatic wave of circular polarization when

$$E(\tau) = E_0 e^{-i\omega_0 \tau}, \tag{4.106}$$

one can find the exact solution of Eq. (4.105). The latter is sought in the form

$$Z(\tau) = Z_0 e^{-i\omega_0 \tau} \tag{4.107}$$

and for the transversal momentum of the particle we obtain the following algebraic equation:

$$\left(1 - \frac{\Omega_0}{\omega_0 \left(1 - n_0 \frac{\mathrm{v}_0}{c}\right) \sqrt{1 \mp \beta^2}}\right) \beta = X, \tag{4.108}$$

where the quantities E_0, Z_0 are expressed in the scale of the parameter ζ:

$$\frac{Z_0}{\zeta} \equiv i\beta; \qquad \frac{eE_0}{mc\omega_0 \zeta} = \frac{\xi_0}{\zeta} \equiv X. \tag{4.109}$$

We will not represent here the exact solution of Eq. (4.108) for β. An interesting nonlinear phenomenon exists in this process which can be found out through the graphical solution of Eq. (4.108). Thus, depending on the ratio of the Larmor and wave frequencies as well as on the initial velocity of the particle (in the case of dielectriclike medium where $\mathrm{v}_0 \lessgtr c/n_0$) the solution of Eq. (4.108) is a single-valued or multivalent that essentially changes the interaction behavior of the particle with a strong EM wave at the nonlinear cyclotron resonance in a medium. Hence, we will consider separately the cases $\Omega_0 \geqslant \omega_0'$ and $\Omega_0 < \omega_0'$ at $\mathrm{v}_0 < c/n_0$ where

$$\omega_0' = \omega_0 \left(1 - n_0 \frac{\mathrm{v}_0}{c}\right) \tag{4.110}$$

is the Doppler-shifted frequency of the wave for the initial velocity of the particle. If $\mathrm{v}_0 > c/n_0$ the effects considered here will take place with the opposite circular polarization of the wave ($\omega_0 \to -\omega_0$) or in the opposite direction of the uniform magnetic field ($\mathbf{H}_0 \to -\mathbf{H}_0$).

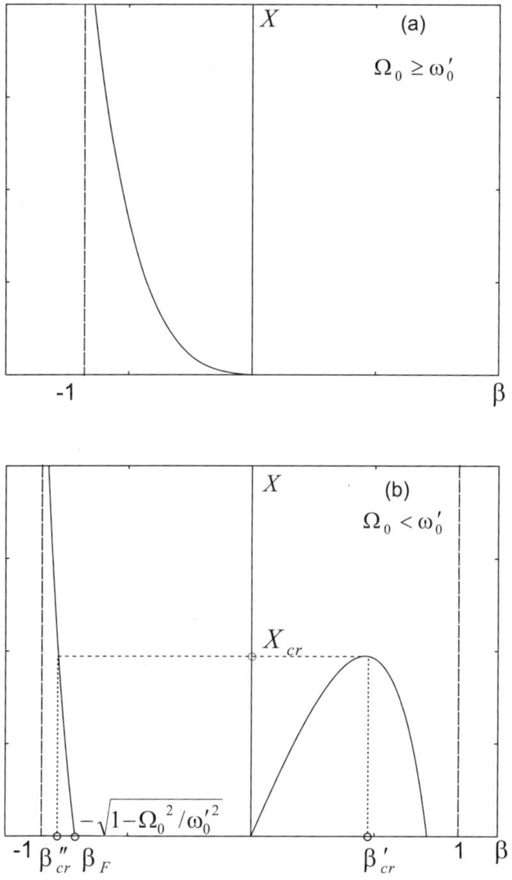

Fig. 4.2. Dependence of normalized transversal momentum β on the normalized EM wave amplitude X at $n_0 > 1$.

Consider first the case of a medium with refractive index $n_0 > 1$ (sign "$-$" under the root) in Eq. (4.108). We will turn on the EM wave adiabatically and draw the graphic of dependence of the particle transversal momentum on the wave intensity $\beta(X)$. For the case $\Omega_0 \geqslant \omega_0'$ the latter is illustrated in Fig. 4.2a. As is seen from this graphic with the increase of the wave intensity the transversal momentum of the particle increases in the field (consequently the energy as well) and vice versa: with the decrease of the wave intensity it decreases in the field and after the passing of the wave ($X = 0$) the transversal momentum becomes zero ($\beta = 0$), i.e., the particle momentum-energy remain unchanged: $p = p_0$ and $\mathcal{E} = \mathcal{E}_0$.

With the increase of the transversal momentum the longitudinal velocity of the particle increases as well, but in contrast to the case $\mathbf{H}_0 = 0$ it always remains smaller than the wave phase velocity if initially the wave overtakes the particle ($\mathrm{v}_0 < c/n_0$) and larger if the particle overtakes the wave ($\mathrm{v}_0 > c/n_0$). For this reason the particle "reflection" phenomenon vanishes in the presence of a uniform magnetic field. Indeed, as is seen from Eq. (4.108) for an arbitrary finite value of X we have $\beta < 1$ and from Eq. (4.100) it follows that the longitudinal velocity of the particle in the field $\mathrm{v}_x < c/n_0$ if $\mathrm{v}_0 < c/n_0$ and $\mathrm{v}_x > c/n_0$ if $\mathrm{v}_0 > c/n_0$. The value $\beta = 1$ may be reached only at $X = \infty$ when the root in Eq. (4.100) becomes zero and $\mathrm{v}_x = c/n_0$. So, the expression under the roots in Eqs. (4.100), (4.101) cannot become zero for finite intensities of the EM wave and, consequently, the root cannot change its sign. According to the latter in Eqs. (4.100), (4.101) before the roots only the sign "−" has been taken so as to satisfy the initial condition.

Consider now the case $\Omega_0 < \omega_0'$. The graphic of dependence of the particle transversal momentum on the wave intensity $\beta(X)$ in this case is illustrated in Fig. 4.2b. As is seen from this graphic $\beta(X)$ is already a multivalent function: for wave intensities smaller than the value corresponding to the maximum point of the curve $\beta(X)$ three values of the particle transversal momentum exist for each value of the wave intensity. At the maximum point, which will be called a critical one, the wave intensity has the value

$$X_{cr} = \left[1 - \left(\frac{\Omega_0}{\omega_0'}\right)^{2/3}\right]^{3/2} . \tag{4.111}$$

There are two values β_{cr}' and β_{cr}'' which correspond to critical intensity (4.111). The first one, β_{cr}', is the value of the parameter β corresponding to particle transversal momentum at the maximum point of the curve $\beta(X)$. From the extremum condition of Eq. (4.108) for β_{cr}' we have

$$\beta_{cr}' = \left[1 - \left(\frac{\Omega_0}{\omega_0'}\right)^{2/3}\right]^{1/2} . \tag{4.112}$$

The second critical value for the parameter β corresponding to critical intensity X_{cr} is situated on the left-hand side branch of the curve $\beta(X)$. To determine its value one needs the analytic solution $\beta = \beta(X)$ of Eq. (4.108), but there is no necessity here to present the bulk expression for β_{cr}''.

We shall decide on that branch of the curve $\beta(X)$ which corresponds to real motion of the particle. Up to the critical point the particle transversal momentum can be changed on that branch which corresponds to initial condition $\beta = 0$ at $X = 0$. On this branch the particle momentum increases with the increase of the wave intensity and vice versa. It is evident that with further increase of the field the particle cannot be situated on the right-hand side from the critical point. Hence, it should pass to the left-hand side branch of the curve $\beta(X)$. Indeed, it is easy to see that the critical point is an unstable state for the particle, while all states on the left-hand side branch of the curve $\beta(X)$ are stable and at the critical point the particle changes instantaneously its transversal momentum and passes by jumping to that branch. The further variation of the particle transversal momentum occurs already on this branch. Note that the instantaneity here is related to the fact that the solution of Eq. (4.105) has been found for the monochromatic wave. It is clear that the momentum change actually occurs during finite time. This jump variation of the particle momentum (energy) is due to the induced resonant absorption of energy from the wave at the critical point because of which the particle state at this point becomes unstable and it leaves the resonance point for a stable state that corresponds to the transversal momentum β''_{cr} on the left-hand side branch of the curve $\beta(X)$. Indeed, if one draws a graphic of the dependence of the particle transversal momentum on the ratio of the Larmor and wave frequencies Ω_0/ω'_0 for a certain intensity of the wave (Fig. 4.3), then it will be seen from the graphic $\beta(\Omega_0/\omega'_0)$ that the cyclotron resonance in the strong EM wave field takes place at the critical point with the satisfaction of the condition $\Omega_0 < \omega'_0$. The latter means that to reach the cyclotron resonance in a medium, in contrast to vacuum autoresonance it is necessary to be initially under the resonance condition, since due to the effect of the strong wave field in a medium with refractive index $n_0 > 1$ the Larmor frequency increases in the field and then reaches the resonance value. In vacuum the cyclotron resonance proceeds at $\Omega_0 = \omega'_0$ which survives infinitely, because of which the energy of the particle turns to infinity. Thus, from Eq. (4.108) in this case ($n_0 = 1$) for the particle transversal momentum we have

$$\beta = \frac{X}{1 - \frac{\Omega_0}{\omega'_0}}, \tag{4.113}$$

which diverges (consequently the energy as well) at $\Omega_0 = \omega'_0$. As is seen from Fig. 4.3 this divergence vanishes in a medium.

With the further increase of the field ($X > X_{cr}$) the transversal momentum of the particle will continuously increase on the left-hand side branch of the curve $\beta(X)$ and tend to value -1 at $X \to \infty$. With the decrease of the field the transversal momentum decreases on this branch and at $X = X_{cr}$ already has only the value β''_{cr} since the value β'_{cr} corresponds to the unstable state at the resonance point and now there is no reason for inverse transition from the stable state to the unstable one. With the further decrease of the field the transversal momentum decreases, but as is seen from Fig. 4.2 after the interaction ($X = 0$) the particle does not return to the initial state ($\beta = 0$ at $X = 0$) and remains with the final transversal momentum

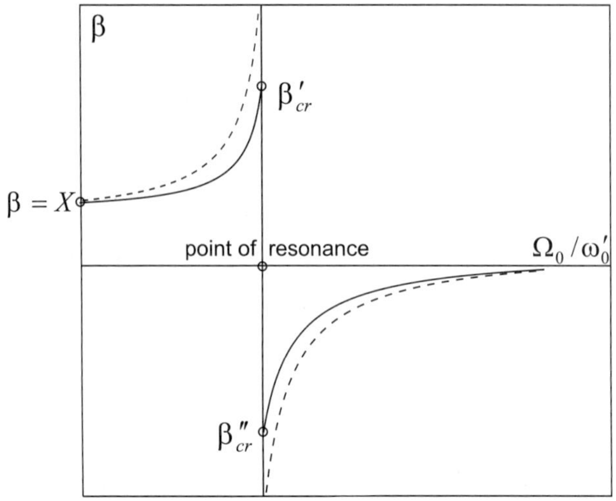

Fig. 4.3. Dependence of normalized transversal momentum on parameter $\Omega_0/\omega_0' < 1$ at $n_0 > 1$.

$$\beta_F = -\left[1 - \left(\frac{\Omega_0}{\omega_0'}\right)^2\right]^{1/2} . \tag{4.114}$$

This is a nonlinear phenomenon of charged particle hysteresis in the cyclotron resonance with a strong EM wave in a medium at intensities exceeding the threshold value (4.111).

The longitudinal velocity of the particle corresponding to the value β_F (4.114) is

$$\mathrm{v}_x = cn_0 \frac{1 - \frac{\mathrm{v}_0}{cn_0} - \left(1 - n_0\frac{\mathrm{v}_0}{c}\right)\frac{\Omega_0}{\omega_0'}}{n_0^2\left(1 - \frac{\mathrm{v}_0}{cn_0}\right) - \left(1 - n_0\frac{\mathrm{v}_0}{c}\right)\frac{\Omega_0}{\omega_0'}} . \tag{4.115}$$

The energy acquired by the particle due to hysteresis is given by

$$\mathcal{E} = \mathcal{E}_0\left[1 + \frac{\left(1 - n_0\frac{\mathrm{v}_0}{c}\right)\left(1 - \frac{\Omega_0}{\omega_0'}\right)}{n_0^2 - 1}\right] . \tag{4.116}$$

If the wave intensity is smaller than the critical value (4.111) the energy of the particle oscillates in the field and after the interaction remains unchanged.

Equation (4.116) determines the particle acceleration due to a strong transversal EM wave at the cyclotron resonance with the powerful static magnetic field in a gaseous medium ($n_0 - 1 << 1$). Because of the latter one can achieve the cyclotron resonance using optical (laser) radiation in a medium with the refractive index $n_0 > 1$, since the Doppler shift for a wave frequency $1 - n_0 \mathrm{v}_0/c$ (see Eq. (4.110)) in this case may be arbitrarily small in contrast to vacuum, where the cyclotron resonance for the existing powerful static magnetic fields is possible only in the radio-frequency domain. On the other hand, the application of powerful laser radiation for large acceleration of the particles in gaseous media is confined by the ionization threshold of the medium.

Consider now the case of a plasmous medium ($n_0 < 1$). In Eq. (4.108) this case should take the sign "+" under the root at which the confinement for the particle transversal momentum, existing in a dielectriclike medium, vanishes. In addition, the above-considered behavior of the cyclotron resonance in a plasmous medium takes place with the inverse relation between the initial Larmor and wave frequencies Ω_0/ω_0'. Thus, at $\Omega_0 \leqslant \omega_0'$ with the increase of the wave intensity the transversal momentum of the particle increases in the field and vice versa: with the decrease of the wave intensity it decreases in the field and after the passing of the wave ($X = 0$) the transversal momentum becomes zero ($\beta = 0$), i.e., the particle momentum-energy remain unchanged: $p = p_0$ and $\mathcal{E} = \mathcal{E}_0$. The nonlinear phenomenon of particle hysteresis in a plasmous medium takes place at $\Omega_0 > \omega_0'$, since in a medium with refractive index $n_0 < 1$ the Larmor frequency decreases in the field and then becomes equal to the resonance value. The graphic of dependence of the particle transversal momentum on the wave intensity $\beta(X)$ in this case is illustrated in Fig. 4.4. As is seen from this graphic, in contrast to the case of dielectriclike media the parameter β in the plasmas increases with no limit at the increase of the field. The latter allows the large acceleration of the particles achieved by the current superstrong laser fields of relativistic intensities ($\xi > 1$) due to this phenomenon of hysteresis in the plasmas. The final transversal momentum of the particle as a result of the hysteresis in this case is

$$\beta_F = \left[\left(\frac{\Omega_0}{\omega_0'}\right)^2 - 1\right]^{1/2}, \tag{4.117}$$

the final energy of which will be determined by the same equation (4.116) since both the numerator and denominator of the fraction in the expression analogous to Eq. (4.116) for the particle energy in a plasma change sign.

Note an interesting effect at the cyclotron resonance in a medium as well. At $\Omega_0 = \omega_0'$ no matter how weak the EM wave field is — $\xi_0 << \zeta$ (that is, $\xi_0 << 1$ even for $\zeta \sim 1$) — from Eq. (4.108) it follows that

$$|\beta| \simeq \left(\frac{2\xi_0}{\zeta}\right)^{1/3}, \tag{4.118}$$

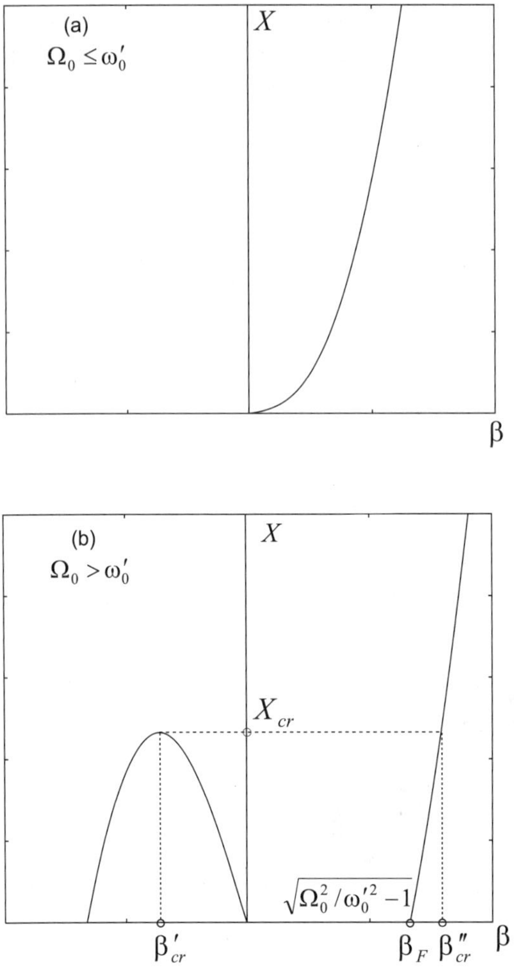

Fig. 4.4. Dependence of normalized transversal momentum β on the normalized EM wave amplitude X at $n_0 < 1$.

that is, an essential nonlinearity ($\sim \xi_0^{1/3} >> \xi_0$) arises in a case where one would expect a linear dependence on the field according to linear theory. It is the consequence of nonlinear cyclotron resonance the width of which is large enough in this case:

$$\triangle\omega \simeq 2^{-1/3} \cdot \left(\frac{\xi_0}{\zeta}\right)^{2/3} \omega_0'. \tag{4.119}$$

4.5 High Harmonics Radiation at Cyclotron Resonance

The considered phenomena at the cyclotron resonance in vacuum and in a medium will resonantly enhance the efficiency of charged particle radiation in the presence of a uniform magnetic field with respect to Compton radiation in the strong wave field. Hence, here we will consider the radiation of a charged particle in the field of a strong monochromatic EM wave in the presence of a uniform magnetic field directed along the wave propagation direction in the scope of the classical theory. We will analyze the case of circular polarization of the incident wave and when the initial velocity of the particle is parallel to the wave propagation direction. This case of particle–wave parallel propagation is of certain interest since in this case the interaction length with the actual laser beams is maximal which is especially important for the problem of high harmonic generation.

To determine the radiation energy at the cyclotron resonance in vacuum and in a medium we will consider the general case of radiation in a medium and then we will move to the vacuum case substituting the refractive index of a medium $n_0 = n(\omega) = 1$ in the ultimate equation for radiation energy. The latter is given by Eq. (2.50) where the kinematic quantities $\mathbf{v}\,(t)$ and $\mathbf{r} = \mathbf{r}\,(t)$ for the cyclotron resonance in a medium will be defined by Eqs. (4.100), (4.101), and (4.108). If in the considered case

$$p_y^2(\tau) + p_z^2(\tau) = p_\perp^2 = \text{const},$$

then the longitudinal velocity and the energy of the particle in the field

$$\mathrm{v}_x = \text{const}; \qquad \mathcal{E} = \text{const}, \tag{4.120}$$

and from Eqs. (4.104), (4.107), and (4.109) for the transversal components of the particle momentum we will have

$$\mathrm{v}_y\,(t) = \frac{mc^3\zeta\beta}{\mathcal{E}} \sin\omega_0 \left(1 - n_0\frac{\mathrm{v}_x}{c}\right) t,$$

$$\mathrm{v}_z\,(t) = \frac{mc^3\zeta\beta}{\mathcal{E}} \cos\omega_0 \left(1 - n_0\frac{\mathrm{v}_x}{c}\right) t. \tag{4.121}$$

The particle law of motion $\mathbf{r} = \mathbf{r}\,(t)$ corresponding to Eqs. (4.120) and (4.121) is

$$x\,(t) = \mathrm{v}_x t,$$

$$y\,(t) = -\frac{mc^3\zeta\beta}{\mathcal{E}\omega_0\left(1 - n_0\frac{\mathrm{v}_x}{c}\right)} \cos\omega_0 \left(1 - n_0\frac{\mathrm{v}_x}{c}\right) t, \tag{4.122}$$

$$z(t) = \frac{mc^3\zeta\beta}{\mathcal{E}\omega_0\left(1 - n_0\frac{v_x}{c}\right)} \sin\omega_0\left(1 - n_0\frac{v_x}{c}\right)t.$$

Substituting Eqs. (4.120)–(4.122) into Eq. (2.50) and integrating over t, the following ultimate equation for the spectral power of the particle radiation at the cyclotron resonance in a medium is obtained:

$$dP_{\mathbf{k}} = \frac{e^2 n(\omega)\omega^2}{2\pi c^3} v_\perp^2 \sum_{s=-\infty}^{\infty} \delta\left(\omega\left(1 - n(\omega)\frac{v_x}{c}\cos\vartheta\right) - s\omega_0\left(1 - n_0\frac{v_x}{c}\right)\right)$$

$$\times\left[\left(\frac{n^2(\omega)v_x^2 - c^2}{n^2(\omega)v_\perp^2} + \left(\frac{s}{\alpha}\right)^2\right) J_s^2(\alpha) + J_s'^2(\alpha)\right] d\omega dO. \tag{4.123}$$

Here

$$v_\perp = \frac{mc^3\zeta\beta}{\mathcal{E}} \tag{4.124}$$

is the amplitude of the transversal velocity of the particle in the field, and the argument of the Bessel function α is

$$\alpha = n(\omega)\frac{mc^2\omega\zeta\beta}{\mathcal{E}\omega_0\left(1 - n_0\frac{v_x}{c}\right)} \sin\vartheta. \tag{4.125}$$

Noting that

$$\frac{n^2(\omega)v^2 - c^2}{n^2(\omega)v_\perp^2} = -\frac{1}{(\zeta\beta)^2}\left[1 - \frac{\mathcal{E}^2}{m^2c^4}\frac{n^2(\omega) - 1}{n^2(\omega)}\right]$$

Eq. (4.123) may be written in the form

$$dP_{\mathbf{k}} = \frac{e^2 n(\omega)\omega^2}{2\pi c\left|1 - n(\omega)\frac{v_x}{c}\cos\vartheta\right|}\left(\zeta\beta\frac{mc^2}{\mathcal{E}}\right)^2$$

$$\times\sum_{s=-\infty}^{\infty}\left\{\left[\left(\frac{s}{\alpha}\right)^2 - 1 - \frac{1}{(\zeta\beta)^2}\left(1 - \frac{\mathcal{E}^2}{m^2c^4}\frac{n^2(\omega) - 1}{n^2(\omega)}\right)\right] J_s^2(\alpha) + J_s'^2(\alpha)\right\}$$

$$\times\delta\left(\omega - s\omega_0\frac{1 - n_0\frac{v_x}{c}}{1 - n(\omega)\frac{v_x}{c}\cos\vartheta}\right) d\omega dO. \tag{4.126}$$

Consider first the case of vacuum. If $n_0 = n(\omega) = 1$ when the autoresonance phenomenon takes place, parameters (4.124) and (4.125) become

$$\mathrm{v}_{\perp} = \frac{mc^3}{\mathcal{E}} \frac{\xi_0}{1 - \frac{\Omega_0}{\omega_0'}}, \qquad \alpha = \frac{\omega mc^2}{\omega_0 \Lambda} \frac{\xi_0}{1 - \frac{\Omega_0}{\omega_0'}} \sin\vartheta,$$

where Λ is the integral of motion in vacuum (1.10) and $\omega_0' = \omega_0 (1 - \mathrm{v}_0/c)$ is the Doppler-shifted frequency of the incident strong wave for the initial velocity of the particle. Then from Eq. (4.126) for the radiation power in vacuum we obtain

$$dP_{\mathbf{k}} = \frac{e^2}{2\pi c}\left(\frac{mc^2}{\mathcal{E}}\right)^2 \frac{\omega^2}{1 - \frac{\mathrm{v}_x}{c}\cos\vartheta} \frac{\xi_0^2}{\left(1 - \frac{\Omega_0}{\omega_0'}\right)^2} \sum_{s=1}^{\infty} \delta\left(\omega - s\omega_0 \frac{1 - \frac{\mathrm{v}_x}{c}}{1 - \frac{\mathrm{v}_x}{c}\cos\vartheta}\right)$$

$$\times \left[\left\{\left(\frac{s}{\alpha}\right)^2 - 1 - \frac{\left(1 - \frac{\Omega_0}{\omega_0'}\right)^2}{\xi_0^2}\right\} J_s^2(\alpha) + J_s'^2(\alpha)\right] d\omega dO. \qquad (4.127)$$

Note that in Eq. (4.127) the term $s = 0$ corresponds to $\omega = 0$ (according to the δ-function) for which the radiation power is zero, so that the summation proceeds from $s = 1$. The $s = 0$ harmonic arises in a dielectriclike medium which corresponds to Cherenkov radiation. Concerning the terms with the negative s in the sum (4.127) then those are zero in vacuum according to the argument of the δ-function taking into account that $\omega_0, \omega > 0$.

In the absence of a static magnetic field ($\Omega_0 = 0$) Eq. (4.127) coincides with the equation for the spectral power of nonlinear Compton radiation (1.61). Comparison of Eq. (4.127) with the latter shows that the radiation power at the cyclotron resonance in vacuum resonantly enhances with the parameter of nonlinearity $\xi_0/(1 - \Omega_0/\omega_0')$ instead of the parameter of nonlinearity ξ_0 for nonlinear Compton radiation. Hence, we will not repeat the analysis of the conditions for revelation of nonlinearities in the considered process that is the radiation of high harmonics, which has been done for nonlinear Compton radiation and the substitution of the strong wave intensity parameter $\xi_0 \to \xi_0/(1 - \Omega_0/\omega_0')$ only should be made.

Consider now the radiation in a medium at the nonlinear cyclotron resonance. In this case the Doppler factor $1 - n_0\mathrm{v}_0/c$ may be as positive as well as negative — anomalous Doppler effect at $n_0 > 1$. However, as has been shown in the previous section, for the anomalous Doppler effect the considered process of cyclotron resonance will take place at the opposite circular polarization of the incident strong wave. Hence, we also assume here $\mathrm{v}_0 < c/n_0$ at which Eq. (4.110) has a meaning. In addition, since for $\mathrm{v}_0 < c/n_0$ the longitudinal velocity in the field always remains smaller than the wave phase velocity ($\mathrm{v}_x < c/n_0$), then the Doppler factor $1 - n_0\mathrm{v}_x/c > 0$ as well.

Taking into account Eqs. (4.124), (4.125), and (4.102) as well as using the δ-function, which expresses the radiation spectrum of the process, the equation for radiation power (4.123) may be written in the form

$$dP_{\mathbf{k}} = \frac{e^2 n(\omega)\omega^2}{2\pi c} \frac{1}{\left|1 - n(\omega)\frac{\mathrm{v}_x}{c}\cos\vartheta\right|} \sum_{s=-\infty}^{\infty} \delta\left(\omega - s\omega_0 \frac{1 - n_0\frac{\mathrm{v}_x}{c}}{1 - n(\omega)\frac{\mathrm{v}_x}{c}\cos\vartheta}\right)$$

$$\times \left[\left(\frac{n(\omega)\frac{\mathrm{v}_x}{c} - \cos\vartheta}{n(\omega)\sin\vartheta}\right)^2 J_s^2(\alpha) + \frac{\mathrm{v}_\perp^2}{c^2} J_s'^2(\alpha)\right] d\omega dO, \tag{4.128}$$

where the argument of the Bessel function is

$$\alpha = n(\omega)\frac{\omega}{\omega_0}\frac{\sin\vartheta}{\sqrt{|n_0^2 - 1|}}\frac{\beta}{\sqrt{1 \mp \beta^2}}. \tag{4.129}$$

Concerning the terms with the negative s in Eq. (4.128), note that according to the argument of the δ function the harmonics with $s < 0$ correspond to the anomalous Doppler effect for radiated frequencies (as for the fundamental frequency $1 - n_0\mathrm{v}_x/c > 0$) which is possible due to the dispersion of the medium, if

$$1 - n(\omega)\frac{\mathrm{v}_x}{c}\cos\vartheta < 0,$$

i.e., the harmonics with $s < 0$ may be radiated inside the Cherenkov cone.

Arising from Eq. (4.108) one can express the argument of the Bessel function via the parameter of the cyclotron resonance Ω_0/ω_0'

$$\alpha = n(\omega)\frac{\omega}{\omega_0'}\frac{mc^2}{\mathcal{E}_0}\frac{\xi_0}{\sqrt{1 \mp \beta^2} - \frac{\Omega_0}{\omega_0'}}\sin\vartheta, \tag{4.130}$$

which evidences the resonant enhancement of the parameter of nonlinearity and, consequently, the intensity of high harmonics radiation ($\alpha \sim s >> 1$). If $\beta^2 << 1$, which corresponds to linear cyclotron resonance, from Eq. (4.130) we see that the radiation power in a medium resonantly enhances with the parameter of nonlinearity $\xi_0/\left(1 - \Omega_0/\omega_0'\right)$ as in the case of vacuum.

The radiation of high harmonics at the nonlinear cyclotron resonance in a medium arises for the wave intensities in the area close to the critical value for electron hysteresis phenomenon (4.111). Corresponding to this intensity the transversal momentum of the particle β in Eq. (4.130) should be substituted by the critical value β_{cr}' from Eq. (4.112). In the other case of particle–wave nonlinear interaction at the cyclotron resonance in a medium that takes place at $\Omega_0 = \omega_0'$ and $\xi_0 << \zeta$ (see Eq. (4.118)), the transversal momentum of the particle β in Eq. (4.130) should be substituted from Eq. (4.118).

Bibliography

A.V. Gaponov, M.A. Miller, Zh. Éksp. Teor. Fiz. **34**, 242 (1958)
M.A. Miller, Izv. VUZov, Radiofizika **1**, 110 (1958) [in Russian]
Ya.B. Fainberg, V.I. Kurilko, Zh. Tekh. Fiz. **29**, 935 (1959) [in Russian]
M.I. Petelin, Izv. VUZov, Radiofizika **4**, 455 (1961) [in Russian]
A.A. Andronov, M.I. Petelin, V.V. Zheleznyakov, Izv. VUZov, Radiofizika **7**, 251 (1961) [in Russian]
M.A. Miller, Izv. VUZov, Radiofizika **5**, 929 (1962) [in Russian]
B.G. Eremin, M.A. Miller, Izv. VUZov, Radiofizika **5**, 1151 (1962) [in Russian]
V.Ya. Davidovsky, Zh. Éksp. Teor. Fiz. **43**, 886 (1962)
A.A. Kolomensky, A.N. Lebedev, Zh. Éksp. Teor. Fiz. **44**, 261 (1963)
V.S. Voronin, A.A. Kolomensky, Zh. Éksp. Teor. Fiz. **47**, 1528 (1964)
A.I. Nikishov, V.I. Ritus, Zh. Éksp. Teor. Fiz. **64**, 776 (1964)
C.S. Roberts, S.J. Buchsbaum, Phys.Rev. A **135**, 381 (1964)
P.J. Redmond, Math. Phys. **6**, 1163 (1965)
V.P. Oleinik, Zh. Éksp. Teor. Fiz. **52**, 1049 (1967)
V.P. Oleinik, Zh. Éksp. Teor. Fiz. **53**, 1997 (1967)
V.M. Haroutunian, H.K. Avetissian, Izv. Akad. Nauk Arm. SSR Ser. Fiz. **9**, 110 (1974) [in Russian]
H.K. Avetissian, Izv. Akad. Nauk Arm. SSR Ser. Fiz. **10**, 3 (1975) [in Russian]
Yu.A. Andreev, V.Ya. Davidovsky, Zh. Tekh. Fiz. **45**, 3 (1975) [in Russian]
Yu.A. Andreev, V.Ya. Davidovsky, Zh. Tekh. Fiz. **46**, 413 (1976) [in Russian]
Yu.A. Andreev, V.Ya. Davidovsky, V.N. Danilenko, Zh. Tekh. Fiz. **46**, 2380 (1976) [in Russian]
Yu.A. Andreev, V.Ya. Davidovsky, V.N. Danilenko, Zh. Tekh. Fiz. **48**, 2184 (1978) [in Russian]
I.M. Ternov, V.R. Khalilov, V.N. Rodionov: Interaction of Charged Particles with Strong Electromagnetic Field (Mosk. Gos. Univ., Moscow 1982) [in Russian]
A.A. Sokolov, I.M. Ternov: Relativistic Electron (Nauka, Moscow 1983) [in Russian]
H.K. Avetissian, K.Z. Hatsagortsian, Zh. Tekh. Fiz. **54**, 2347 (1984) [in Russian]
R.M. Robb, Phys. Rev. E **50**, 3345 (1994)
G.S. Nusinovich, P.E. Latham, O. Dumbrajs, Phys. Rev. E **52**, 998 (1995)
S.J. Cooke, A.W.Cross, W. He, A.D.R. Phelps, Phys. Rev. Lett. **77**, 4836 (1996)
V.L. Bratman, A.D.R. Phelps, A.V. Savilov, Physics of Plasmas **4**, 2285 (1997)
B.W.J. McNeil, G.R.M. Robb, A.D.R. Phelps, J. Phys. D **30**, 1688 (1997)
P. Aitken et al., J. Phys. D **30**, 2482 (1997)
N.S. Ginzburg et al., Phys. Rev. Lett. **78**, 2365 (1997)

P. Aitken, B.W.J. McNeil, G.R.M. Robb, A.D.R. Phelps, Phys. Rev. E **59**, 1152 (1999)
N.S. Ginzburg et al., IEEE Trans. Plasma Sci. **27**, 462 (1999)
Y.I. Salamin, F.H.M. Faisal, C. H. Keitel, Phys. Rev. A **62**, 53809 (2000)
H.K. Avetissian, G.F. Mkrtchian, M.G. Poghosyan, Zh. Éksp. Teor. Fiz. **99**, 290 (2004)

5 Nonlinear Dynamics of Induced Compton and Undulator Processes

In this chapter we will consider the interaction of charged particles with superstrong radiation fields of relativistic intensities in induced coherent processes in vacuum where there is no restriction on the field intensity taking place at the induced Cherenkov interaction in dielectriclike media. Those are the induced Compton and undulator processes.

In the presence of a second wave of different frequency, the Compton scattering, as well as spontaneous undulator radiation in the external EM wave field acquire induced character. Because of its coherent nature (as the Cherenkov one) these induced processes have the same peculiarity and, consequently, the nonlinear interaction of charged particles with the mentioned fields leads to analogous threshold phenomena of particle "reflection" and capture by the plane EM waves in vacuum.

On the other hand, it is clear that the second wave in the induced Compton process or the undulator field perform the role of the third body for the real radiation/absorption of photons by the free electrons in vacuum. Hence, irrespective of revelation of new phenomena the consideration of nonlinear dynamics of induced Compton and undulator processes in current superstrong laser fields is of great interest, especially from the point of view of FEL and laser accelerators. Further, the significance of the undulator (wiggler) is great enough as the unique version of the current FEL and expected X-ray laser due to its large coherent length and effective power of the static magnetic field for relativistic particles.

To achieve relatively large coherent lengths in the induced Compton process we will consider the case of counterpropagating waves.

5.1 Interaction of Charged Particles with Superstrong Counterpropagating Waves of Different Frequencies

Consider the classical dynamics of a charged particle at the interaction with two counterpropagating (along the axis OX) plane EM waves having arbitrary electric field strengths $\mathbf{E}_1\left(t-\frac{x}{c}\right)$ and $\mathbf{E}_2\left(t+\frac{x}{c}\right)$ in vacuum. The relativistic equation of motion in components is written as

$$\frac{dp_x}{dt} = \frac{e}{c}(\mathbf{vE}_1 - \mathbf{vE}_2), \tag{5.1}$$

$$\frac{dp_y}{dt} = e\left(1 - \frac{\mathrm{v}_x}{c}\right)E_{1y} + e\left(1 + \frac{\mathrm{v}_x}{c}\right)E_{2y},$$

$$\frac{dp_z}{dt} = e\left(1 - \frac{\mathrm{v}_x}{c}\right)E_{1z} + e\left(1 + \frac{\mathrm{v}_x}{c}\right)E_{2z}. \tag{5.2}$$

This set of equations allows exact solution when the particle initial velocity is directed along the axis OX and the waves are monochromatic with circular polarization:

$$\mathbf{E}_1(x,t) = \left\{0, E_1\cos\omega_1\left(t - \frac{x}{c}\right), E_1\sin\omega_1\left(t - \frac{x}{c}\right)\right\},$$

$$\mathbf{E}_2(x,t) = \left\{0, E_2\cos\omega_2\left(t + \frac{x}{c}\right), E_2\sin\omega_2\left(t + \frac{x}{c}\right)\right\}. \tag{5.3}$$

From Eq. (5.2) in the field (5.3) we obtain

$$p_y = \frac{eE_1}{\omega_1}\sin\omega_1\left(t - \frac{x}{c}\right) + \frac{eE_2}{\omega_2}\sin\omega_2\left(t + \frac{x}{c}\right),$$

$$p_z = -\frac{eE_1}{\omega_1}\cos\omega_1\left(t - \frac{x}{c}\right) - \frac{eE_2}{\omega_2}\cos\omega_2\left(t + \frac{x}{c}\right) \tag{5.4}$$

(the waves are turned on and turned off adiabatically at $t \to \mp\infty$).

For the integration of Eq. (5.1) we will use the equation for the particle energy exchange in the field

$$\frac{d\mathcal{E}}{dt} = e(\mathbf{vE}_1 + \mathbf{vE}_2). \tag{5.5}$$

Thus, defining the particle transverse velocity in the field by Eqs. (5.4), from Eqs. (5.1) and (5.5) we obtain the following integral of motion in the induced Compton process:

$$\mathcal{E} - c\frac{\omega_1 - \omega_2}{\omega_1 + \omega_2}p_x = \text{const}. \tag{5.6}$$

The latter together with Eq. (5.4) determines the particle energy in the field

$$\mathcal{E} = \frac{\mathcal{E}_0}{n_1^2 - 1}\left\{n_1^2\left(1 - \frac{\mathrm{v}_0}{cn_1}\right) \mp \left[\left(1 - n_1\frac{\mathrm{v}_0}{c}\right)^2 - (n_1^2 - 1)\left(\frac{mc^2}{\mathcal{E}_0}\right)^2\right.\right.$$

$$\times \left[\xi_1^2 + \xi_2^2 + 2\xi_1\xi_2 \cos\left(\omega_1 - \omega_2\right)\left(t - n_1 \frac{x}{c}\right)\right]\Bigg]^{1/2} \Bigg\}. \tag{5.7}$$

The parameter n_1 included in Eq. (5.7) is

$$n_1 = \frac{\omega_1 + \omega_2}{|\omega_1 - \omega_2|}, \tag{5.8}$$

and the parameters $\xi_{1,2} \equiv eE_{1,2}/mc\omega_{1,2}$.

As is seen from Eq. (5.7) due to the effective interaction of the particle with the counterpropagating waves a slowed traveling wave in vacuum arises. The parameter n_1 denotes the refractive index of this interference wave and since $n_1 > 1$ (see Eq. (5.8)) the phase velocity of the effective traveling wave $\mathrm{v}_{ph} = c/n_1 < c$. Then the expression under the root in Eq. (5.7) evidences the peculiarity in the interaction dynamics like the induced Cherenkov one that causes the analogous threshold phenomena of particle "reflection" and capture by the interference wave in the induced Compton process. Hence, omitting the same procedure related to bypass of the multivalence and complexity of Eq. (5.7), which has been made in detail for the analogous expression in the Cherenkov process, we will present the final results for particle "reflection" and capture by the effective interference wave in the induced Compton process. The threshold value of the "reflection" phenomenon or the critical field for nonlinear Compton resonance is

$$\xi_{cr}\left(\omega_{1,2}\right) \equiv \left(\xi_1 + \xi_2\right)_{cr} = \frac{\mathcal{E}_0}{mc^2} \frac{\left|\omega_1\left(1 - \frac{\mathrm{v}_0}{c}\right) - \omega_2\left(1 + \frac{\mathrm{v}_0}{c}\right)\right|}{2\sqrt{\omega_1\omega_2}}. \tag{5.9}$$

If one knows the longitudinal velocity v_x of the particle in the field, then it is easy to see that $\xi_{cr}\left(\omega_{1,2}\right)$ is the value of the total intensity of counterpropagating waves at which v_x becomes equal to the phase velocity of the effective interference wave: $\mathrm{v}_x = \mathrm{v}_{ph} = c/n_1$ irrespective of the magnitude of particle initial velocity v_0. The latter is the condition of coherency of induced Compton process

$$\omega_1\left(1 - \frac{\mathrm{v}_x}{c}\right) = \omega_2\left(1 + \frac{\mathrm{v}_x}{c}\right). \tag{5.10}$$

Under condition (5.10) the nonlinear resonance in the field of counterpropagating waves of different frequencies occurs and because of induced Compton radiation/absorption the particle velocity becomes smaller or larger than the phase velocity of the interference wave and the particle leaves the slowed effective wave. In the rest frame of the latter the particle swoops on the motionless barrier (if $\xi_1 + \xi_2 > \xi_{cr}\left(\omega_{1,2}\right)$) and the elastic reflection occurs. In the laboratory frame it corresponds to inelastic "reflection" and from Eq. (5.7) for particle energy after the "reflection" ($\xi_{1,2} \to 0$ adiabatically at $t \to +\infty$)

we have

$$\mathcal{E} = \mathcal{E}_0 \frac{\omega_1^2 \left(1 - \frac{\mathrm{v}_0}{c}\right) + \omega_2^2 \left(1 + \frac{\mathrm{v}_0}{c}\right)}{2\omega_1\omega_2}. \tag{5.11}$$

From this equation it follows that the energy of the particle with the initial velocity $\mathrm{v}_0 = c\left|\omega_1 - \omega_2\right| / \left(\omega_1 + \omega_2\right)$ corresponding to the resonance value of the induced Compton process does not change after the interaction ($\mathcal{E} = \mathcal{E}_0$). For such particle $\xi_{cr}\left(\omega_{1,2}\right) = 0$, i.e., it cannot enter the field: $\xi_1 = \xi_2 = 0$. The particle with the initial velocity $\mathrm{v}_0 > c\left|\omega_1 - \omega_2\right| / \left(\omega_1 + \omega_2\right)$ after the "reflection" is decelerated, while at $\mathrm{v}_0 < c\left|\omega_1 - \omega_2\right| / \left(\omega_1 + \omega_2\right)$ it is accelerated because of direct and inverse induced Compton processes. At the acceleration the particle absorbs photons from the wave of frequency ω_1 and coherently radiates into the wave of frequency ω_2 if $\omega_1 > \omega_2$ and at the deceleration the inverse process takes place. Hence, at the particle acceleration the amplification of the wave of a smaller frequency holds, while at the deceleration the wave of a larger frequency is amplified.

In the case of $\omega_1 = \omega_2 \equiv \omega$ the refractive index of the interference wave $n_1 = \infty$ and nonlinear interaction of the particle with the strong standing wave occurs. It is evident that in this case the process is elastic: $\mathcal{E} = \mathcal{E}_0 = \mathrm{const}$ (see Eq. (5.11)) and for the longitudinal momentum of the particle in the field we have

$$p_x = \pm\sqrt{p_0^2 - m^2c^2\left(\xi_1^2 + \xi_2^2 + 2\xi_1\xi_2\cos\frac{2\omega}{c}x\right)}. \tag{5.12}$$

From this equation it is seen that at $\xi_1 + \xi_2 > \xi_{cr}\left(\omega\right) = \left|p_0\right| / mc$ the standing wave becomes a potential barrier for the particle and elastic reflection occurs: the root changes its sign and $p_x = -p_0$ (if $\xi_1 + \xi_2 < \xi_{cr}\left(\omega\right)$ we have $p_x = p_0$).

Consider now the nonlinear dynamics of a particle with the arbitrary direction of velocity $\mathbf{v}_0$ initially situated in the field of counterpropagating waves (internal particle). It is clear that at the wave intensities $\xi_1 + \xi_2 > \xi_{cr}\left(\omega_{1,2}\right)$ when the "reflection" of an external particle from the slowed traveling wave holds, an internal particle under the specified conditions may be captured by the such slowed wave. Consequently, one needs to define the conditions for the particle capture by the effective field in the induced Compton process.

Let a particle with velocity $\mathbf{v}_0$ be situated in the initial phases $\phi_{10} = \omega_1(t_0 - x_0/c)$ and $\phi_{20} = \omega_2(t_0 + x_0/c)$ of linearly polarized along the axis OY counterpropagating waves (in Eq. (5.3) $E_{1z} = E_{2z} = 0$, so the coordinate z is free and one can assume $\mathrm{v}_{0z} = 0$). The solution of Eqs. (5.1) and (5.2) under these initial conditions for the particle momentum in the field gives

$$p_x = p_{0x} + \frac{n_1^2}{n_1^2 - 1}\frac{\mathcal{E}_0}{c}\left\{1 - n_1\frac{\mathrm{v}_{0x}}{c} \mp \left[\left(1 - n_1\frac{\mathrm{v}_{0x}}{c}\right)^2\right.\right.$$

$$
- \left(n_1^2 - 1\right) \left(\frac{mc^2}{\mathcal{E}_0}\right)^2 \left[\frac{1}{2}\left(\xi_1^2 + \xi_2^2\right) + \left(\xi_1 \sin\phi_{10} + \xi_2 \sin\phi_{20}\right) \left(\xi_1 \sin\phi_{10} \right.\right.
$$

$$
\left.\left.\left. + \xi_2 \sin\phi_{20} - 2\frac{P_{0y}}{mc}\right) + \xi_1 \xi_2 \cos\left(\phi_1 - \phi_2\right)\right]\right]^{1/2}\Bigg\}, \qquad (5.13)
$$

$$
p_y = p_{0y} + mc\xi_1 \left(\sin\phi_1 - \sin\phi_{10}\right) + mc\xi_2 \left(\sin\phi_2 - \sin\phi_{20}\right), \qquad (5.14)
$$

where

$$
\phi_1 - \phi_2 = \left(\omega_1 - \omega_2\right)\left(t - n_1 \frac{x}{c}\right).
$$

In the derivation of Eq. (5.13) the averaging over fast oscillations of separate waves with respect to the interference wave (in the intrinsic frame of which only a static magnetic field acts on the particle) in Eqs. (5.1) and (5.5) has been made. Physically it corresponds to time averaging of noncoherent interaction with separate waves in relation to coherent interaction due to induced Compton resonance. In this approximation the integral of motion (5.6) remains applicable and with Eq. (5.13) it determines the energy of the particle at the coherent interaction with the counterpropagating waves of different frequencies.

The equilibrated phases for the particle capture in this process correspond to extrema of the interference wave and the motion of the particle is stable in the phases

$$
\left(\phi_1 - \phi_2\right)_s = \left(\omega_1 - \omega_2\right)\left(t - n_1 \frac{x}{c}\right)_s = \pi\left(2k + 1\right); \quad k = 0, \pm 1, \ldots. \qquad (5.15)
$$

Equation (5.15) shows that the particle situated in the equilibrated phases moves with the velocity

$$
\mathrm{v}_{xs} = c\left(\omega_1 - \omega_2\right) / \left(\omega_1 + \omega_2\right).
$$

Let the particle initial longitudinal velocity be equilibrated: $\mathrm{v}_{0x} = \mathrm{v}_{xs}$. If $p_{0y} = 0$ as well, then the analysis of Eq. (5.13) shows that the capture of such particle is possible at $\xi_1 = \xi_2$ ($eE_1/\omega_1 = eE_2/\omega_2$, i.e., the waves should transfer to the particle equal momenta) and $\left(\phi_1 - \phi_2\right)_0 = \pi\left(2k + 1\right) = \left(\phi_1 - \phi_2\right)_s$. From Eq. (5.14) for equilibrated transverse momentum in this case we have $p_{ys} = p_{0y} = 0$. If $\mathrm{v}_{0x} = \mathrm{v}_{xs} + \triangle\mathrm{v}$ and $p_{0y} = 0$, then we have the following condition for the particle capture:

$$
|\triangle\mathrm{v}| < \frac{c}{n_1}\frac{mc^2}{\mathcal{E}_0}\xi\sqrt{\left(n_1^2 - 1\right)\left[2 + \left(\sin\phi_{10} + \sin\phi_{20}\right)^2\right]}, \qquad (5.16)
$$

from which one can define the tolerance for divergences of initial phases and velocity of a nonequilibrium particle. On the other hand, condition (5.16) defines the threshold value of the wave intensities for the capture of a nonequilibrium particle, which coincides with the critical intensity for the "reflection" of an external particle (5.9) at $\xi_1 = \xi_2 \equiv \xi$ and $\phi_{10} = \phi_{20} = 0$ (coefficient $\sqrt{2}$ arises because of different polarization of the waves).

Now let $\mathrm{v}_{0x} = \mathrm{v}_{xs}$ but $p_{0y} \neq 0$. If at that $(\phi_1 - \phi_2)_0 \neq \pi\,(2k+1)$, then the motion of the particle will be stable at the condition

$$p_{0y}\,(\sin\phi_{10} + \sin\phi_{20}) > 0; \qquad \frac{|p_{0y}|}{mc\xi} > 1. \tag{5.17}$$

The condition for the capture in this case is $|p_{0y}|\,/mc\xi < 3/2$, which with the condition of stability (5.17) strictly restricts the transverse momentum of the particle. Meanwhile the conditions of stability and capture in the minimums of the interference wave $(\phi_1 - \phi_2)_0 = \pi\,(2k+1)$ are automatically satisfied. Hence, these phases are equilibrated at the arbitrary transverse momentum of the particle ($p_{0y} = p_{ys}$).

If the particle initial velocity differs from the equilibrated one ($\mathrm{v}_{0x} \neq \mathrm{v}_{xs}$) and $p_{0y} \neq 0$, the tolerance for the capture of a nonequilibrium particle is defined analogously to condition (5.16).

5.2 Interaction of Charged Particles with Superstrong Wave in a Wiggler

Consider the nonlinear dynamics of a charged particle at the interaction with a strong EM wave in a magnetic undulator. Let a particle with an initial velocity $\mathrm{v}_0 = \mathrm{v}_{0x}$ enter into a magnetic undulator with circularly polarized field

$$\mathbf{H}(x) = \left\{0, -H\cos\frac{2\pi}{l}x, H\sin\frac{2\pi}{l}x\right\} \tag{5.18}$$

(l is the space period or step of an undulator) along the axis of which propagates a plane monochromatic EM wave of circular polarization with the electric field strength

$$\mathbf{E}(x,t) = \left\{0, E_0\sin\omega_0(t-\frac{x}{c}), E_0\cos\omega_0(t-\frac{x}{c})\right\}. \tag{5.19}$$

The equation of motion of the particle in the fields (5.18) and (5.19) in components is written as

$$\frac{dp_x}{dt} = \frac{e}{c}E_0\left[\mathrm{v}_y\sin\omega_0(t-\frac{x}{c}) + \mathrm{v}_z\cos\omega_0(t-\frac{x}{c})\right]$$

$$+\frac{e}{c}H\left[\mathrm{v}_y\sin\frac{2\pi}{l}x+\mathrm{v}_z\cos\frac{2\pi}{l}x\right], \tag{5.20}$$

$$\frac{dp_y}{dt}=eE_0\left(1-\frac{\mathrm{v}_x}{c}\right)\sin\omega_0(t-\frac{x}{c})-e\frac{\mathrm{v}_x}{c}H\sin\frac{2\pi}{l}x,$$

$$\frac{dp_z}{dt}=eE_0\left(1-\frac{\mathrm{v}_x}{c}\right)\cos\omega_0(t-\frac{x}{c})-e\frac{\mathrm{v}_x}{c}H\cos\frac{2\pi}{l}x. \tag{5.21}$$

Integration of Eqs. (5.21) under the assumed initial conditions (at $t=-\infty$ the particle has only longitudinal velocity, i.e., $p_{0y}=p_{0z}=0$) gives

$$p_y=-\frac{eE_0}{\omega_0}\cos\omega_0(t-\frac{x}{c})+\frac{elH}{2\pi c}\cos\frac{2\pi}{l}x,$$

$$p_z=\frac{eE_0}{\omega_0}\sin\omega_0(t-\frac{x}{c})-\frac{elH}{2\pi c}\sin\frac{2\pi}{l}x. \tag{5.22}$$

The integration of Eq. (5.20) is made analogously to the integration of Eq. (5.1). Using the equation for the particle energy exchange in the field

$$\frac{d\mathcal{E}}{dt}=eE_0\left[\mathrm{v}_y\sin\omega_0(t-\frac{x}{c})+\mathrm{v}_z\cos\omega_0(t-\frac{x}{c})\right], \tag{5.23}$$

with the help of Eqs. (5.1), (5.22), and (5.23) we obtain the integral of motion in the induced undulator process

$$\mathcal{E}-\frac{c}{1+\frac{\lambda}{l}}p_x=\text{const}. \tag{5.24}$$

Equations (5.22) and (5.24) determine the particle energy

$$\mathcal{E}=\frac{\mathcal{E}_0}{n_2^2-1}\left\{n_2^2\left(1-\frac{\mathrm{v}_0}{cn_2}\right)\mp\left[\left(1-n_2\frac{\mathrm{v}_0}{c}\right)^2-(n_2^2-1)\left(\frac{mc^2}{\mathcal{E}_0}\right)^2\right.\right.$$

$$\left.\left.\times\left[\xi_0^2+\xi_H^2-2\xi_0\xi_H\cos\omega_0(t-n_2\frac{x}{c})\right]\right]^{1/2}\right\} \tag{5.25}$$

in the field of a strong EM wave in the magnetic undulator, which is characterized by relativistic parameter

$$\xi_H=\frac{elH}{2\pi mc^2} \tag{5.26}$$

(for large magnitudes of undulator field strength H and space period l when $\xi_H > 1$ such undulator is called a wiggler).

From Eq. (5.25) it follows that at the particle–wave nonlinear resonance interaction in the undulator an effective slowed traveling wave is formed as in the induced Compton process. The parameter

$$n_2 = 1 + \frac{\lambda}{l} \tag{5.27}$$

is the refractive index of this slowed wave, which causes the analogous threshold phenomenon of particle "reflection"% in the induced undulator process. The effective critical field at which the nonlinear resonance and then the particle "reflection" take place in the undulator, is

$$\xi_{cr}\left(\frac{\lambda}{l}\right) \equiv (\xi_0 + \xi_H)_{cr} = \frac{\left|1 - \left(1 + \frac{\lambda}{l}\right)\frac{\mathrm{v}_0}{c}\right|}{\sqrt{\frac{2\lambda}{l}\left(1 + \frac{\lambda}{2l}\right)}}\frac{\mathcal{E}_0}{mc^2}. \tag{5.28}$$

At this value of the resulting field the longitudinal velocity of the particle v_x reaches the resonant value in the field at which the condition of coherency in the undulator

$$\frac{2\pi}{l}\mathrm{v}_x = \omega_0\left(1 - \frac{\mathrm{v}_x}{c}\right) \tag{5.29}$$

is satisfied. The latter has a simple physical explanation in the intrinsic frame of the particle. In this frame of reference the static magnetic field (5.18) becomes a traveling EM wave with the frequency

$$\omega = \frac{2\pi}{l}\frac{\mathrm{v}_x}{\sqrt{1 - \frac{\mathrm{v}_x^2}{c^2}}}$$

and phase velocity $\mathrm{v}_{ph} = \mathrm{v}_x$. For coherent interaction process this frequency must coincide with the Doppler-shifted frequency of stimulated wave.

The energy of the particle after the "reflection" (in Eq. (5.25) $\xi_0 = \xi_H = 0$ at the sign "+" before the root) is

$$\mathcal{E} = \mathcal{E}_0\left[1 + \frac{1 - \left(1 + \frac{\lambda}{l}\right)\frac{\mathrm{v}_0}{c}}{\frac{\lambda}{l}\left(1 + \frac{\lambda}{2l}\right)}\right]. \tag{5.30}$$

From this equation it follows that the particle with the initial velocity $\mathrm{v}_0 < c/(1 + \lambda/l)$ after the "reflection" accelerates, while at $\mathrm{v}_0 > c/(1 + \lambda/l)$ it decelerates because of induced undulator radiation.

If a particle is initially situated in the field, under the certain conditions it may be captured by the slowed-in-the-undulator effective wave. We shall define those conditions.

Let a particle with the velocity $\mathbf{v}_0$ be situated in the initial phases $\phi_{10} = \omega_0(t_0 - x_0/c)$ and $\phi_{20} = 2\pi x_0/l$ of a linearly polarized EM wave and undulator field

$$E_y(x,t) = -E_0 \cos\omega_0(t - \frac{x}{c}); \quad H_z(x) = H\cos\frac{2\pi}{l}x. \tag{5.31}$$

The solution of Eqs. (5.1) and (5.2) under these initial conditions for the particle momentum in the field gives

$$p_x = p_{0x} + \frac{n_2}{n_2^2 - 1}\frac{\mathcal{E}_0}{c}\left\{1 - n_2\frac{\mathrm{v}_{0x}}{c} \mp \left[\left(1 - n_2\frac{\mathrm{v}_{0x}}{c}\right)^2\right.\right.$$

$$- \left(n_2^2 - 1\right)\left(\frac{mc^2}{\mathcal{E}_0}\right)^2\left[\frac{1}{2}\left(\xi_0^2 + \xi_H^2\right) + \left(\xi_0 \sin\phi_{10} + \xi_H \sin\phi_{20}\right)\right.$$

$$\left.\left.\left.\times\left(\xi_0 \sin\phi_{10} + \xi_H \sin\phi_{20} - 2\frac{P_{0y}}{mc}\right) + \xi_0\xi_H \cos\omega_0(t - n_2\frac{x}{c})\right]\right]^{1/2}\right\}, \tag{5.32}$$

$$p_y = p_{0y} + mc\xi_0\left[\sin\omega_0(t - \frac{x}{c}) - \sin\phi_{10}\right]$$

$$+ mc\xi_H\left(\sin\frac{2\pi}{l}x - \sin\phi_{20}\right). \tag{5.33}$$

Note that at the derivation of Eq. (5.32) in Eqs. (5.20) and (5.23) the time averaging of noncoherent interaction with respect to coherent interaction has been made. In this approximation the integral of motion (5.24) remains applicable and with Eq. (5.32) determines the energy of the particle at the coherent interaction with the strong EM wave in a wiggler.

The equilibrated phases for the particle capture correspond to extrema of slowed-in-the-undulator effective wave and the motion of the particle is stable in the phases

$$\phi_s = \omega_0\left[t - \left(1 + \frac{\lambda}{l}\right)\frac{x}{c}\right]_s = \pi\left(2k+1\right); \quad k = 0, \pm1, \ldots. \tag{5.34}$$

From Eq. (5.34) one can define the particle velocity in the equilibrated phase: $\mathrm{v}_{xs} = c/(1 + \lambda/l)$. If the initial velocity of the particle $\mathrm{v}_{0x} = \mathrm{v}_{xs}$ and $p_{0y} = 0$ the capture of such particle is possible at $\xi_0 = \xi_H$ that is $\lambda E_0 = lH$, i.e., the strong wave and wiggler field should transfer to the particle equal momenta and $\phi_{10} - \phi_{20} = \phi_s$ (at that $p_{ys} = 0$). If the initial velocity of the particle differs from the equilibrated one ($\mathrm{v}_{0x} \neq \mathrm{v}_{xs}$) and $p_{0y} = 0$ the tolerance for the capture of nonequilibrium particles is defined analogously to condition (5.16) in the induced Compton process. If $p_{0y} \neq 0$, then as in the case of counterpropagating waves the phases $\phi_0 = \pi\,(2k+1)$ automatically are equilibrated for the arbitrary p_{0y} ($p_{0y} = p_{ys}$). In the other cases the conditions for particle capture by the effective slowed wave in the regime of stable motion in the wiggler are defined as for those in the induced Compton interaction.

The "reflection" phenomenon of charged particles from a plane EM wave, as was shown in the induced Cherenkov process, may be used for monochromatization of the particle beams. Note that the considered vacuum versions of this phenomenon are more preferable for this goal taking into account the influence of negative effects of the multiple scattering and ionization losses in a medium. On the other hand, the refractive index of the effective slowed waves in vacuum n_1 or n_2 in corresponding induced Compton and undulator processes may be varied choosing the appropriate frequencies of counterpropagating waves or wiggler step. In particular, for monochromatization of particle beams with moderate or low energies via the induced Cherenkov process one needs a refractive index of a medium $n_0 - 1 \sim 1$ that corresponds to solid states. Meanwhile, such values of effective refractive index may be reached in the induced Compton process at the frequencies $\omega_1 \sim \omega_2$ of the counterpropagating waves. However, we will not consider here the possibility of particle beam monochromatization on the basis of the vacuum versions of "reflection" phenomenon since the principle of conversion of energetic or angular spreads is the same. To study the subject in more detail we refer the reader to original papers listed in the bibliography of this chapter.

5.3 Inelastic Diffraction Scattering on a Moving Phase Lattice

Consider now the quantum dynamics of a particle coherent interaction with the counterpropagating waves of different frequencies in the induced Compton process. Neglecting the spin interaction (with the same justification that has been made in the above-considered processes) we will derive from the Klein–Gordon equation in the field of quasimonochromatic waves with the vector potentials $\mathbf{A}_1(t - x/c)$ and $\mathbf{A}_2(t + x/c)$ which is written as

$$-\hbar^2 \frac{\partial^2 \Psi}{\partial t^2} = \left\{ -\hbar^2 c^2 \nabla^2 + m^2 c^4 + e^2 \left[\mathbf{A}_1(t - \frac{x}{c}) + \mathbf{A}_2(t + \frac{x}{c}) \right]^2 \right.$$

$$+2ie\hbar c\left[\mathbf{A}_1(t-\frac{x}{c})+\mathbf{A}_2(t+\frac{x}{c})\right]\nabla\Big\}\Psi. \tag{5.35}$$

As we saw in the classical consideration of the dynamics of the induced Compton process the effective interaction occurs with the slowed interference wave. At the intensities of the waves $\xi_1+\xi_2<\xi_{cr}(\omega_{1,2})$ when the particle can penetrate into the interference wave the latter will stand for a phase lattice for the particle (at the satisfaction of the condition of coherency (5.10)) and the coherent scattering will occur as for the diffraction effect on a crystal lattice. However, in contrast to diffraction on a motionless lattice (elastic scattering) the diffraction scattering on the moving phase lattice has inelastic character. To determine this quantum effect we will solve Eq. (5.35) in the eikonal approximation by the particle wave function (3.91) corresponding to multiphoton processes in strong fields. In accordance with the latter the solution of Eq. (5.35) for the waves of linear polarizations (along the axis OY)

$$\mathbf{A}_1(t-x/c)=\mathbf{A}_1(t)\cos\omega_1(t-x/c),$$

$$\mathbf{A}_2(t+x/c)=\mathbf{A}_2(t)\cos\omega_2(t+x/c)$$

we look for in the form Eq. (3.91) and for the slowly varying function $f(x,t)$ (see Eq. (3.92)) we obtain the following equation:

$$\frac{\partial f}{\partial t}+\mathrm{v}_{0x}\frac{\partial f}{\partial x}=\Big\{-\frac{ie^2}{2\hbar\mathcal{E}_0}\Big[A_1^2(t)\cos^2\omega_1(t-\frac{x}{c})+A_2^2(t)\cos^2\omega_2(t+\frac{x}{c})$$

$$+A_1(t)A_2(t)\cos(\omega_1+\omega_2)\left(t-\frac{\omega_1-\omega_2}{\omega_1+\omega_2}\frac{x}{c}\right)$$

$$+A_1(t)A_2(t)\cos(\omega_1-\omega_2)\left(t-\frac{\omega_1+\omega_2}{\omega_1-\omega_2}\frac{x}{c}\right)\Big]$$

$$+\frac{iecp_{0y}}{\hbar\mathcal{E}_0}\left[A_1(t)\cos\omega_1(t-\frac{x}{c})+A_2(t)\cos\omega_2(t+\frac{x}{c})\right]\Big\}f(x,t). \tag{5.36}$$

As is seen from Eq. (5.36) at the interaction with the counterpropagating waves of different frequencies two interference waves are formed — third and fourth terms on the right-hand side — which propagate with the phase velocities

$$\mathrm{v}_{ph}=c\frac{\omega_1+\omega_2}{|\omega_1-\omega_2|}>c$$

and

$$\mathrm{v}_{ph} = c\frac{|\omega_1 - \omega_2|}{\omega_1 + \omega_2} < c,$$

respectively. It is clear that the interaction of the particle with the wave propagating with the phase velocity $\mathrm{v}_{ph} > c$, as well as with the incident separate waves propagating in the vacuum with the phase velocity c (remaining four terms on the right-hand side of Eq. (5.36)), cannot be coherent. These terms correspond to noncoherent scattering of the particle in the separate wave fields which vanish after the interaction. Coherent interaction in this process occurs with the slowed interference wave (fourth term), in accordance with the classical results (see Eqs. (5.8) and (5.10)).

For the integration of Eq. (5.36) we will pass to characteristic coordinates $\tau' = t - x/\mathrm{v}_{0x}$ and $\eta' = t$. Then, if one directs the particle velocity $\mathbf{v}_0$ at the angle ϑ_0 with respect to the waves' propagation axis providing the condition of coherency of the induced Compton process (resonance between the waves' Doppler-shifted frequencies) for the free-particle velocity

$$\mathrm{v}_0 \cos\vartheta_0 = c\frac{|\omega_1 - \omega_2|}{\omega_1 + \omega_2}, \tag{5.37}$$

the traveling interference wave in this frame of coordinates becomes a standing phase lattice over the coordinate τ' and diffraction scattering of the particle occurs. From Eq. (5.36) for the amplitude of the scattered particle wave function we obtain

$$f(\tau') = \exp\left\{-\frac{ie^2}{2\hbar\mathcal{E}_0}\cos(\omega_1 - \omega_2)\tau'\int_{\eta_1}^{\eta_2} A_1(\eta')A_2(\eta')d\eta'\right\}, \tag{5.38}$$

where η_1 and η_2 are the moments of the particle entrance into the field and exit, respectively.

If one expands the exponential (5.38) into a series by Bessel functions and returns again to coordinates x, t with the help of Eq. (3.91) for the total wave function we will have

$$\Psi(\mathbf{r}, t) = \sqrt{\frac{N_0}{2\mathcal{E}_0}}\exp\left[\frac{i}{\hbar}(p_0 \sin\vartheta_0)\,y\right]\sum_{s=-\infty}^{+\infty}(-i)^s J_s(\alpha)$$

$$\times\exp\left\{\frac{i}{\hbar}\left[p_0\cos\vartheta_0 + s\hbar\frac{\omega_1 + \omega_2}{c}\right]x - \frac{i}{\hbar}\left[\mathcal{E}_0 + s\hbar(\omega_1 - \omega_2)\right]t\right\}, \tag{5.39}$$

where the argument of the Bessel function

$$\alpha = \frac{e^2c^2}{2\hbar\mathcal{E}_0\omega_1\omega_2}\int_{t_1}^{t_2} E_1(\eta')E_2(\eta')d\eta' \tag{5.40}$$

(E_1 and E_2 are the amplitudes of the waves' electric field strengths).

Equation (5.39) shows that the diffraction scattering of the particles in the field of counterpropagating waves of different frequencies is inelastic. Due to the induced Compton effect the particle absorbs s photons from the one wave and coherently radiates s photons into the other wave and vice versa (resonance between the Doppler-shifted frequencies in the intrinsic frame of the particle), i.e., the conservation of the number of photons in the induced Compton process takes place in contrast to spontaneous Compton effect in the strong wave field where after the multiphoton absorption a single photon is emitted. However, because of the different photon energies the scattering process is inelastic. From Eq. (5.39) for the change of the particle energy-momentum we have

$$\Delta\mathcal{E} = s\hbar\left(\omega_1 - \omega_2\right); \ \Delta p_x = s\hbar\left(\omega_1 + \omega_2\right)/c\ ; \ \Delta p_y = 0; \ s = 0, \pm 1, \ldots. \tag{5.41}$$

The probability of inelastic diffraction scattering is

$$W_s = J_s^2\left[\frac{e^2c^2}{2\hbar\omega_1\omega_2\mathcal{E}_0}\int_{t_1}^{t_2} E_1(\eta')E_2(\eta')d\eta'\right]. \tag{5.42}$$

According to the condition of eikonal approximation (3.92): $|\Delta p| << p_0$ and $|\Delta\mathcal{E}| << \mathcal{E}_0$ from Eq. (5.41) we have the condition of applicability of the obtained results: $|s|\,\hbar\left(\omega_1 + \omega_2\right)/c << p_0$.

In the case of monochromatic waves

$$W_s = J_s^2\left(\frac{e^2c^2E_1E_2t_0}{2\hbar\mathcal{E}_0\omega_1\omega_2}\right), \tag{5.43}$$

where $t_0 = t_2 - t_1$ is the time duration of the particle motion in the interference wave ($l_c = \mathrm{v}_0t_0\cos\vartheta_0$ is the coherent length of the process). For the actual values of the parameters including in Eq. (5.43) the argument of the Bessel function $\alpha >> 1$, consequently the most probable number of absorbed/radiated photons

$$\overline{s} \simeq \frac{1}{2}\xi_1\xi_2\frac{mc^2}{\mathcal{E}_0}\frac{mc^2}{\hbar}t_0. \tag{5.44}$$

The energetic width of the main diffraction maximums

$$\Gamma(\overline{s}) \simeq \overline{s}^{1/3}\hbar\left(\omega_1 - \omega_2\right)$$

and since $\overline{s} >> 1$ then

$$\Gamma(\overline{s}) << |\mathcal{E} - \mathcal{E}_0|\,.$$

The scattering angles of s-photon diffraction on the counterpropagating waves are

$$\tan\vartheta_s = \frac{s\hbar\left(\omega_1+\omega_2\right)\sin\vartheta_0}{cp_0 + s\hbar\left(\omega_1+\omega_2\right)\cos\vartheta_0}\,; \qquad s = 0, \pm 1, \ldots . \tag{5.45}$$

As in the Cherenkov process at the inelastic diffraction there is an asymmetry in the angular distribution of the scattered particle: $|\vartheta_{-s}| > \vartheta_s$, i.e., the main diffraction maximums are situated at the different angles with respect to the direction of particle initial motion. However, since $|s|\,\hbar\left(\omega_1+\omega_2\right)/c << p_0$ this asymmetry can be neglected, i.e., $|\vartheta_{-s}| \simeq \vartheta_s$ and the scattering angles of the main diffraction maximums will be determined by the equation

$$\vartheta_{\pm\overline{s}} = \pm\overline{s}\frac{\hbar\left(\omega_1+\omega_2\right)}{cp_0}\sin\vartheta_0 . \tag{5.46}$$

In the case of counterpropagating waves of equal frequencies ($\omega_1 = \omega_2 \equiv \omega$) the phase velocity of the interference wave $\mathrm{v}_{ph} = 0$ and the coherent scattering on the motionless phase lattice takes place, which is elastic: $\Delta\mathcal{E} = 0$ and $\Delta p_x = 2s\hbar\omega/c$. This is the known Kapitza–Dirac effect for electron diffraction on a standing wave (in the one-photon approximation for the weak waves). As follows from Eq. (5.37) the coherent scattering in this case is possible at the incident angle $\vartheta_0 = \pi/2$, i.e., if the particle velocity is perpendicular to the axis of waves' propagation, to exclude the Doppler shift of waves frequencies because of its counterpropagation (a longitudinal component of the particle velocity will result in different Doppler shifts of equal laboratory frequencies because of different wave vectors $\mathbf{k}$ and $-\mathbf{k}$ of counterpropagating waves and, consequently, will violate the resonance between the waves).

5.4 Inelastic Diffraction Scattering on a Traveling Wave in an Undulator

Charged particles diffraction scattering is also possible on a plane EM wave propagating in vacuum if the interaction proceeds in an undulator. As the diffraction effect is the result of particle coherent interaction with the periodic wave field the effective field in the undulator should be smaller than the threshold value of "reflection" phenomenon: $\xi_0 + \xi_H < \xi_{cr}\left(\lambda/l\right)$ (to prohibit the nonlinear resonance in the field at which the periodic EM field becomes a potential barrier for the particle and coherent interaction with the periodic wave field impossible). Under this condition we will solve the relativistic quantum equation of motion

$$-\hbar^2\frac{\partial^2\Psi}{\partial t^2} = \left\{-\hbar^2c^2\nabla^2 + m^2c^4 + e^2\left[\mathbf{A}_1(t-\frac{x}{c}) + \mathbf{A}_2(x)\right]^2\right.$$

$$+2ie\hbar c\left[\mathbf{A}_1(t-\frac{x}{c})+\mathbf{A}_2(x)\right]\nabla\Big\}\Psi, \tag{5.47}$$

where $\mathbf{A}_1(t-x/c)$ is the vector potential of the quasimonochromatic EM wave and $\mathbf{A}_2(x)$ is the vector potential of the undulator magnetic field. For the linear undulator

$$H_z(x)=H\cos\frac{2\pi}{l}x$$

the vector potential will be described by the equation

$$A_{2y}(x)=\frac{lH}{2\pi}\sin\frac{2\pi}{l}x,$$

and correspondingly the EM wave will be assumed linearly polarized along the axis OY

$$A_{1y}(t-x/c)=A(t)\sin\omega_0(t-x/c).$$

To determine the multiphoton diffraction effect Eq. (5.47) will be solved again in the eikonal approximation. In accordance with the latter we present the solution of Eq. (5.47) in the form of Eq. (3.91). Then taking into account the condition (3.92) for the slowly varying function $f(x,t)$ we obtain the equation

$$\frac{\partial f}{\partial t}+\mathrm{v}_{0x}\frac{\partial f}{\partial x}=\Bigg\{-\frac{ie^2}{2\hbar\mathcal{E}_0}\Bigg[A^2(t)\sin^2\omega_0(t-\frac{x}{c})+\frac{l^2H^2}{4\pi^2}\sin^2\frac{2\pi}{l}x$$

$$+\frac{lH}{2\pi}A(t)\cos\omega_0\left(t-\left(1+\frac{\lambda}{l}\right)\frac{x}{c}\right)-\frac{lH}{2\pi}A(t)\cos\omega_0\left(t-\left(1-\frac{\lambda}{l}\right)\frac{x}{c}\right)\Bigg]$$

$$+\frac{iecp_{0y}}{\hbar\mathcal{E}_0}\left[A(t)\sin\omega_0(t-\frac{x}{c})+\frac{lH}{2\pi}\sin\frac{2\pi}{l}x\right]\Bigg\}f(x,t). \tag{5.48}$$

As is seen from Eq. (5.48) under the induced interaction in the undulator, traveling waves propagating with the phase velocities $\mathrm{v}_{ph}=c/\left(1+\lambda/l\right)<c$ and $\mathrm{v}_{ph}=c/\left(1-\lambda/l\right)>c$ arise. We will not repeat here the analogous interpretation of the terms in Eq. (5.48) which correspond to interaction of the particle with the waves propagating with the phase velocities $\mathrm{v}_{ph}\gtrless c$ that has been done for the above-considered induced Compton process. Note only that coherent interaction in this process occurs with the slowed interference wave propagating with the phase velocity $\mathrm{v}_{ph}=c/\left(1+\lambda/l\right)<c$ (third term on the right-hand side of Eq. (5.48)), in accordance with the classical results for the induced interaction in the magnetic undulator (see Eqs. (5.27) and (5.29)).

The integration of Eq. (5.48) is simple if we pass to characteristic coordinates $\tau' = t - x/\mathrm{v}_{0x}$ and $\eta' = t$. Then, if one directs the particle velocity $\mathbf{v}_0$ at the angle ϑ_0 with respect to the wave propagation direction (undulator axis) thus providing the condition of coherency in the undulator for the free-particle velocity

$$\mathrm{v}_0 \cos\vartheta_0 = \frac{c}{1+\frac{\lambda}{l}}, \tag{5.49}$$

the slowed traveling wave in this frame of coordinates becomes a motionless phase lattice (over the coordinate τ') and diffraction scattering of the particle occurs. For the amplitude of the scattered particle wave function we obtain

$$f(\tau') = \exp\left\{-\frac{ie^2 lH}{4\pi\hbar\mathcal{E}_0}\cos\omega_0\tau' \int_{\eta_1}^{\eta_2} A(\eta')d\eta'\right\}, \tag{5.50}$$

where η_1 and η_2 are the moments of the particle entrance into the undulator and exit, respectively.

Expanding the exponential in Eq. (5.50) into a series by Bessel functions with the help of Eq. (3.91) for the final wave function of the scattered particle we will have

$$\Psi(\mathbf{r},t) = \sqrt{\frac{N_0}{2\mathcal{E}_0}} \exp\left[\frac{i}{\hbar}(p_0\sin\vartheta_0)\,y\right] \sum_{s=-\infty}^{+\infty} (-i)^s J_s(\alpha)$$

$$\times \exp\left\{\frac{i}{\hbar}\left[p_0\cos\vartheta_0 + s\hbar\frac{\omega_0}{c}\left(1+\frac{\lambda}{l}\right)\right]x - \frac{i}{\hbar}(\mathcal{E}_0 + s\hbar\omega_0)\,t\right\}, \tag{5.51}$$

where the argument of the Bessel function

$$\alpha = \frac{e^2 lH}{4\pi\hbar\mathcal{E}_0}\int_{t_1}^{t_2} A(\eta')d\eta'. \tag{5.52}$$

The expression for the particle wave function (5.51) shows that the initial plane wave of the free particle as a result of the induced undulator effect is expanded into the envelope of plane waves with all possible numbers of absorbed and emitted photons — the inelastic diffraction scattering occurs. The energy and momentum of the particle after the scattering are

$$\mathcal{E} = \mathcal{E}_0 + s\hbar\omega_0; \quad p_x = p_0\cos\vartheta_0 + \left(1+\frac{\lambda}{l}\right)\frac{s\hbar\omega_0}{c};$$

$$p_y = \mathrm{const}; \quad s = 0, \pm1, \ldots . \tag{5.53}$$

According to the condition of eikonal approximation (3.92) $s\hbar\omega_0 << \mathcal{E}_0$.

The probability of inelastic diffraction scattering in the undulator is

$$W_s = J_s^2 \left[\frac{e^2 l H}{4\pi\hbar\mathcal{E}_0} \int_{t_1}^{t_2} A(\eta')d\eta' \right] . \tag{5.54}$$

If the incident strong EM wave is monochromatic, the probability of this process is

$$W_s = J_s^2 \left(\frac{e^2 c E_0 l H}{4\pi\hbar\omega_0\mathcal{E}_0} t_0 \right) , \tag{5.55}$$

where $t_0 = t_2 - t_1$ is the time duration of the particle motion in the undulator, and E_0 is the amplitude of the electric field strength of stimulating wave.

For the actual values of the parameters the argument of the Bessel function $\alpha >> 1$, consequently the inelastic diffraction scattering in the undulator is essentially multiphoton as in the Cherenkov and Compton processes. The main diffraction maximums correspond to the most probable number of absorbed/radiated photons

$$\bar{s} \simeq \xi_0 \frac{mc^2}{\mathcal{E}_0} \frac{elH}{4\pi\hbar} t_0 \tag{5.56}$$

with the energetic width $\Gamma(\bar{s}) \simeq \bar{s}^{1/3}\hbar\omega_0$.

The scattering angles of s-photon diffraction in the undulator are

$$\tan\vartheta_s = \frac{s\hbar\omega_0 \left(1 + \frac{\lambda}{l}\right) \sin\vartheta_0}{cp_0 + s\hbar\omega_0 \left(1 + \frac{\lambda}{l}\right) \cos\vartheta_0} ; \qquad s = 0, \pm 1, \ldots . \tag{5.57}$$

The main diffraction maximums are situated at the angles (taking into account the condition of applied eikonal approximation)

$$\vartheta_{\pm\bar{s}} = \pm \frac{\left(1 + \frac{\lambda}{l}\right) \bar{s}\hbar\omega_0}{cp_0} \sin\vartheta_0 , \tag{5.58}$$

with respect to the direction of the particle initial motion.

5.5 Quantum Modulation of Particle Beam in Induced Compton Process

Consider the effect of a particle beam quantum modulation at the interaction with the counterpropagating waves of different frequencies and intensities smaller than the threshold value for nonlinear Compton resonance or the critical value of the particle "reflection" phenomenon (5.9) (since the quantum modulation of the particle state is the result of coherent interaction with the periodic wave field, while at values larger than the critical one the latter becomes a potential barrier for the particle).

Neglecting the spin interaction the quantum equation of motion (5.35) for the plane waves of circular polarization

$$\mathbf{A}_1 = \{0, A_1 \cos\omega_1 \left(t - \frac{x}{c}\right), A_1 \sin\omega_1 \left(t - \frac{x}{c}\right)\},$$

$$\mathbf{A}_2 = \{0, A_2 \cos\omega_2 \left(t + \frac{x}{c}\right), A_2 \sin\omega_2 \left(t + \frac{x}{c}\right)\}$$

may be presented in the form

$$\hbar^2 c^2 \Delta\Psi - \hbar^2 \frac{\partial^2 \Psi}{\partial t^2} = \left\{ e^2 \left(A_1^2 + A_2^2\right) + m^2 c^4 + 2ie\hbar c \left[\mathbf{A}_1 \left(t - \frac{x}{c}\right)\right.\right.$$

$$\left.\left. + \mathbf{A}_2 \left(t + \frac{x}{c}\right)\right] \nabla + 2e^2 A_1 A_2 \cos\left(\omega_1 - \omega_2\right) \left(t - \frac{\omega_1 + \omega_2}{\omega_1 - \omega_2} \frac{x}{c}\right) \right\} \Psi. \quad (5.59)$$

If the initial velocity of the particle is directed along the axis of wave propagation ($\mathbf{p}_{0\perp} = 0$) the noncoherent interaction with the separate waves $\sim A_1$ and A_2 vanishes and we have the equation

$$\hbar^2 c^2 \Delta\Psi - \hbar^2 \frac{\partial^2 \Psi}{\partial t^2} = \left\{ e^2 \left(A_1^2 + A_2^2\right) + m^2 c^4 \right.$$

$$\left. + 2e^2 A_1 A_2 \cos\left(\omega_1 - \omega_2\right) \left(t - \frac{\omega_1 + \omega_2}{\omega_1 - \omega_2} \frac{x}{c}\right) \right\} \Psi, \quad (5.60)$$

which describes the coherent interaction with the slowed interference wave of frequency $\omega_1 - \omega_2$ (corresponding to Compton resonance between the counterpropagating waves) and constant renormalization of the particle mass in the field because of the intensity effect of strong waves $\sim A_1^2 + A_2^2$. To determine the effect of quantum modulation at the harmonics of the fundamental frequency $\omega_1 - \omega_2$ the problem will be solved in the approximation of perturbation theory (besides, the wave intensities should be smaller than the critical value in the induced Compton process). It is found this renormalization in the field is rather small and since it vanishes after the interaction as well, we will omit this term. Then one needs to take into account the quantum recoil which has been vanished by consideration of the diffraction effect on the basis of eikonal-type wave function, when the second-order derivatives of the wave function have been neglected. Hence, we will keep the second-order derivatives in Eq. (5.59) and solve it within perturbation theory by the wave function. Then the solution of Eq. (5.60) is sought by the series of harmonics of the fundamental frequency $\omega_1 - \omega_2$:

$$\Psi(\mathbf{r}, t) = \sqrt{\frac{N_0}{2\mathcal{E}_0}} \exp\left[\frac{i}{\hbar}(p_0 x - \mathcal{E}_0 t)\right]$$

$$\times \sum_{s=-\infty}^{+\infty} \Psi_s \exp\left[is(\omega_1 - \omega_2)\left(t - \frac{\omega_1 + \omega_2}{\omega_1 - \omega_2}\frac{x}{c}\right)\right]. \qquad (5.61)$$

(for N_0 particles per unit volume) corresponding to s-photon absorption by the particle from the wave of frequency ω_2 and s-photon coherent radiation into the wave of frequency ω_1 and vice-versa (induced Compton effect with the conservation of the number of interacting photons). Substituting the wave function (5.61) into Eq. (5.60) we obtain the following recurrent equation for the amplitudes Ψ_s:

$$\left[4\hbar^2 s^2 \omega_1 \omega_2 + 2\mathcal{E}_0 s\hbar\left(\omega_1 - \omega_2 - (\omega_1 + \omega_2)\frac{\mathrm{v}_0}{c}\right)\right]\Psi_s$$

$$= -e^2 A_1 A_2 \left[\Psi_{s-1} + \Psi_{s+1}\right]. \qquad (5.62)$$

Equation (5.62) will be solved in the approximation of perturbation theory by the wave function:

$$|\Psi_{\pm 1}| << |\Psi_0|\,; \qquad |\Psi_{\pm 2}| << |\Psi_{\pm 1}|\,, \ldots .$$

Thus, for the amplitude of the particles' wave function corresponding to absorption of s photons of frequency ω_2 and induced radiation of s photons of frequency ω_1 we obtain

$$\Psi_s = \frac{(-1)^s}{s!}\left(\frac{e^2 A_1 A_2}{2\hbar\mathcal{E}_0}\right)^s \prod_{s_1=1}^{s} \frac{1}{\omega_1 - \omega_2 - (\omega_1 + \omega_2)\frac{\mathrm{v}_0}{c} + 2s_1\frac{\hbar\omega_1\omega_2}{\mathcal{E}_0}}, \qquad (5.63)$$

and for the inverse process (absorption of s photons of frequency ω_1 and induced radiation of s photons of frequency ω_2):

$$\Psi_{-s} = \frac{1}{s!}\left(\frac{e^2 A_1 A_2}{2\hbar\mathcal{E}_0}\right)^s \prod_{s_1=1}^{s} \frac{1}{\omega_1 - \omega_2 - (\omega_1 + \omega_2)\frac{\mathrm{v}_0}{c} - 2s_1\frac{\hbar\omega_1\omega_2}{\mathcal{E}_0}}. \qquad (5.64)$$

Hence, for the total wave function of the particles after the interaction we have the equation

$$\Psi(\mathbf{r},t) = \sqrt{\frac{N_0}{2\mathcal{E}_0}} \left\{ 1 + \sum_{s=1}^{\infty} \frac{1}{s!} \left(\frac{e^2 A_1 A_2}{2\hbar\mathcal{E}_0} \right)^s \right.$$

$$\times \left[\prod_{s_1=1}^{s} \frac{(-1)^s \exp\left[is(\omega_1 - \omega_2)\left(t - \frac{\omega_1+\omega_2}{\omega_1-\omega_2}\frac{x}{c} \right) \right]}{\omega_1 - \omega_2 - (\omega_1+\omega_2)\frac{\text{v}_0}{c} + 2s_1 \frac{\hbar\omega_1\omega_2}{\mathcal{E}_0}} \right.$$

$$\left. \left. + \prod_{s_1=1}^{s} \frac{\exp\left[-is(\omega_1 - \omega_2)\left(t - \frac{\omega_1+\omega_2}{\omega_1-\omega_2}\frac{x}{c} \right) \right]}{\omega_1 - \omega_2 - (\omega_1+\omega_2)\frac{\text{v}_0}{c} - 2s_1 \frac{\hbar\omega_1\omega_2}{\mathcal{E}_0}} \right] \right\} e^{\frac{i}{\hbar}(p_0 x - \mathcal{E}_0 t)}. \tag{5.65}$$

Here the dimensionless parameter of one-photon absorption-radiation is the small parameter of applied perturbation theory

$$\frac{e^2 A_1 A_2}{2\hbar\mathcal{E}_0 \left| \omega_1 - \omega_2 - (\omega_1+\omega_2)\frac{\text{v}_0}{c} \pm 2\frac{\hbar\omega_1\omega_2}{\mathcal{E}_0} \right|} << 1. \tag{5.66}$$

The denominators in Eq. (5.65) become zero at the fulfillment of exact resonance (with the quantum recoil $2\hbar\omega_1\omega_2/\mathcal{E}_0$) corresponding to the conservation law for the induced Compton process

$$\omega_1 = \omega_2 \frac{1 + \frac{\text{v}_0}{c}}{1 - \frac{\text{v}_0}{c} \pm 2s\frac{\hbar\omega_2}{\mathcal{E}_0}}. \tag{5.67}$$

In this case, perturbation theory is not applicable and consideration must be given to secular perturbation theory.

Corresponding to wave function (5.65) the current density of the particles after the interaction will be expressed by the equation

$$\mathbf{j}(t,x) = \mathbf{j}_0 \left\{ 1 + 2\sum_{s=1}^{\infty} \frac{1}{s!} \left(\frac{e^2 A_1 A_2}{2\hbar\mathcal{E}_0} \right)^s \right.$$

$$\times \left[\prod_{s_1=1}^{s} \frac{(-1)^s}{\omega_1 - \omega_2 - (\omega_1+\omega_2)\frac{\text{v}_0}{c} + 2s_1 \frac{\hbar\omega_1\omega_2}{\mathcal{E}_0}} \right.$$

$$\left. + \prod_{s_1=1}^{s} \frac{1}{\omega_1 - \omega_2 - (\omega_1+\omega_2)\frac{\text{v}_0}{c} - 2s_1 \frac{\hbar\omega_1\omega_2}{\mathcal{E}_0}} \right]$$

$$\times \cos\left[s(\omega_1-\omega_2)\left(t-\frac{\omega_1+\omega_2}{\omega_1-\omega_2}\frac{x}{c}\right)\right]$$

$$+2\sum_{s=1}^{\infty}\sum_{s'=1}^{\infty}\frac{(-1)^s}{s!s'!}\left(\frac{e^2A_1A_2}{2\hbar\mathcal{E}_0}\right)^{s+s'}\cos\left[(s+s')(\omega_1-\omega_2)\left(t-\frac{\omega_1+\omega_2}{\omega_1-\omega_2}\frac{x}{c}\right)\right]$$

$$\times\prod_{s_1=1}^{s}\prod_{s_2=1}^{s'}\frac{1}{\omega_1-\omega_2-(\omega_1+\omega_2)\frac{\mathrm{v}_0}{c}+2s_1\frac{\hbar\omega_1\omega_2}{\mathcal{E}_0}}$$

$$\left.\times\frac{1}{\omega_1-\omega_2-(\omega_1+\omega_2)\frac{\mathrm{v}_0}{c}-2s_2\frac{\hbar\omega_1\omega_2}{\mathcal{E}_0}}\right\}, \tag{5.68}$$

where $\mathbf{j}_0 = \text{const}$ is the initial current density of the particles.

We present in explicit form the expression of modulated current density of the particles for the first three harmonics

$$\mathbf{j}(t,x)=\mathbf{j}_0\left\{1+B(\omega_{1,2})\cos(\omega_1-\omega_2)\left(t-\frac{\omega_1+\omega_2}{\omega_1-\omega_2}\frac{x}{c}\right)\right.$$

$$+\frac{3}{4}B^2(\omega_{1,2})\cos 2(\omega_1-\omega_2)\left(t-\frac{\omega_1+\omega_2}{\omega_1-\omega_2}\frac{x}{c}\right)$$

$$+\frac{5}{8}B^3(\omega_{1,2})\cos 3(\omega_1-\omega_2)\left(t-\frac{\omega_1+\omega_2}{\omega_1-\omega_2}\frac{x}{c}\right)+\cdots, \tag{5.69}$$

where the modulation depth at the fundamental frequency $\omega_1-\omega_2$

$$B(\omega_{1,2})=\frac{\xi_1\xi_2}{\xi_{cr}^2(\omega_{1,2})} \tag{5.70}$$

is represented by the parameter of critical field (5.9) in the induced Compton process. As was mentioned above for quantum modulation of the particle state at the harmonics of interference wave the intensity of the latter should be smaller than the threshold value of nonlinear resonance in the field or the critical value in the induced Compton process. Equation (5.70) shows that this requirement ($\xi_1\xi_2<\xi_{cr}^2(\omega_{1,2})$) holds in any case since in accordance with perturbation theory (condition (5.66)) $\xi_1\xi_2 << \xi_{cr}^2(\omega_{1,2})$. Note that for the representation of modulation depth in the form of Eq. (5.70) it was assumed that the quantum recoil is smaller than the Compton resonance width because of nonmonochromaticity of actual particle beams.

5.6 Quantum Modulation of Particle Beam in the Undulator

If in the induced Compton process the particles' quantum modulation takes place at the difference of frequencies (and harmonics) of two waves, the induced interaction in the undulator leads to particles' quantum modulation at the stimulating wave frequency and its harmonics. The latter is similar to Cherenkov modulation, but it is important that in this case the modulation takes place in the vacuum.

The quantum equation of motion of the particle (5.47) in the undulator with circular polarization of the magnetic field in the presence of a plane monochromatic EM wave of circular polarization with vector potentials respectively

$$\mathbf{A}_2(x) = \left\{0, -\frac{lH}{2\pi}\cos\frac{2\pi}{l}x, \frac{lH}{2\pi}\sin\frac{2\pi}{l}x\right\},$$

$$\mathbf{A}_1(x,t) = \left\{0, A_0\cos\omega_0(t-\frac{x}{c}), -A_0\sin\omega_0(t-\frac{x}{c})\right\}$$

is written as

$$\hbar^2c^2\Delta\Psi - \hbar^2\frac{\partial^2\Psi}{\partial t^2} = \left\{e^2\left(A_0^2+\frac{l^2H^2}{4\pi^2}\right) + m^2c^4 + 2ie\hbar c\left[\mathbf{A}_1\left(t-\frac{x}{c}\right)\right.\right.$$

$$\left.\left.+\mathbf{A}_2\left(x\right)\right]\nabla - e^2\frac{lH}{\pi}A_0\cos\omega_0\left(t-\left(1+\frac{\lambda}{l}\right)\frac{x}{c}\right)\right\}\Psi. \tag{5.71}$$

The coherent interaction in this process which leads to particles' quantum modulation proceeds with the effective slowed wave $\sim HA_0$ (last term on the right-hand side of Eq. (5.71)). If the free-particle initial velocity is directed along the undulator axis ($\mathbf{p}_{0\perp}=0$) the noncoherent interaction with the EM wave $\sim A_1$ and magnetic field of the undulator $\sim A_2$ vanishes and we have the equation

$$\hbar^2c^2\Delta\Psi - \hbar^2\frac{\partial^2\Psi}{\partial t^2} = \left\{e^2\left(A_0^2+\frac{l^2H^2}{4\pi^2}\right) + m^2c^4\right.$$

$$\left.-e^2\frac{lH}{\pi}A_0\cos\omega_0\left(t-\left(1+\frac{\lambda}{l}\right)\frac{x}{c}\right)\right\}\Psi, \tag{5.72}$$

which describes the particle coherent interaction with the effective slowed wave in the undulator and constant renormalization of the particle mass in the field due to the intensity effect of strong wave $\sim A_0^2$ and powerful magnetic field of the wiggler $\sim H^2l^2$. With the same justification made at the solution of this problem in the induced Compton process these constant terms will be neglected and the solution of Eq. (5.72) will be sought in the form

$$\Psi(\mathbf{r},t)=\sqrt{\frac{N_0}{2\mathcal{E}_0}}\exp\left[\frac{i}{\hbar}\left(p_0x-\mathcal{E}_0t\right)\right]$$

$$\times\sum_{s=-\infty}^{+\infty}\Psi_s\exp\left[is\omega_0\left(t-\left(1+\frac{\lambda}{l}\right)\frac{x}{c}\right)\right]. \tag{5.73}$$

Substituting the wave function (5.73) into Eq. (5.72) we obtain the recurrent equation for the amplitudes Ψ_s corresponding to s-photon induced absorption by the particle from the effective slowed wave ($s<0$) and induced undulator radiation ($s>0$)

$$\left[\frac{2\pi c\hbar}{l\mathcal{E}_0}\left(1+\frac{\lambda}{2l}\right)s^2+s\left(1-\left(1+\frac{\lambda}{l}\right)\frac{\mathrm{v}_0}{c}\right)\right]\Psi_s$$

$$=\frac{e^2lHA_0}{4\pi\mathcal{E}_0\hbar\omega_0}\left[\Psi_{s-1}+\Psi_{s+1}\right], \tag{5.74}$$

which will be solved in the approximation of perturbation theory by the wave function:

$$|\Psi_{\pm1}|<<|\Psi_0|;\qquad |\Psi_{\pm2}|<<|\Psi_{\pm1}|,\ldots.$$

For the amplitude of the particle wave function corresponding to s-photon induced radiation we obtain

$$\Psi_s=\frac{1}{s!}\left(\frac{e^2lHA_0}{4\pi\mathcal{E}_0\hbar\omega_0}\right)^s\prod_{s_1=1}^{s}\frac{1}{1-\left(1+\frac{\lambda}{l}\right)\frac{\mathrm{v}_0}{c}+2s_1\frac{\pi c\hbar}{l\mathcal{E}_0}\left(1+\frac{\lambda}{2l}\right)}, \tag{5.75}$$

and for s-photon absorption

$$\Psi_{-s}=\frac{(-1)^s}{s!}\left(\frac{e^2lHA_0}{4\pi\mathcal{E}_0\hbar\omega_0}\right)^s\prod_{s_1=1}^{s}\frac{1}{1-\left(1+\frac{\lambda}{l}\right)\frac{\mathrm{v}_0}{c}-2s_1\frac{\pi c\hbar}{l\mathcal{E}_0}\left(1+\frac{\lambda}{2l}\right)}. \tag{5.76}$$

Hence, for total wave function of the particles after the interaction we have

$$\Psi(\mathbf{r},t)=\sqrt{\frac{N_0}{2\mathcal{E}_0}}\left\{1+\sum_{s=1}^{\infty}\frac{1}{s!}\left(\frac{e^2lHA_0}{4\pi\mathcal{E}_0\hbar\omega_0}\right)^s\right.$$

$$\times \left[\prod_{s_1=1}^{s} \frac{\exp\left[is\omega_0\left(t-\left(1+\frac{\lambda}{l}\right)\frac{x}{c}\right)\right]}{1-\left(1+\frac{\lambda}{l}\right)\frac{\mathrm{v}_0}{c}+2s_1\frac{\pi c\hbar}{l\mathcal{E}_0}\left(1+\frac{\lambda}{2l}\right)}\right.$$

$$\left.\left. + \prod_{s_1=1}^{s} \frac{(-1)^s\exp\left[-is\omega_0\left(t-\left(1+\frac{\lambda}{l}\right)\frac{x}{c}\right)\right]}{1-\left(1+\frac{\lambda}{l}\right)\frac{\mathrm{v}_0}{c}-2s_1\frac{\pi c\hbar}{l\mathcal{E}_0}\left(1+\frac{\lambda}{2l}\right)}\right]\right\} e^{\frac{i}{\hbar}(p_0x-\mathcal{E}_0t)}. \tag{5.77}$$

The small parameter of applied perturbation theory (dimensionless parameter of induced one-photon absorption-radiation in the undulator) is

$$\frac{e^2lHA_0}{4\pi\mathcal{E}_0\hbar\omega_0\left|1-\left(1+\frac{\lambda}{l}\right)\frac{\mathrm{v}_0}{c}\pm 2\frac{\pi c\hbar}{l\mathcal{E}_0}\left(1+\frac{\lambda}{2l}\right)\right|} << 1. \tag{5.78}$$

The denominators in Eq. (5.77) become zero at the fulfillment of exact resonance (with the quantum recoil) between the EM wave and undulator fields

$$\frac{\lambda}{l} = \frac{c}{\mathrm{v}_0} - 1 \pm 2s\frac{\pi\hbar c^2}{l\mathcal{E}_0\mathrm{v}_0}\left(1+\frac{\lambda}{2l}\right), \tag{5.79}$$

for which the perturbation theory is not applicable and the consideration should be made in the scope of secular perturbation theory.

With the help of the wave function (5.77) for the current density of the particles after the interaction we obtain the equation

$$\mathbf{j}(t,x) = \mathbf{j}_0\left\{1+2\sum_{s=1}^{\infty}\frac{1}{s!}\left(\frac{e^2lHA_0}{4\pi\mathcal{E}_0\hbar\omega_0}\right)^s\right.$$

$$\times\left[\prod_{s_1=1}^{s}\frac{1}{1-\left(1+\frac{\lambda}{l}\right)\frac{\mathrm{v}_0}{c}+2s_1\frac{\pi c\hbar}{l\mathcal{E}_0}\left(1+\frac{\lambda}{2l}\right)}\right.$$

$$\left.+\prod_{s_1=1}^{s}\frac{(-1)^s}{1-\left(1+\frac{\lambda}{l}\right)\frac{\mathrm{v}_0}{c}-2s_1\frac{\pi c\hbar}{l\mathcal{E}_0}\left(1+\frac{\lambda}{2l}\right)}\right]\times\cos\left[s\omega_0\left(t-\left(1+\frac{\lambda}{l}\right)\frac{x}{c}\right)\right]$$

$$+2\sum_{s=1}^{\infty}\sum_{s'=1}^{\infty}\frac{(-1)^{s'}}{s!s'!}\left(\frac{e^2lHA_0}{4\pi\mathcal{E}_0\hbar\omega_0}\right)^{s+s'}\cos\left[(s+s')\,\omega_0\left(t-\left(1+\frac{\lambda}{l}\right)\frac{x}{c}\right)\right]$$

$$\times\prod_{s_1=1}^{s}\prod_{s_2=1}^{s'}\frac{1}{1-\left(1+\frac{\lambda}{l}\right)\frac{\mathrm{v}_0}{c}+2s_1\frac{\pi c\hbar}{l\mathcal{E}_0}\left(1+\frac{\lambda}{2l}\right)}$$

$$\times \frac{1}{1-\left(1+\frac{\lambda}{l}\right)\frac{\mathrm{v}_0}{c}-2s_2\frac{\pi c\hbar}{l\mathcal{E}_0}\left(1+\frac{\lambda}{2l}\right)}\Bigg\}. \tag{5.80}$$

From Eq. (5.80) for the modulation at the fundamental frequency of the stimulating wave we have

$$\mathbf{j}_1(t,x)=\mathbf{j}_0\left\{1-B(\lambda/l)\cos\omega_0\left(t-\left(1+\frac{\lambda}{l}\right)\frac{x}{c}\right)\right\}, \tag{5.81}$$

where the modulation depth

$$B(\lambda/l)=2\xi_0\xi_H\left(\frac{mc^2}{\mathcal{E}_0}\right)^2\frac{\frac{\lambda}{l}\left(1+\frac{\lambda}{2l}\right)}{\left[1-\left(1+\frac{\lambda}{l}\right)\frac{\mathrm{v}_0}{c}\right]^2-\frac{4\pi^2c^2\hbar^2}{l^2\mathcal{E}_0^2}\left(1+\frac{\lambda}{2l}\right)^2}. \tag{5.82}$$

The depth of quantum modulation can be represented by the parameter of critical field (5.28) in the induced undulator process. As the resonance width because of nonmonochromaticity of actual particle beams is rather larger than the quantum recoil, then neglecting the latter, for the modulation depth we will have

$$B(\lambda/l)=\frac{\xi_0\xi_H}{\xi_{cr}^2(\lambda/l)}. \tag{5.83}$$

In accordance with perturbation theory the modulation depth $B(\lambda/l) \ll 1$ (condition (5.78)) and Eq. (5.83) shows that $\xi_0\xi_H < \xi_{cr}^2(\lambda/l)$, i.e., the effective field in the undulator for the considered regime of coherent interaction holds under the threshold of nonlinear resonance or critical value in the undulator (above which the quantum modulation of particles, as well as the above-considered diffraction scattering, do not proceed).

Bibliography

P.L. Kapitsa, P.A.M. Dirac, Proc. Cambridge Philos. Soc. **29**, 297 (1933)
J.H. Eberly, Phys. Rev. Lett. **15**, 91 (1965)
L.S. Bartell, H.B. Thomson, R.R. Roskos, Phys. Rev. Lett. **143**, 851 (1965)
H. Schwarz, H.A. Tourtellotte, W.W. Gaertner, Phys. Lett. **19**, 202 (1965)
V.S. Letokhov, Usp. Fiz. Nauk **88**, 396 (1966)
M.V. Fedorov, Zh. Éksp. Teor. Fiz. **52**, 1434 (1967)
S. Takede, T.J. Matsui, Phys. Soc. Japan **25**, 1202 (1968)
L.S. Bartell, R.R. Roskos, H.B. Thomson, Phys. Rev. **166**, 1494 (1968)
M.M. Nieto, Am. J. Phys. **37**, 162 (1969)
R. Gush, H.P. Gush, Phys. Rev. D **3**, 1712 (1971)
R.B. Palmer, J. Appl. Phys. **43**, 3014 (1972)

F. Ehlotzky, Opt. Commun. **10**, 175 (1974)
V.M. Haroutunian, H.K. Avetissian, Phys. Lett. A **51**, 320 (1975)
V.M. Haroutunian, H.K. Avetissian, Phys. Lett. A **59**, 115 (1976)
D.F. Alferov, Yu.A. Bashmakov, E.G. Bessonov, Zh. Tekh. Fiz. **46**, 2392 (1976) [in Russian]
H. Al Abawi, F.A. Horf, P. Meystree, Phys. Rev. A **16**, 666 (1977)
H.K. Avetissian, Phys. Lett. A **67**, 101 (1978)
A.A. Kolomensky, A.N. Lebedev, Kvant. Electron. (Moscow) **7**, 1543 (1978) [in Russian]
D.F. Alferov, Yu.A. Bashmakov, E.G. Bessonov, Zh. Tekh. Fiz. **48**, 1592 (1978)
D.F. Alferov, Yu.A. Bashmakov, E.G. Bessonov, Zh. Tekh. Fiz. **48**, 1598 (1978)
H.K. Avetissian, A.A. Jivanian, R.G. Petrossian, Phys. Lett. A **66**, 161 (1978)
H.K. Avetissian, Zh. Tekh. Fiz. **49**, 2118 (1979) [in Russian]
J.K. McIver, M.V. Fedorov, Zh. Éksp. Teor. Fiz. **76**, 1996 (1979)
P.G. Jukov et al., Zh. Éksp. Teor. Fiz. **76**, 2065 (1979)
V.L. Bratman et al., Zh. Éksp. Teor. Fiz. **76**, 930 (1979)
D.F. Alferov, E.G. Bessonov, Zh. Éksp. Teor. Fiz. **49**, 777 (1979)
A.N. Didenko et al., Zh. Éksp. Teor. Fiz. **76**, 1919 (1979)
T.G. Kuper, G.T. Moore, M.O. Scully, Opt. Commun. **34**, 117 (1980)
W.B. Colson, S.B. Segall, J. Appl. Phys. **22**, 219 (1980)
W.B. Colson, S.K. Ride, Phys. Lett. A **76**, 379 (1980)
R. Bonifacio, M.O. Scully, Opt. Commun. **32**, 291 (1980)
A. Bambini, R. Bonifacio, S. Stenholm, Opt. Commun. **2**, 306 (1980)
T. Taguchi, K. Mima, T. Mochizuki, Phys. Rev. Lett. **46**, 824 (1981)
H.K. Avetissian, A.A. Jivanian, R.G. Petrossian, Pis'ma Zh. Éksp. Teor. Fiz. **34**, 561 (1981)
H.K. Avetissian, A.A. Jivanian, R.G. Petrossian, Phys. Lett. A **9**, 449 (1981)
H.K. Avetissian, A.A. Jivanian, R.G. Petrossian, Phys. Lett. A **5**, 263 (1981)
M.V. Fedorov: Electron in a Strong Light Field (Nauka, Moscow 1991) [in Russian]
Y.I. Salamin, C. H. Keitel, J. Phys. B **33**, 5057 (2000)
Y.I. Salamin, C.H. Keitel, F.H.M. Faisal, J. Phys. A **34**, 2819 (2001)
Y. I. Salamin, G. R. Mocken, C. H. Keitel, Phys. Rev. E **67**, 016501 (2003)
H.K. Avetissian, G.F. Mkrtchian: Quantum Signatures in X-ray Compton Laser. 26th International FEL Conference, Trieste, Italy (2004)

6 Induced Nonstationary Transition Process

How will the nonstationarity of a medium reflect on the process of charged particle interaction with strong laser radiation?

In the current laser fields of ultrashort pulse duration and relativistic intensities any medium turns instantaneously (on a time span much smaller than one wave cycle) into a plasma, that is, abrupt change of the medium properties, particularly the dielectric permittivity, occurs in time.

On the other hand, with the abrupt change in time of the dielectric permittivity of a medium, charged particle radiation occurs similar to transition radiation on the boundary of two media with different dielectric permittivity.

In the presence of an external EM radiation field this nonstationary transition process acquires induced character and the inverse process of radiation absorption by a charged particle is actualized, particularly in plasmas where in the stationary states the radiation or absorption of quanta of a transversal EM radiation field (monochromatic radiation such as a laser one) by a free particle cannot proceed.

With the abrupt change in time of the medium dielectric permittivity the production of hard quanta of relativistic energies from the laser radiation is possible and, consequently, electron–positron pair creation in nonstationary plasma of common densities is available. Meanwhile, for electron–positron pair production in a stationary plasma (a medium should be plasmalike for this process) by a γ*-quantum a superdense plasma with electron densities greater than* $10^{34}\,cm^{-3}$ *is necessary. Such superdense matter exists in astrophysical objects (in the core of neutron stars — pulsars), leading to special interest in the processes of electron–positron pair production and annihilation in superdense plasma. On the other hand, the matter in the astrophysical objects may also be in a strongly nonstationary state.*

Hence, it is important to study the induced nonstationary transition process in the strong EM radiation field in a medium with an arbitrary dielectric permittivity changing abruptly in time.

6.1 Effect of Abrupt Temporal Variation of Dielectric Permittivity of a Medium

In the investigation of charged particle interaction with strong EM radiation in a medium, overall it was supposed that the electromagnetic properties of the latter, i.e., the dielectric (ε_0) and magnetic (μ_0) permittivities and, consequently, refractive index n_0, are not changed in the field and the medium being initially in the stationary state maintains its electromagnetic characteristics $n_0 = \sqrt{\varepsilon_0 \mu_0} = \text{const}$.

Consider now how the nonstationarity of a medium will reflect on the process of charged particle interaction with strong EM radiation. From the physical point of view it is clear that the effects that arise here because of the nonstationarity of a medium will be essential at the abrupt temporal change of the dielectric permittivity (as it is generally assumed the magnetic permittivity of the medium will be taken as $\mu_0 = 1$). Under the abrupt change of ε here we mean its change at the time $\Delta t << 2\pi/\omega$, where ω is the characteristic frequency because of the nonstationarity of a medium (then radiation frequency by a charged particle in this process). Such abrupt change of the dielectric permittivity occurs with the propagation of ultrashort laser pulses of relativistic intensities in a medium when the tunneling ionization of atoms on a time span smaller than a few femtoseconds/attoseconds occurs and the medium instantaneously becomes a plasma.

Let a charged particle with constant initial velocity $\mathbf{v}_0$ move in a spatially homogeneous and isotropic medium whose dielectric permittivity ε changes abruptly at the time from a value ε_1 to ε_2

$$\varepsilon = \begin{cases} \varepsilon_1, & t < 0, \\ \varepsilon_2, & t > 0, \end{cases} \tag{6.1}$$

and let a strong EM wave propagate in this medium. To determine the electromagnetic field in that type of nonstationary medium one should solve the macroscopic Maxwell equations

$$\text{rot}\mathbf{H}(\mathbf{r}, t) = \frac{1}{c}\frac{\partial \mathbf{D}(\mathbf{r}, t)}{\partial t} + \frac{4\pi}{c}\mathbf{J}(\mathbf{r}, t), \tag{6.2}$$

$$\text{rot}\mathbf{E}(\mathbf{r}, t) = -\frac{1}{c}\frac{\partial \mathbf{B}(\mathbf{r}, t)}{\partial t} \tag{6.3}$$

for $t < 0$ and for $t > 0$, then the obtained solutions should be laced at the instant of time $t = 0$. At the discontinuity of the dielectric permittivity (in general, properties of the medium) only the derivatives of the physical quantities can have large values. Hence, the conditions of the lacing can be obtained by the integration of the Maxwell equations (6.2) and (6.3) over t in the arbitrary small region including the instant of time $t = 0$ at which the stepwise discontinuity of the dielectric permittivity (6.1) occurs. The latter means that the integration should be made between the moments $t_1 = -\Delta t$ and $t_2 = \Delta t$ and then one should take the limit $\Delta t \to 0$. Taking into account that the quantities rot$\mathbf{H}$, rot$\mathbf{E}$, and $\mathbf{J}$ are finite, after this procedure we obtain

$$\mathbf{D}(\mathbf{r}, t)|_{t=-0} = \mathbf{D}(\mathbf{r}, t)|_{t=+0},$$

$$\mathbf{B}(\mathbf{r}, t)|_{t=-0} = \mathbf{B}(\mathbf{r}, t)|_{t=+0}.$$

These equations can be written in terms of electric and magnetic field strengths with the help of the constitutive equations

$$\mathbf{D}(\mathbf{r}, t) = \varepsilon(t)\,\mathbf{E}(\mathbf{r}, t); \qquad \mathbf{B}(\mathbf{r}, t) = \mathbf{H}(\mathbf{r}, t),$$

which yield to "boundary conditions"

$$\varepsilon_1\, \mathbf{E}(\mathbf{r}, t)|_{t=-0} = \varepsilon_2\, \mathbf{E}(\mathbf{r}, t)|_{t=+0}, \tag{6.4}$$

$$\mathbf{H}(\mathbf{r}, t)|_{t=-0} = \mathbf{H}(\mathbf{r}, t)|_{t=+0}. \tag{6.5}$$

Under the conditions (6.4) and (6.5) the charged particle radiation will occur in the nonstationary medium similar to transition radiation on the boundary of two media with different dielectric permittivity. This spontaneous radiation field can be obtained from the Maxwell equations (6.2), (6.3) with the corresponding current density of a charged particle $\mathbf{J}(\mathbf{r}, t)$ under the conditions (6.4) and (6.5). However, we will not describe here the spontaneous nonstationary transition radiation effect and refer the reader interested in this process to the original work presented in the bibliography of this chapter. We will consider the induced nonstationary transition process in the external EM wave field. For the latter one needs also to clear up the question of how the change of the dielectric permittivity (6.1) of the medium affects the external monochromatic wave.

If a plane monochromatic wave of frequency ω_0, wave vector $\mathbf{k}_0$, and electric field amplitude $\mathbf{E}_0$ propagates in a medium with the mentioned properties, then at $t < 0$ when $\varepsilon = \varepsilon_1$

$$\mathbf{E}(\mathbf{r}, t) = \mathbf{E}_0 e^{i(\omega_0 t - \mathbf{k}_0 \mathbf{r})} + \text{c.c.}; \quad t < 0 \tag{6.6}$$

and at $t > 0$ when $\varepsilon = \varepsilon_2$ there are two waves — transmitted and reflected:

$$\mathbf{E}(\mathbf{r}, t) = \mathbf{E}_1 e^{i(\omega_1 t - \mathbf{k}_1 \mathbf{r})} + \mathbf{E}_2 e^{i(-\omega_2 t - \mathbf{k}_2 \mathbf{r})} + \text{c.c.}; \quad t > 0. \tag{6.7}$$

Here ω_1, $\mathbf{k}_1$, $\mathbf{E}_1$ and ω_2, $\mathbf{k}_2$, $\mathbf{E}_2$ are the frequencies, wave vectors, and amplitudes of the electric fields of the transmitted and reflected waves, respectively. Since the medium is assumed to be spatially homogeneous, for the wave vectors the condition takes place:

$$\mathbf{k}_0 = \mathbf{k}_1 = \mathbf{k}_2 = \text{const}, \tag{6.8}$$

and the nonstationarity of the medium leads to a change of frequency. From the condition for the wave vectors (6.8) follows the relations between the frequencies of the incident, transmitted, and reflected waves:

$$\omega_0\sqrt{\varepsilon_1} = \omega_1\sqrt{\varepsilon_2} = \omega_2\sqrt{\varepsilon_2}. \tag{6.9}$$

Let the wave propagate along the axis OX with the vector of electric field amplitude $\mathbf{E}_0$ directed along the OY axis. Then using conditions (6.4), (6.5) and Maxwell equations (6.2), (6.3) for the field (6.6), (6.7) in the case of the wave linear polarization, for the amplitudes of the electric field of the transmitted and reflected waves we obtain

$$E_1 = \frac{\sqrt{\varepsilon_1}(\sqrt{\varepsilon_1} + \sqrt{\varepsilon_2})}{2\varepsilon_2} E_0, \tag{6.10}$$

$$E_2 = \frac{\sqrt{\varepsilon_1}(\sqrt{\varepsilon_1} - \sqrt{\varepsilon_2})}{2\varepsilon_2} E_0. \tag{6.11}$$

Equations (6.10), (6.11) with the analogous equations for the magnetic strengths, and Eqs. (6.8), (6.9) determine the electromagnetic fields of the transmitted and reflected waves at the propagation of a plane monochromatic EM wave in a medium the dielectric permittivity of which changes abruptly at the time.

6.2 Classical Description of Induced Nonstationary Transition Process

As was mentioned above in the presence of an external EM radiation field the nonstationary transition process acquires induced character and the interaction of a charged particle with the incident plane monochromatic wave in a medium will proceed with the actual energy change and the acceleration of the particles or induced coherent radiation will take place. It is of special interest, in particular, in plasmas where for the stationary states the real energy change between a charged particle and a transversal EM wave cannot proceed because of the violation of the conservation law of energy-momentum for the absorption/emission of quanta in the field of a plane monochromatic wave by a free charged particle. Hence, we will study the classical and quantum dynamics of the induced nonstationary transition process in the external wave field on the basis of relativistic equations of motion for a charged particle.

Consider first the classical dynamics of the particle–wave interaction in a medium with the abrupt temporal change of the dielectric permittivity. Then, the initial monochromatic wave is transformed into a continuous wave spectrum (in general, finite since the change of ε actually occurs in finite time). This spectrum of frequencies (ω) depends on the time during which the electromagnetic properties of the medium are changed. If the characteristic time $\tau << 2\pi/\omega$, then the abrupt temporal change of the dielectric permittivity can be described by the stepwise function ε (6.1).

With the stepwise discontinuity of the dielectric permittivity (6.1) the initial monochromatic wave (of linear polarization) is transformed into a spectrum that can be found via Fourier transformation over t

$$E_y\left(x,t\right) = \int_{-\infty}^{\infty} E_y\left(x,\omega\right) e^{i\omega t} d\omega. \tag{6.12}$$

Then for the field (6.6), (6.7) the Fourier transform $E_y\left(x,\omega\right)$ may be presented in the form

$$\begin{aligned} E_y\left(x,\omega\right) &= \frac{e^{-ik_0x}}{2\pi}\left\{E_0\int_{-\infty}^{0} e^{\epsilon t}e^{i(\omega_0-\omega)t}dt + E_1\int_0^{\infty} e^{-\epsilon t}e^{i(\omega_1-\omega)t}dt\right. \\ &\left. + E_2\int_0^{\infty} e^{-\epsilon t}e^{-i(\omega_1+\omega)t}dt\right\} + \frac{e^{ik_0x}}{2\pi}\left\{E_0\int_{-\infty}^{0} e^{\epsilon t}e^{-i(\omega_0+\omega)t}dt\right. \\ &\left. + E_1\int_0^{\infty} e^{-\epsilon t}e^{-i(\omega_1+\omega)t}dt + E_2\int_0^{\infty} e^{-\epsilon t}e^{i(\omega_1-\omega)t}dt\right\}, \end{aligned} \tag{6.13}$$

where we have introduced an arbitrarily small damping factor $\epsilon \to 0$ to switch on/off adiabatically the wave at $t = \mp\infty$. After the integration in Eq. (6.13) for the Fourier transform of the field we obtain

$$\begin{aligned} E_y\left(x,\omega\right) &= \frac{e^{-ik_0x}}{2\pi i}\left\{\frac{E_2}{\omega+\omega_1-i\epsilon} + \frac{E_1}{\omega-\omega_1-i\epsilon} - \frac{E_0}{\omega-\omega_0+i\epsilon}\right\} \\ &+ \frac{e^{ik_0x}}{2\pi i}\left\{\frac{E_2}{\omega-\omega_1-i\epsilon} + \frac{E_1}{\omega+\omega_1-i\epsilon} - \frac{E_0}{\omega+\omega_0+i\epsilon}\right\}. \end{aligned} \tag{6.14}$$

The infinitesimal quantity $i\epsilon$ in the poles of Eq. (6.14) indicates the path that should be chosen at the integration over ω (at the inverse Fourier transformation as well). Taking into account Eqs. (6.9), (6.10), and (6.11) for the

$E_y(x,\omega)$ we will have

$$E_y(x,\omega) = E(\omega)e^{-ik_0 x} - E(-\omega)e^{ik_0 x}, \tag{6.15}$$

where

$$E(\omega) = \frac{E_0}{2\pi i}\left(\frac{\varepsilon_1}{\varepsilon_2} - 1\right)\frac{\omega^2}{(\omega - \omega_0)\left(\omega^2 - \omega_0^2\frac{\varepsilon_1}{\varepsilon_2}\right)}. \tag{6.16}$$

Here we have omitted the infinitesimal $i\epsilon$ bearing in mind the role of the poles bypass.

The analogous equations can be obtained for the magnetic field strength:

$$H_z(x,\omega) = H(\omega)e^{-ik_0 x} - H(-\omega)e^{ik_0 x}, \tag{6.17}$$

$$H(\omega) = \frac{\sqrt{\varepsilon_1}\omega_0}{\omega}E(\omega).$$

Now the problem of the particle–wave interaction in a nonstationary medium with the abrupt temporal change of the dielectric permittivity reduces to the particle interaction with the EM field possesing the spectral components (6.15), (6.17). Consequently, the relativistic classical equations of motion of the particle take the form

$$\frac{dp_x}{dt} = \frac{e}{c}\mathrm{v}_y\int_{-\infty}^{\infty}\left[H(\omega)e^{-ik_0 x} - H(-\omega)e^{ik_0 x}\right]e^{i\omega t}d\omega, \tag{6.18}$$

$$\frac{dp_y}{dt} = e\int_{-\infty}^{\infty}\left[E(\omega)e^{-ik_0 x} - E(-\omega)e^{ik_0 x}\right]e^{i\omega t}d\omega$$

$$-\frac{e}{c}\mathrm{v}_x\int_{-\infty}^{\infty}\left[H(\omega)e^{-ik_0 x} - H(-\omega)e^{ik_0 x}\right]e^{i\omega t}d\omega, \tag{6.19}$$

$$\frac{dp_z}{dt} = 0. \tag{6.20}$$

The energy change of the particle is given by the equation

$$\frac{d\mathcal{E}}{dt} = e\mathrm{v}_y\int_{-\infty}^{\infty}\left[E(\omega)e^{-ik_0 x} - E(-\omega)e^{ik_0 x}\right]e^{i\omega t}d\omega. \tag{6.21}$$

The equations of motion (6.18)–(6.20) can be presented in the form

$$\frac{dp_x}{dt} = -i\frac{e}{c}k_0 \int_{-\infty}^{\infty} \mathrm{v}_y F(\omega, x, t) d\omega, \tag{6.22}$$

$$\frac{dp_y}{dt} = i\frac{e}{c} \int_{-\infty}^{\infty} (k_0 \mathrm{v}_x - \omega) F(\omega, x, t) d\omega \ , \tag{6.23}$$

$$\frac{dp_z}{dt} = 0, \tag{6.24}$$

where the kernel in the integrals (6.22), (6.23)

$$F(\omega, x, t) = A(\omega) \exp\left[i(\omega t - k_0 x)\right] - A^*(\omega) \exp\left[-i(\omega t - k_0 x)\right],$$

and

$$A(\omega) = \frac{cE_0}{2\pi}\left(\frac{\varepsilon_1}{\varepsilon_2} - 1\right) \frac{\omega}{(\omega - \omega_0)\left(\omega^2 - \omega_0^2 \frac{\varepsilon_1}{\varepsilon_2}\right)} \tag{6.25}$$

is the spectral amplitude of the vector potential of the field (6.12).

We shall solve the set of equations (6.22)–(6.24) in the approximation of the perturbation theory by the field. The parameter of the perturbation theory is $\xi_0 = eE_0/mc\omega_0 << 1$. As long as the particle motion along the z axis remains free we can choose the initial velocity of the particle in the xy plane: $\mathbf{v}_0 = \{\mathrm{v}_0 \cos\theta, \mathrm{v}_0 \sin\theta, 0\}$. According to perturbation theory

$$\mathbf{p} = \mathbf{p}_0 + \Delta\mathbf{p}; \quad |\Delta\mathbf{p}| << |\mathbf{p}_0| \, ,$$

and from the Eqs. (6.22), (6.23) in first-order approximation by ξ_0 (keeping only the uniform part of motion $x(t) = x_0 + \mathrm{v}_{0x} t$ on the right-hand side of the equations) for the changes of the particle momentum in the field $\Delta\mathbf{p}$ we will obtain the following equations:

$$\frac{d\Delta p_x}{dt} = -i\frac{e}{c}k_0 \int_{-\infty}^{\infty} \mathrm{v}_{0y} F(\omega, x_0 + \mathrm{v}_{0x} t, t) d\omega, \tag{6.26}$$

$$\frac{d\Delta p_y}{dt} = i\frac{e}{c} \int_{-\infty}^{\infty} (k_0 \mathrm{v}_{0x} - \omega) F(\omega, x_0 + \mathrm{v}_{0x} t, t) d\omega \ . \tag{6.27}$$

Integrating Eqs. (6.26) and (6.27) over t from $-\infty$ to $+\infty$ we obtain in first-order approximation by ξ_0 the following expressions for the particle momentum change after the interaction:

$$\Delta p_x = -i\frac{2\pi e k_0}{c} \mathrm{v}_{0y} \int_{-\infty}^{\infty} \left[A(\omega) e^{-ik_0 x_0}\right.$$

$$-A^{*}(\omega)\, e^{ik_0x_0}\big]\,\delta(\omega - k_0 \mathrm{v}_{0x})\, d\omega, \tag{6.28}$$

$$\Delta p_y = i\frac{2\pi e}{c}\int_{-\infty}^{\infty}(k_0\mathrm{v}_{0x}-\omega)\big[A(\omega)\,e^{-ik_0x_0}$$

$$-A^{*}(\omega)\, e^{ik_0x_0}\big]\,\delta(\omega - k_0 \mathrm{v}_{0x})\, d\omega. \tag{6.29}$$

The δ-function in these expressions defines the condition of induced radiation/absorption by a free charged particle in the field of a transversal monochromatic EM wave under the nonstationary transition process:

$$\omega - \mathbf{k}_0\mathbf{v}_0 = 0. \tag{6.30}$$

Integrating in the same way Eqs. (6.21) and taking into account Eq. (6.30) for the particle momentum and energy changes after the interaction we obtain the following ultimate formulas:

$$\triangle p_y = \triangle p_z = 0, \qquad \triangle p_x = \frac{\triangle\mathcal{E}}{\mathrm{v}_0\cos\theta}, \tag{6.31}$$

$$\triangle\mathcal{E}{=}2mc^2\xi_0\frac{\mathrm{v}_0^3}{c^3}(\varepsilon_1-\varepsilon_2)\frac{\sin\theta\cos^2\theta}{\left(1-\sqrt{\varepsilon_1}\frac{\mathrm{v}_0}{c}\cos\theta\right)\left(1-\varepsilon_2\frac{\mathrm{v}_0^2}{c^2}\cos^2\theta\right)}$$

$$\times\sin\left(\omega_0\sqrt{\varepsilon_1}\frac{\mathrm{v}_0\cos\theta}{c}t_0\right). \tag{6.32}$$

Here t_0 is the instant of time corresponding to the initial phase of the particle in the external EM wave. Note that Eq. (6.32) besides the induced nonstationary transition process describes generally the induced Cherenkov effect as well (see the denominator) if a medium initially (at $t < 0$) was dielectriclike (in principle, it includes also the Cherenkov effect at $t > 0$ if $\varepsilon_2 > 1$, but for actual physical cases we assume that the stepwise discontinuity of ε (6.1) may be realistic at the abrupt transformation of a dielectriclike medium into a plasma for which $\varepsilon_2 < 1$ and the induced Cherenkov effect is excluded).

As is seen from Eq. (6.32) depending on the initial phase

$$\Phi_0 = \omega_0 t_0\sqrt{\varepsilon_1}\,(\mathrm{v}_0/c)\cos\theta$$

the particle is either accelerated after the interaction or is decelerated radiating coherently into the wave. This real energy exchange is due to the direct and inverse induced nonstationary transition effect. In the case of a particle beam, various particles situated initially in the diverse phases Φ_0 will acquire or lose different energies in the field and the particles' free drift after the interaction will result in bunching of an initially homogeneous particle beam.

6.3 Quantum Description of Multiphoton Interaction

Consider now the quantum dynamics of the induced nonstationary transition process. Quantitative analysis of Eqs. (6.31) and (6.32) shows that the classical energy exchange of a particle with strong EM radiation in a nonstationary medium as a result of the induced nonstationary transition effect corresponds to absorption and emission of a large number of photons. On the basis of the quantum theory such multiphoton process can be described by the quasiclassical-type wave function neglecting, in fact, the quantum recoil at the absorption/emission of photons by the particle. The latter corresponds to a slowly varying wave function for which the derivatives of the second order of the particle wave function can be neglected with respect to the first order ones that have been made in the consideration of the multiphoton processes in the previous chapters. The role of the particle spin is inessential here, hence by neglecting the spin interaction the Dirac equation in quadratic form is written as the Klein–Gordon equation (3.30) for the particle in the specified EM field. Assuming the same geometry as in Section 6.1 the latter takes the form

$$-\hbar^2 \frac{\partial^2 \Psi}{\partial t^2} = \left[-\hbar^2 c^2 \nabla^2 + 2iec\hbar \, \nabla_y \, A_y(x,t) + e^2 A_y^2(x,t) + m^2 c^4\right] \Psi, \quad (6.33)$$

where

$$A_y(x,t) = \int_{-\infty}^{\infty} \left[A(\omega) e^{-ik_0 x} + A(-\omega) e^{ik_0 x}\right] e^{i\omega t} d\omega \quad (6.34)$$

is the vector potential of the field (6.12) expressed via the spectral amplitude $A(\omega)$ (6.25).

Equation (6.33) will be solved in the mentioned approximation by the particle wave function

$$\Psi(\mathbf{r},t) = \sqrt{\frac{N_0}{2\mathcal{E}_0}} f(x,t) \exp\left[\frac{i}{\hbar}\left(\mathbf{p}_0 \mathbf{r} - \mathcal{E}_0 t\right)\right], \quad (6.35)$$

where $f(x,t)$ is a slowly varying function with respect to the free-particle wave function (see Section 3.5). Taking into account the conditions (3.92) and Eq. (6.35) from Eq. (6.33) for $f(x,t)$ we will obtain the differential equation of the first order:

$$\frac{\partial f}{\partial t} + \mathrm{v}_{0x} \frac{\partial f}{\partial x} = \frac{i}{2\hbar \mathcal{E}_0} \left[2ecp_{0y} A_y(x,t) + e^2 A^2(x,t)\right] f(x,t). \quad (6.36)$$

The conditions (3.92) correspond to a small change of the momentum and energy of the electron in the field compared with the initial values $\Delta p << p_0$ and $\Delta\mathcal{E} << \mathcal{E}_0$, that is, the approximation made in the classical consideration, where the intensity of the EM wave is restricted by the condition $\xi_0 << 1$. Then for actual values of parameters $p_{0y}/mc >> \xi_0$ and the last term $\sim A^2$ in Eq. (6.36) will be neglected.

Passing from x, t to characteristic coordinates $\tau' = t - x/\mathrm{v}_{0x}$, $\eta' = t$ and integrating Eq. (6.36) we obtain

$$f\left(\tau', \eta'\right) = \exp\left\{\frac{ie\mathrm{v}_{0y}}{\hbar c}\int_{-\infty}^{\eta'} A_y(\mathrm{v}_{0x}\left(\eta'' - \tau'\right), \eta'')d\eta''\right\}. \tag{6.37}$$

Then after the interaction ($\eta' \to +\infty$) taking into account Eq. (6.34) we obtain

$$f\left(\tau\right) = \exp\left\{\frac{i4\pi e\mathrm{v}_{0y}}{\hbar c} A\left(\omega_0\sqrt{\varepsilon_1}\frac{\mathrm{v}_{0x}}{c}\right)\cos\left(\omega_0\sqrt{\varepsilon_1}\frac{\mathrm{v}_{0x}}{c}\tau\right)\right\}. \tag{6.38}$$

The spectral amplitude in Eq. (6.38) is determined by Eq. (6.25):

$$A\left(\omega_0\sqrt{\varepsilon_1}\frac{\mathrm{v}_{0x}}{c}\right) = \frac{E_0}{2\pi\omega_0^2}\frac{\varepsilon_1 - \varepsilon_2}{\sqrt{\varepsilon_1}}\frac{\mathrm{v}_0\cos\theta}{\left(\sqrt{\varepsilon_1}\frac{\mathrm{v}_0}{c}\cos\theta - 1\right)\left(\varepsilon_2\frac{\mathrm{v}_0^2}{c^2}\cos^2\theta - 1\right)}. \tag{6.39}$$

Returning to coordinates x, t and expanding the exponential (6.38) into a series by the Bessel functions and taking into account Eq. (6.39) for the total wave function (6.35) we will have

$$\Psi\left(\mathbf{r}, t\right) = \sqrt{\frac{N_0}{2\mathcal{E}_0}}\exp\left[\frac{i}{\hbar}p_{0y}y\right]\sum_{s=-\infty}^{+\infty} i^s J_s\left(\alpha\right)$$

$$\times\exp\left\{\frac{i}{\hbar}\left[p_{0x} - s\hbar\sqrt{\varepsilon_1}\frac{\omega_0}{c}\right]x - \frac{i}{\hbar}\left[\mathcal{E}_0 - s\hbar\omega_0\sqrt{\varepsilon_1}\frac{\mathrm{v}_0}{c}\cos\theta\right]t\right\}, \tag{6.40}$$

where the argument of the Bessel function is

$$\alpha = 2\xi_0\frac{m\mathrm{v}_0^2}{\hbar\omega_0}\frac{\varepsilon_1 - \varepsilon_2}{\sqrt{\varepsilon_1}}\frac{\sin\theta\cos\theta}{\left(1 - \sqrt{\varepsilon_1}\frac{\mathrm{v}_0}{c}\cos\theta\right)\left(1 - \varepsilon_2\frac{\mathrm{v}_0^2}{c^2}\cos^2\theta\right)}. \tag{6.41}$$

As is seen from Eq. (6.40), due to the induced nonstationary transition effect the particle absorbs or emits s photons, as a result of which the momentum and energy after the interaction are changed:

$$\triangle p_x = s\hbar\frac{\omega_0}{c}\sqrt{\varepsilon_1}, \quad \triangle p_y = 0, \quad \triangle\mathcal{E} = s\hbar\omega_0\sqrt{\varepsilon_1}\frac{\mathrm{v}_0}{c}\cos\theta. \tag{6.42}$$

The probability of the induced s-photon process is

$$W_s = J_s^2\left(\frac{2\xi_0 m\mathrm{v}_0^2\left(\varepsilon_1-\varepsilon_2\right)\sin\theta\cos\theta}{\hbar\omega_0\sqrt{\varepsilon_1}\left(1-\sqrt{\varepsilon_1}\frac{\mathrm{v}_0}{c}\cos\theta\right)\left(1-\varepsilon_2\frac{\mathrm{v}_0^2}{c^2}\cos^2\theta\right)}\right). \tag{6.43}$$

The comparison of the expression for α with the amplitude of the classical change of the particle momentum $(\triangle p_x)_{\max}$ (6.31) and energy $(\triangle\mathcal{E})_{\max}$ (6.32) shows that

$$\alpha = \frac{(\triangle p_x)_{\max}}{\hbar k_0}, \tag{6.44}$$

in accordance with the correspondence principle ($s \sim \alpha >> 1$).

At the small value of α or small number of photons s when the interaction has entirely quantum character it is necessary to take into account the quantum recoil as well. It is especially important in this process, because at the abrupt temporal variation of the dielectric permittivity the hard quanta in the spectrum of the initial radiation arise. We will solve for this purpose Eq. (6.33) keeping also the derivatives of the second order of the particle wave function for a single-photon absorption or emission. Correspondingly, in first-order approximation of the perturbation theory from Eq. (6.33) we have the following equation for the particle wave function at the single-photon interaction with the field (6.35) in the nonstationary transition process:

$$\frac{\partial^2\Psi_1}{\partial x^2} - \frac{1}{c^2}\frac{\partial^2\Psi_1}{\partial t^2} - \frac{1}{\hbar^2c^2}\left(m^2c^4 + c^2p_{0y}^2\right)\Psi_1$$

$$= -2\frac{ep_{0y}}{c\hbar^2}\left[A_y(t)\,e^{-ik_0x} + A_y^*(t)\,e^{ik_0x}\right]\Psi_0, \tag{6.45}$$

where

$$\Psi_0(\mathbf{r},t) = \sqrt{\frac{N_0}{2\mathcal{E}_0}}\exp\left[\frac{i}{\hbar}\left(\mathbf{p}_0\mathbf{r} - \mathcal{E}_0t\right)\right] \tag{6.46}$$

is the initial wave function of the particle (normalized on N_0 particles per unit volume). The solution of Eq. (6.45) is sought in the form

$$\Psi_1(\mathbf{r},t) = \left[\Phi_1(t)e^{-ik_0x} + \Phi_2(t)e^{ik_0x}\right]\exp\left[\frac{i}{\hbar}\left(\mathbf{p}_0\mathbf{r} - \mathcal{E}_0t\right)\right]. \tag{6.47}$$

Substituting Eq. (6.47) in Eq. (6.45) for the functions $\Phi_1(t)$ and $\Phi_2(t)$ we obtain the equations:

$$\frac{d^2\Phi_1}{dt^2} - 2i\frac{\mathcal{E}_0}{\hbar}\frac{d\Phi_1}{dt} - c^2 k_0\left(2\frac{p_{0x}}{\hbar} - k_0\right)\Phi_1 = 2\sqrt{\frac{N_0}{\mathcal{E}_0}}\frac{ecp_{0y}}{\hbar^2}A_y(t), \tag{6.48}$$

$$\frac{d^2\Phi_2}{dt^2} - 2i\frac{\mathcal{E}_0}{\hbar}\frac{d\Phi_2}{dt} + c^2 k_0\left(2\frac{p_{0x}}{\hbar} + k_0\right)\Phi_2 = 2\sqrt{\frac{N_0}{\mathcal{E}_0}}\frac{ecp_{0y}}{\hbar^2}A_y^*(t). \tag{6.49}$$

The solution of Eq. (6.48) is

$$\Phi_1(t) = -2i\sqrt{\frac{N_0}{\mathcal{E}_0}}\frac{ecp_{0y}}{\hbar^2(\Omega_1 - \Omega_2)}$$

$$\times\left[e^{i\Omega_1 t}\int_{-\infty}^{t} e^{-i\Omega_1 t'}A_y(t')\,dt' - e^{i\Omega_2 t}\int_{-\infty}^{t} e^{-i\Omega_2 t'}A_y(t')\,dt'\right], \tag{6.50}$$

where the characteristic frequencies Ω_1 and Ω_2 are given by the expressions

$$\Omega_{1,2} = \frac{\mathcal{E}_0}{\hbar} \mp \left[\left(\frac{\mathcal{E}_0}{\hbar} - \omega_0\sqrt{\varepsilon_1}\frac{\mathrm{v}_{0x}}{c}\right)^2 + \omega_0^2\varepsilon_1\left(1 - \frac{\mathrm{v}_{0x}^2}{c^2}\right)\right]^{1/2} \tag{6.51}$$

with the signs "$\mp$" correspondingly.

Passing from $A_y(t)$ to the Fourier component of the field we obtain for $\Phi_1(t)$ after the interaction $(t \to +\infty)$

$$\Phi_1(t) = -4i\sqrt{\frac{N_0}{\mathcal{E}_0}}\frac{\pi ecp_{0y}}{\hbar^2(\Omega_1 - \Omega_2)}\left[A(\Omega_1)e^{i\Omega_1 t} - A(\Omega_2)e^{i\Omega_2 t}\right], \tag{6.52}$$

where the spectral amplitudes of the wave vector potential $A(\Omega_1)$ and $A(\Omega_2)$ are determined by Eq. (6.25).

Solving Eq. (6.49) in an analogous way for the function $\Phi_2(t)$ we obtain

$$\Phi_2(t) = -4i\sqrt{\frac{N_0}{\mathcal{E}_0}}\frac{\pi ecp_{0y}}{\hbar^2(\Omega_1' - \Omega_2')}\left[A^*(-\Omega_1')e^{i\Omega_1' t} - A^*(-\Omega_2')e^{i\Omega_2' t}\right], \tag{6.53}$$

with the characteristic frequencies

$$\Omega_{1,2}' = \frac{\mathcal{E}_0}{\hbar} \mp \left[\left(\frac{\mathcal{E}_0}{\hbar} + \omega_0\sqrt{\varepsilon_1}\frac{\mathrm{v}_{0x}}{c}\right)^2 + \omega_0^2\varepsilon_1\left(1 - \frac{\mathrm{v}_{0x}^2}{c^2}\right)\right]^{1/2}. \tag{6.54}$$

Equations (6.51) and (6.54) correspond to the energy-momentum conservation law for a particle in the induced nonstationary transition process: the particle can emit only the photons with frequencies $\Omega_{1,2}$ and absorb photons with frequencies $\Omega'_{1,2}$. As long as $\mathcal{E}_0/\hbar >> \omega_0\sqrt{\varepsilon_1}\mathrm{v}_{0x}/c$ for the frequencies of a strong coherent radiation field we expand the square roots in Eqs. (6.51), (6.54) in a series and retain only the small terms of first order. We then obtain for the radiation frequencies:

$$\Omega_1 \simeq \omega_0\sqrt{\varepsilon_1}\frac{\mathrm{v}_{0x}}{c} - \varepsilon_1\frac{\hbar\omega_0^2}{2\mathcal{E}_0}\left(1-\frac{\mathrm{v}_{0x}^2}{c^2}\right),$$

$$\Omega_2 \simeq 2\frac{\mathcal{E}_0}{\hbar} - \omega_0\sqrt{\varepsilon_1}\frac{\mathrm{v}_{0x}}{c} + \varepsilon_1\frac{\hbar\omega_0^2}{2\mathcal{E}_0}\left(1-\frac{\mathrm{v}_{0x}^2}{c^2}\right) \tag{6.55}$$

and for the absorption frequencies:

$$\Omega'_1 \simeq -\omega_0\sqrt{\varepsilon_1}\frac{\mathrm{v}_{0x}}{c} - \varepsilon_1\frac{\hbar\omega_0^2}{2\mathcal{E}_0}\left(1-\frac{\mathrm{v}_{0x}^2}{c^2}\right),$$

$$\Omega'_2 \simeq 2\frac{\mathcal{E}_0}{\hbar} + \omega_0\sqrt{\varepsilon_1}\frac{\mathrm{v}_{0x}}{c} + \varepsilon_1\frac{\hbar\omega_0^2}{2\mathcal{E}_0}\left(1-\frac{\mathrm{v}_{0x}^2}{c^2}\right). \tag{6.56}$$

These expressions show that the emission of a photon with frequency Ω_2 and absorption with frequency Ω'_2 has a clearly quantum character, and its probability, as is seen from Eq. (6.25), depends on the change of the dielectric permittivity of the medium $\varepsilon_1 - \varepsilon_2$. We therefore consider two cases: $\varepsilon_1/\varepsilon_2 \lesssim 1$ and $\varepsilon_1/\varepsilon_2 >> 1$.

If $\varepsilon_1/\varepsilon_2 \lesssim 1$ we get from Eq. (6.25)

$$A(\Omega_2) \simeq A\left(2\frac{\mathcal{E}_0}{\hbar}\right) << A(\Omega_1) \simeq A\left(\omega_0\sqrt{\varepsilon_1}\frac{\mathrm{v}_{0x}}{c}\right), \tag{6.57}$$

so that in this case we can neglect in Eqs. (6.52) and (6.53) the pure quantum process of emission and absorption of hard quanta $\Omega_2 \simeq 2\mathcal{E}_0/h$. Then for the amplitudes of the particle wave function $\Phi_1(t)$ and $\Phi_2(t)$ we will have correspondingly

$$\Phi_{1,2}(t) = i\sqrt{\frac{N_0}{\mathcal{E}_0}\frac{e\mathrm{v}_0^2 E_0}{\hbar\omega_0^2 c}}\frac{\varepsilon_1-\varepsilon_2}{\sqrt{\varepsilon_1}}\frac{\sin\theta\cos\theta}{\left(1-\sqrt{\varepsilon_1}\frac{\mathrm{v}_0}{c}\cos\theta\right)\left(1-\varepsilon_2\frac{\mathrm{v}_0^2}{c^2}\cos^2\theta\right)}$$

$$\times\exp\left\{i\omega_0\left[\pm\sqrt{\varepsilon_1}\frac{\mathrm{v}_0}{c}\cos\theta - \frac{\varepsilon_1\hbar\omega_0}{2\mathcal{E}_0}\left(1-\frac{\mathrm{v}_0^2}{c^2}\cos^2\theta\right)\right]t\right\} \tag{6.58}$$

with the signs "±" correspondingly. Equation (6.58) with Eq. (6.47) determines the particle's wave function after the single-photon interaction with the field (6.35) in the nonstationary transition process. In this case ($\varepsilon_1/\varepsilon_2 \lesssim 1$) we obtain for the current density ($\sim |\Psi_0 + \Psi_1|^2$) of the particles after the interaction

$$\mathbf{j}(x,t) = \mathbf{j}_0 \left\{ 1 + 2\alpha \sin \left[\varepsilon_1 \frac{\hbar\omega_0^2}{2\mathcal{E}_0} \left(1 - \frac{\mathrm{v}_0^2}{c^2} \cos^2\theta \right) t \right] \right.$$

$$\left. \times \cos \left[\omega_0 \sqrt{\varepsilon_1} \frac{\mathrm{v}_0 \cos\theta}{c} \left(t - \frac{x}{\mathrm{v}_0 \cos\theta} \right) \right] \right\}, \qquad (6.59)$$

where $\mathbf{j}_0 = \text{const}$ is the particle's initial current density and α is defined by Eq. (6.41) or (6.44). As is seen from Eq. (6.59) as a result of the stimulated absorption and emission of the photons of frequency

$$\Omega_1 = \omega_0 \sqrt{\varepsilon_1} \frac{\mathrm{v}_0}{c} \cos\theta$$

the quantum modulation of the particle's probability density and, consequently, current density at this frequency occurs with a depth $\Gamma_1 = 2\alpha$. Also, in contrast to the effect of quantum modulation in coherent processes considered in previous chapters, the pure temporal modulation here takes place as well that is caused by the nonstationarity of the medium. The period of this temporal modulation is

$$T_1 = \frac{4\pi\mathcal{E}_0}{\hbar\omega_0^2 \varepsilon_1 \left(1 - \frac{\mathrm{v}_0^2}{c^2} \cos^2\theta \right)}.$$

If we derive the particle's wave function in the next orders of perturbation theory, then we obtain the modulation at higher harmonics of the wave frequency. The modulation depth at the s-th harmonic will be $\Gamma_s \sim \Gamma_1^s$.

For $\varepsilon_1/\varepsilon_2 >> 1$, it is necessary to also take into account in Eqs. (6.52), (6.53) the pure quantum process of emission and absorption of hard quanta $\Omega_2 \simeq 2\mathcal{E}_0/\hbar$. The spectral amplitude of the wave vector potential $A(\Omega_2)$ at such frequencies is

$$A(\Omega_2) \simeq \frac{cE_0}{8\pi} \frac{\varepsilon_1}{\varepsilon_2} \left(\frac{\mathcal{E}_0^2}{\hbar^2} - \frac{\varepsilon_1}{\varepsilon_2} \frac{\omega_0^2}{4} \right)^{-1}. \qquad (6.60)$$

In an analogous way for the particles current density after the interaction we will have

$$\mathbf{j}(x,t) = \mathbf{j}_0 \left\{ 1 + \Gamma_1 \sin \left[\varepsilon_1 \frac{\hbar\omega_0^2}{2\mathcal{E}_0} \left(1 - \frac{\mathrm{v}_0^2}{c^2} \cos^2\theta \right) t \right] \right.$$

$$\times \cos\left[\omega_0\sqrt{\varepsilon_1}\frac{v_0\cos\theta}{c}\left(t-\frac{x}{v_0\cos\theta}\right)\right]$$

$$+\Gamma_2\sin\left(2\frac{\mathcal{E}_0}{\hbar}t\right)\cos\left[\omega_0\sqrt{\varepsilon_1}\frac{v_0\cos\theta}{c}\left(t+\frac{x}{v_0\cos\theta}\right)\right]\Bigg\}, \qquad (6.61)$$

where $\Gamma_1 = 2\alpha$, and the modulation depth Γ_2 due to the absorption-emission of hard quanta Ω_2 is

$$\Gamma_2 = \xi\frac{mv_0c\hbar\omega_0}{\mathcal{E}_0^2}\frac{\varepsilon_1}{\varepsilon_2}\frac{\sin\theta}{1-\frac{\varepsilon_1}{\varepsilon_2}\left(\frac{\hbar\omega}{2\mathcal{E}_0}\right)^2}. \qquad (6.62)$$

The period of temporal modulation in this case is $T_2 = \pi\hbar/\mathcal{E}_0$.

As the modulated particle beam radiates coherently this mechanism can be of interest in astrophysics where the radiating matter may be in a strongly nonstationary state.

6.4 Electron–Positron Pair Production by a γ-Quantum in a Medium

The formation of hard γ-quanta of frequencies $\sim \mathcal{E}_0/\hbar$ in the spectrum of a strong monochromatic EM wave propagating in a nonstationary medium, the dielectric permittivity of which abruptly changes in time, makes available the single-photon production of electron–positron (e^-, e^+) pairs from the intense light fields in a nonstationary medium.

In general, the single-photon reaction $\gamma \rightarrow e^- + e^+$ as well as the inverse reaction of the electron–positron annihilation ($e^- + e^+ \rightarrow \gamma$) can proceed in a medium that must be plasmalike (for the satisfaction of conservation laws for these reactions one needs $n(\omega) < 1$). However, as will be shown below, excessively large densities of the plasma in this case are required. Meanwhile, the single-photon production of e^-, e^+ pairs in a nonstationary plasma is possible at ordinary densities. Moreover, this process can proceed in the strong light fields in an arbitrary medium turning abruptly into a plasma (with the temporal variation law of ε (6.1)). Hence, we will consider both single-photon reactions $\gamma \rightleftarrows e^- + e^+$ in a stationary plasma and the production of e^-, e^+ pairs from the intense light beam in a nonstationary medium.

Consider first the production of electron–positron pairs by a γ-quantum and its annihilation in a stationary medium. It is easy to see from the conservation laws of the energy and momentum for the single-photon reactions $\gamma \rightleftarrows e^- + e^+$

$$\hbar\mathbf{k} = \mathbf{p}_1 + \mathbf{p}_2; \qquad \hbar\omega = \mathcal{E}_1 + \mathcal{E}_2 \qquad (6.63)$$

(ω, $\mathbf{k}$ are the γ-quantum frequency and wave vector, $|\mathbf{k}| = n(\omega)\omega/c$, $\mathbf{p}_{1,2}$ and $\mathcal{E}_{1,2}$ are the momenta and energies of the electron and positron, respectively) that the phase velocity of a γ-quantum $\mathrm{v}_{ph} = c/n(\omega)$ must be larger than c, i.e., a medium for these processes must be plasmalike: $n(\omega) < 1$. The latter restricts the energy of a γ-quantum because of the dispersive properties of a medium. Indeed, for the macroscopic meaning of the refractive index of a medium for a γ-quantum at least one particle within a distance of the order of $\lambda/2$ is required (λ is the wavelength of the γ-quantum), that is, the condition $\lambda/2 \gtrsim l$ must be satisfied, where l is the distance between the electrons in a plasma. Therefore, besides the threshold condition that follows from the conservation laws (6.63):

$$\hbar\omega > \frac{2mc^2}{\sqrt{1 - n^2(\omega)}}, \tag{6.64}$$

for the reactions $\gamma \rightleftarrows e^- + e^+$ in a medium the following requirement on the plasma density N/V for a specified frequency ω of a γ-quantum arises:

$$\omega \lesssim \pi \left(\frac{N}{V}\right)^{1/3} \equiv \omega_{\mathrm{lim}}. \tag{6.65}$$

Hence, condition (6.65) determines the lower bound for the density of the medium or the upper bound for the energy of the γ-quantum, while threshold condition (6.64) determines the lower bound for the energy of the γ-quantum to cause the reactions $\gamma \rightleftarrows e^- + e^+$ to proceed in a medium.

From the standpoint of single-photon pair creation and annihilation in plasma, the latter must compensate the longitudinal momentum $\triangle p = [1 - n(\omega)]\hbar\omega/c$ transferred in these processes. Consequently, the characteristic length in the macroscopic description of the dispersion of the medium is the wavelength $\hbar/\triangle p$, which corresponds to the transferred momentum, and the condition necessary for this is $\hbar/\triangle p > (V/N)^{1/3}$. Since $n(\omega) < 1$, this condition is satisfied automatically when condition (6.65) is satisfied.

The plasma densities satisfying conditions (6.64) and (6.65) are at least: $N/V > 10^{33}\mathrm{cm}^{-3}$. Such superdense matter exists only in astrophysical objects, particularly in the core of the neutron stars (pulsars). At these densities the electron component of the superdense plasma is highly degenerate (the dispersion of the transverse electromagnetic waves is determined by electrons). Actually, the degeneracy temperature of the electron component of such plasma is $T_F > 10^{10}$ K. On the other hand, because of neutrino energy losses, the physically attainable temperatures in an equilibrium system are much lower than this: $T << T_F$ and the superdense plasma is fully degenerate.

Since the Fermi energy at the densities $N/V > 10^{33}\mathrm{cm}^{-3}$ is $\mathcal{E}_F > mc^2$ we need the dispersion law of the fully degenerate relativistic plasma. To determine the dispersion relation $n = n(\omega)$ of the latter we shall solve the self-consistent set of Maxwell–Vlasov equations for the transverse monochromatic EM wave in the relativistic collisionless plasma with the distribution function $f(\mathbf{p}, \mathbf{r}, t)$ (we will not consider the ions' motion).

The characteristic equations of $f(\mathbf{p}, \mathbf{r}, t)$ coincide with the single particle equation of motion. The latter has been solved for an arbitrary medium in Section 2.1 and in the case of plasma we have the following solutions in the wave field with the vector potential $\mathbf{A} = \{0, A_0 \cos(\omega t - n(\omega)\omega x/c), 0\}$:

$$p_x = p_{0x} - \frac{n(\omega)}{c(1 - n^2(\omega))}\left\{\mathcal{E}_0 - n(\omega) c p_{0x}\right.$$

$$\left. - \sqrt{(\mathcal{E}_0 - n(\omega) c p_{0x})^2 + (1 - n^2(\omega))\left[e^2 A_y^2 - 2ecp_{0y}A_y\right]}\right\}, \tag{6.66}$$

$$p_y = p_{0y} - \frac{e}{c}A_y; \qquad p_z = p_{0z}, \tag{6.67}$$

and for the energy of the particle in the field:

$$\mathcal{E} = \mathcal{E}_0 - \frac{1}{1 - n^2(\omega)}\left\{\mathcal{E}_0 - n(\omega) c p_{0x}\right.$$

$$\left. - \sqrt{(\mathcal{E}_0 - n(\omega) c p_{0x})^2 + (1 - n^2(\omega))\left[e^2 A_y^2 - 2ecp_{0y}A_y\right]}\right\}. \tag{6.68}$$

The density of the electric current induced in the plasma can be defined by the equation

$$\mathbf{j}(\mathbf{r}, t) = e \int \mathbf{v} f(\mathbf{p}, \mathbf{r}, t)\, d\mathbf{p}, \tag{6.69}$$

where $\mathbf{v} = c^2\mathbf{p}/\mathcal{E}$ is the velocity of the electrons with the distribution function in the field $f(\mathbf{p}, \mathbf{r}, t)$. According to the Liouville theorem for the collisionless plasma we have

$$f(\mathbf{p}, \mathbf{r}, t) = f_0(\mathbf{p}_0, \mathbf{r}_0, t_0) = f_0(p_0), \tag{6.70}$$

since the electrons before the interaction were distributed stationary, uniformly and isotropic.

Defining from Eqs. (6.66)–(6.68) the velocity of the electrons as a function of the $\mathbf{p}_0$, $\mathbf{r}$, and t and then passing from the integration over $\mathbf{p}$ to integration over $\mathbf{p}_0$ (taking into account Eq. (6.70)), Eq. (6.69) may be presented in the form

$$\mathbf{j}(\mathbf{r},t) = ec^2 \int \frac{\mathbf{p}(\mathbf{p}_0,\mathbf{r},t)}{\mathcal{E}(\mathbf{p}_0,\mathbf{r},t)} f_0(p_0)\, J(\mathbf{p}_0,\mathbf{r},t) d\mathbf{p}_0, \tag{6.71}$$

where

$$J(\mathbf{p}_0,\mathbf{r},t) = \frac{\partial(p_x, p_y, p_z)}{\partial(p_{0x}, p_{0y}, p_{0z})}$$

is the Jacobian of transformation. From Eqs. (6.66), (6.67) for the latter we have

$$J(\mathbf{p}_0,\mathbf{r},t) = 1 - \frac{n(\omega)}{1-n^2(\omega)} \left(\frac{cp_{0x}}{\mathcal{E}_0} - n(\omega) \right)$$

$$\times \left[1 - \frac{\mathcal{E}_0 - n(\omega)\, cp_{0x}}{\sqrt{(\mathcal{E}_0 - n(\omega)\, cp_{0x})^2 + (1-n^2(\omega)) \left[e^2 A_y^2 - 2ecp_{0y} A_y \right]}} \right]. \tag{6.72}$$

In the linear approximation by a weak wave field (since it will be applied for a γ-quantum) Eq. (6.72) can be written as follows:

$$J(\mathbf{p}_0,\mathbf{r},t) = 1 + \frac{n(\omega)}{(\mathcal{E}_0 - n(\omega)\, cp_{0x})^2} \left(\frac{cp_{0x}}{\mathcal{E}_0} - n(\omega) \right) ecp_{0y} A_y. \tag{6.73}$$

The components of the electric current density (6.71) in this linear regime of interaction can be expressed in the form

$$j_y(\mathbf{r},t) = ec^2 \int \left\{ \frac{p_{0y}}{\mathcal{E}_0} \left(1 + \frac{(1-n^2(\omega))\, cp_{0y} e A_y}{(\mathcal{E}_0 - n(\omega)\, cp_{0x})^2} \right) - \frac{eA_y}{\mathcal{E}_0} \right\}$$

$$\times f_0(p_0)\, d\mathbf{p}_0, \tag{6.74}$$

$$j_x = j_z = 0. \tag{6.75}$$

Then turning to spherical coordinates in Eq. (6.71)

$$p_{0x} = p_0 \cos\theta; \; p_{0y} = p_0 \sin\theta \cos\varphi; \; p_{0z} = p_0 \sin\theta \sin\varphi,$$

and taking into account that the initial distribution of the electrons in a plasma is isotropic, after the integration in the equation

$$j_y(\mathbf{r},t) = -e^2 c A_y \int \left\{ 1 - \frac{(1-n^2(\omega))\, c^2 p_{0y}^2}{(\mathcal{E}_0 - n(\omega)\, cp_{0x})^2} \right\}$$

$$\times \frac{f_0(p_0)\,p_0^2}{\mathcal{E}_0}\sin\theta d\theta d\varphi dp_0 \tag{6.76}$$

by the angles, for the electric current density induced by a wave field in the plasma we will have

$$j_y(\mathbf{r},t) = -\frac{4\pi e^2 c A_y}{n^2(\omega)}\int \frac{f(p_0)p_0^2}{\mathcal{E}_0}$$

$$\times\left\{1 - \frac{\mathcal{E}_0\left(1-n^2(\omega)\right)}{2n(\omega)\,cp_0}\ln\left\{\frac{\mathcal{E}_0 + n(\omega)\,cp_0}{\mathcal{E}_0 - n(\omega)\,cp_0}\right\}\right\} dp_0. \tag{6.77}$$

The Maxwell equation for the vector potential

$$\left[\nabla^2 - \frac{1}{c^2}\frac{\partial^2}{\partial t^2}\right] A_y(\mathbf{r},t) = -\frac{4\pi}{c} j_y(\mathbf{r},t) \tag{6.78}$$

with the current density (6.77) gives the following equation for the refractive index of a relativistic plasma:

$$n^2(\omega) = 1 - \frac{16\pi^2 e^2 c^2}{n^2(\omega)\,\omega^2}\int \frac{f(p_0)p_0^2}{\mathcal{E}_0}$$

$$\times\left\{1 - \frac{\mathcal{E}_0\left(1-n^2(\omega)\right)}{2n(\omega)\,cp_0}\ln\left\{\frac{\mathcal{E}_0 + n(\omega)\,cp_0}{\mathcal{E}_0 - n(\omega)\,cp_0}\right\}\right\} dp_0. \tag{6.79}$$

Equation (6.79) describes in general the dispersion law of a relativistic plasma for an arbitrary electron distribution function. In principle, it is also valid for a nondegenerate (relativistic and Maxwellian) electron plasma if an equilibrium distribution with temperature $T \gtrsim T_F$ can be realized in nature.

Now consider the production of electron–positron pairs by a γ-quantum in a stationary medium (homogeneous and isotropic) with a refractive index $n(\omega) < 1$ (6.79). As this process is a QED effect of the first order, then using the general rules for constructing the matrix element of a single-vertex $\gamma \to e^- + e^+$ diagram in a dispersive medium the probability amplitude will be written in the form

$$S_{if} = -e\sqrt{\frac{1}{2\omega a_\omega n^2(\omega)}}\int \overline{\psi}_1 \widehat{\epsilon}^{(\lambda)} e^{ikx}\psi_2 d^4x. \tag{6.80}$$

Here

$$a_\omega = 1 + \frac{\omega}{n(\omega)} \frac{dn(\omega)}{d\omega},$$

$k^i(\omega, \mathbf{k})$ is the 4-dimensional wave vector of the photon, quantization volume $V = 1$, $\epsilon^{(\lambda)}$ is the four-dimensional polarization vector of the photon ($\widehat{\epsilon}^{(\lambda)} = \epsilon_\mu^{(\lambda)}\gamma^\mu$), and

$$\psi_1 = u_1(\mathbf{p}_1) e^{i(\mathbf{p}_1\mathbf{r} - \mathcal{E}_1 t)}; \quad \psi_2 = u_2(-\mathbf{p}_2) e^{-i(\mathbf{p}_2\mathbf{r} - \mathcal{E}_2 t)} \tag{6.81}$$

are the free electron and positron wave functions. Here the units $\hbar = c = 1$ are used.

Performing integration in Eq. (6.80) with the wave functions (6.81) by the standard method for the differential probability of the $\gamma \to e^- + e^+$ process per unit time and unit space volume (in the momentum volumes $d\mathbf{p}_1/(2\pi)^3$ of the electrons and $d\mathbf{p}_2/(2\pi)^3$ of the positrons, respectively) we will have

$$dW = \frac{e^2}{8\pi^2 \omega a_\omega n^2(\omega)} \left| \overline{u}_1(\mathbf{p}_1) \widehat{\epsilon}^{(\lambda)} u_2(-\mathbf{p}_2) \right|^2 \delta(\omega - \mathcal{E}_1 - \mathcal{E}_2)$$

$$\times \delta(\mathbf{k} - \mathbf{p}_1 - \mathbf{p}_2)\, d\mathbf{p}_1 d\mathbf{p}_2. \tag{6.82}$$

We will assume that the γ-quantum is nonpolarized and perform averaging by the polarization states of the γ-quantum and summation over the electron and positron spin projections. Then the probability of the e^-, e^+ pair production per unit time is given by the expression

$$W = \frac{e^2}{8\pi^2 a_\omega \omega n^2(\omega)} \int \frac{\mathcal{E}_1\mathcal{E}_2 + m^2 - p_1 p_2 \cos\vartheta_1 \cos\vartheta_2}{\mathcal{E}_1\mathcal{E}_2} \delta(\omega - \mathcal{E}_1 - \mathcal{E}_2)$$

$$\times \delta(\mathbf{k} - \mathbf{p}_1 - \mathbf{p}_2)\, d\mathbf{p}_1 d\mathbf{p}_2, \tag{6.83}$$

where $\vartheta_{1,2}$ is the angle between the vectors $\mathbf{k}$ and $\mathbf{p}_{1,2}$, respectively.

Integrating Eq. (6.83) over the positron momentum $\mathbf{p}_2$ we obtain the following expression for the pair production probability:

$$W = \frac{e^2}{8\pi^2 a_\omega \omega n^2(\omega)} \int \left(1 + \frac{m^2 + p_1 \cos\vartheta_1 (p_1 \cos\vartheta_1 - k)}{\mathcal{E}_1 \sqrt{\mathcal{E}_1^2 + k^2 + k p_1 \cos\vartheta_1}} \right)$$

$$\times \delta\left(\omega - \mathcal{E}_1 - \sqrt{\mathcal{E}_1^2 + k^2 + k p_1 \cos\vartheta_1} \right) d\mathbf{p}_1. \tag{6.84}$$

For the integration over the electron momentum $\mathbf{p}_1$ note that because of azimuthal symmetry

$$d\mathbf{p}_1 = 2\pi p_1 \mathcal{E}_1 d\mathcal{E}_1 \sin\vartheta_1 d\vartheta_1$$

and the integration over ϑ_1 reduces formally to the following replacement in Eq. (6.84):

$$\delta\left(\omega - \mathcal{E}_1 - \sqrt{\mathcal{E}_1^2 + k^2 + kp_1\cos\vartheta_1}\right)\sin\vartheta_1 d\vartheta_1$$

$$\to \frac{\omega - \mathcal{E}_1}{kp_1}\left[H\left(\mathcal{E}_1 - \mathcal{E}_{\min}(\omega)\right) - H\left(\mathcal{E}_1 - \mathcal{E}_{\max}(\omega)\right)\right],$$

where $H(x)$ is the Heaviside function

$$H(x) = \begin{cases} 1, & x \geq 0, \\ 0, & x < 0. \end{cases}$$

After the integration over ϑ_1, Eq. (6.84) becomes

$$W = \frac{e^2}{4\pi a_\omega \omega^2 n^5(\omega)} \int_{\mathcal{E}_{\min}(\omega)}^{\mathcal{E}_{\max}(\omega)} \left[\left(1 - n^2(\omega)\right)\left(\mathcal{E}_1^2 - \omega\mathcal{E}_1\right) + n^2(\omega)m^2 \right.$$
$$\left. + \frac{1 - n^4(\omega)}{4}\omega^2\right] d\mathcal{E}_1. \quad (6.85)$$

The limits of integration over $\mathcal{E}_1 \in [\mathcal{E}_{\min}, \mathcal{E}_{\max}]$ in Eq. (6.85)

$$\mathcal{E}_{\min,\max}(\omega) = \frac{\omega}{2} \mp \frac{n(\omega)}{2}\left[\omega^2 - \frac{4m^2}{1 - n^2(\omega)}\right]^{1/2} \quad (6.86)$$

are determined by the conservation laws for the $\gamma \rightleftarrows e^- + e^+$ processes in a medium (6.63) with the threshold value (6.64). Taking into account Eq. (6.86) after the integration over the electron energy in Eq. (6.85) we obtain the total probability for the single-photon e^-, e^+ pair production in a plasma:

$$W = \frac{e^2 m^2}{6\pi\omega^2 a_\omega n^2(\omega)}\left[\omega^2 - \frac{4m^2}{1 - n^2(\omega)}\right]^{1/2}$$
$$\times \left\{\frac{1}{2}\left(\frac{\omega}{m}\right)^2\left[1 - n^2(\omega)\right] + 1\right\}. \quad (6.87)$$

Equation (6.86) with the dispersion law (6.79) of a relativistic plasma for an arbitrary electron distribution function determine the probability of the electron–positron pair production by a γ-quantum. As the electron component the of superdense plasma required for this process is fully degenerate the Pauli principle must also be taken into account that imposes an additional restriction on the $\gamma \to e^- + e^+$ reaction. The general picture of this process taking into account the conditions (6.64), (6.65) and the Pauli principle will be analyzed together with the electron–positron annihilation process in the next section.

6.5 Annihilation of Electron–Positron Pairs in a Medium

Now we will consider the inverse process of a single-photon annihilation of an electron–positron pair in a stationary plasma. This process is also a QED effect of the first order and the matrix element of a single-vertex $e^- + e^+ \to \gamma$ diagram is the complex conjugate to the $\gamma \to e^- + e$ diagram matrix element:

$$S'_{if} = -e\sqrt{\frac{1}{2\omega a_\omega n^2(\omega)}} \int \overline{\psi}_2 \widehat{\epsilon}^{(\lambda)} e^{-ikx} \psi_1 d^4x. \tag{6.88}$$

The differential probability of the annihilation process per unit time and unit space volume, summed by the polarization states of the created γ -quantum in the momentum volume $d\mathbf{k}/(2\pi)^3$, is given by the expression

$$dW_\gamma = \frac{\pi e^2}{2\omega a_\omega n^2(\omega)} \frac{\mathcal{E}_1\mathcal{E}_2 + m^2 - p_1 p_2 \cos\vartheta_1 \cos\vartheta_2}{\mathcal{E}_1\mathcal{E}_2}$$

$$\times \delta(\omega - \mathcal{E}_1 - \mathcal{E}_2)\, \delta(\mathbf{k} - \mathbf{p}_1 - \mathbf{p}_2)\, d\mathbf{k}. \tag{6.89}$$

Equation (6.89) determines the annihilation probability for a single e^-, e^+ pair in plasma. To obtain the total probability of annihilation of an initial positron with the plasma electrons one must define the probability of annihilation of a positron of specified energy $\mathcal{E}_2$ with the electrons of the medium in the momentum range $\mathbf{p}_1, \mathbf{p}_1 + d\mathbf{p}_1$:

$$W_\gamma = \frac{\pi e^2}{2\omega a_\omega n^2(\omega)} \int f(p_1) \frac{\mathcal{E}_1\mathcal{E}_2 + m^2 - p_1 p_2 \cos\vartheta_1 \cos\vartheta_2}{\mathcal{E}_1\mathcal{E}_2}$$

$$\times \delta(\omega - \mathcal{E}_1 - \mathcal{E}_2)\, \delta(\mathbf{k} - \mathbf{p}_1 - \mathbf{p}_2)\, d\mathbf{k} d\mathbf{p}_1, \tag{6.90}$$

where $f(p_1)$ is the distribution function of the plasma electrons. We first integrate over $\mathbf{k}$ in Eq. (6.90) and then over $\mathbf{p}_1$ taking into account that

$d\mathbf{p}_1 = 2\pi p_1 \mathcal{E}_1 d\mathcal{E}_1 \sin\vartheta d\vartheta$, where ϑ is the angle between the vectors $\mathbf{p}_1$ and $\mathbf{p}_2$. The integration over ϑ reduces formally to the following replacement in Eq. (6.90):

$$\delta\left(\omega - \mathcal{E}_1 - \mathcal{E}_2\right) \sin\vartheta d\vartheta$$

$$\rightarrow \frac{\omega a_\omega n^2(\omega)}{p_1 p_2}\left[H\left(\mathcal{E}_1 - \mathcal{E}_{\min}(\omega)\right) - H\left(\mathcal{E}_1 - \mathcal{E}_{\max}(\omega)\right)\right],$$

where the quantities $\mathcal{E}_{\min(\max)}(\omega)$ are given by Eq. (6.86) and ω must be replaced by $\mathcal{E}_1 + \mathcal{E}_2$ according to conservation law (6.63). Then for the probability of annihilation of a positron (with an energy $\mathcal{E}_2$) with the electrons of the medium we will have

$$W_\gamma = \frac{\pi e^2}{p_2 \mathcal{E}_2} \int f\left(p_1\right) \left\{ m^2 + \left(\mathcal{E}_1 + \mathcal{E}_2\right)^2 \frac{1 - n^4(\omega)}{4n^2(\omega)} - \frac{1 - n^2(\omega)}{n^2(\omega)} \mathcal{E}_1 \mathcal{E}_2 \right\}$$

$$\times \left[H\left(\mathcal{E}_1 - \mathcal{E}_{\min}(\omega)\right) - H\left(\mathcal{E}_1 - \mathcal{E}_{\max}(\omega)\right)\right] d\mathcal{E}_1. \qquad (6.91)$$

In contrast to the pair-production process (its probability can be obtained without resorting to the explicit form of $n(\omega)$), here we must have the explicit form of the function $n = n(\omega)$ in order to be able to integrate over the electron energy $\mathcal{E}_1$ (ω is now a function of $\mathcal{E}_1$, since $\omega = \mathcal{E}_1 + \mathcal{E}_2$).

As the considered processes $\gamma \rightleftarrows e^- + e^+$ are possible in the superdense plasma where the electrons are fully degenerate, then the dispersion law of such relativistic plasma can be obtained substituting the Fermi distribution function for a fully degenerate electron gas

$$f(p_1) = \begin{cases} \frac{1}{4\pi^3}, & p_1 \le p_F \\ 0, & p_1 > p_F \end{cases} \qquad (6.92)$$

in Eq. (6.79), describing in general the dispersion law of a relativistic plasma for an arbitrary distribution function of electrons $f(p_0)$. Here p_F is the boundary Fermi momentum:

$$p_F = \left(3\pi^2 \rho_e\right)^{1/3}, \qquad (6.93)$$

and ρ_e is the electron density of a degenerate Fermi gas.

Integrating in Eq. (6.79) with the distribution function (6.92) over the electron momenta we obtain the following dispersion law of a relativistic degenerate plasma:

$$n^2\left(\omega\right) = 1 - \frac{2e^2}{n^2\left(\omega\right)\pi\omega^2}$$

$$\times \left\{ p_F \mathcal{E}_F - \frac{\mathcal{E}_F^2 - n^2(\omega) p_F^2}{2n(\omega)} \ln \left\{ \frac{\mathcal{E}_F + n(\omega) p_F}{\mathcal{E}_F - n(\omega) p_F} \right\} \right\}, \tag{6.94}$$

where $\mathcal{E}_F$ is the relativistic Fermi energy corresponding to boundary momentum (6.93). Inserting the dimensionless parameter

$$\beta = \frac{n(\omega) p_F}{\mathcal{E}_F}$$

Eq. (6.94) can be written in the form

$$n^2(\omega) = 1 - \frac{2e^2 p_F \mathcal{E}_F}{n^2(\omega) \pi \omega^2} \left\{ 1 - \frac{1-\beta^2}{2\beta} \ln \left\{ \frac{1+\beta}{1-\beta} \right\} \right\}, \tag{6.95}$$

or in the form more convenient for further investigation

$$n^2(\omega) = 1 - \frac{2e^2 p_F^3}{\omega^2 \pi \mathcal{E}_F} \phi(\beta), \tag{6.96}$$

where the function $\phi(\beta)$ is

$$\phi(\beta) = \frac{1}{\beta^2} \left\{ 1 - \frac{1-\beta^2}{2\beta} \ln \frac{1+\beta}{1-\beta} \right\}. \tag{6.97}$$

By analogy with the usual determination of a plasma frequency, from the equation $n(\omega_p) = 0$ we obtain the plasma frequency for a relativistic degenerate one

$$\omega_p = \sqrt{\frac{4e^2 p_F^3}{3\pi \mathcal{E}_F}}. \tag{6.98}$$

The frequency range corresponding to transverse waves that can propagate in a superdense relativistic degenerate plasma — $\omega_p \leq \omega < \infty$ — can then be obtained by varying the refractive index in the range $0 \leq n < 1$. Therefore, we present the dispersion relation (6.96) in the inverted form $\omega = \omega(n)$:

$$\omega^2 = \frac{2e^2}{\pi} \frac{p_F^3}{\mathcal{E}_F} \frac{1}{1-n^2} \phi(\beta). \tag{6.99}$$

The parameter β in Eq. (6.99) then varies in the range $0 \leqslant \beta < p_F/\mathcal{E}_F$. The analysis of the function $\phi(\beta)$, which can be expressed in the form

$$\phi(\beta) = 2 \sum_{s=1}^{\infty} \frac{\beta^{2s-2}}{4s^2 - 1},$$

shows that throughout the physically admissible range $0 \le \beta < 1$ (for superdense ultrarelativistic plasma $p_F/\mathcal{E}_F \sim 1$) the function $\phi(\beta)$ varies monotonically between the values 2/3 and 1.

The problem now reduces to the determination of the range of variation of the energies of electrons that actually participate in the annihilation process taking account of conditions (6.64), (6.65) and $\mathcal{E}_1 \le \mathcal{E}_F$ for the annihilation process. The situation may be clarified by defining this region graphically. Figure 6.1 shows the $\mathcal{E}_{\min(\max)}(\omega)$ curves and the lines corresponding to frequencies $\omega = \omega_{\lim} = (\pi/3)^{1/3} p_F$ (see Eq. (6.65)) and $\omega = \omega_{\max} = \mathcal{E}_F + \mathcal{E}_2$. The energies of the particles and γ-quantum can vary within the region $ABCA$, and the limits of integration with respect to the electron energy $\mathcal{E}_{1\,\min}$ and $\mathcal{E}_{1\,\max}$ are determined by the points at which the $\mathcal{E}_1 = \omega - \mathcal{E}_2$ line cuts the boundaries of this region.

Evaluating the integral in Eq. (6.91) with the dispersion law (6.99) we obtain a bulky expression for the total probability of the annihilation process. However, for the admissible values of $n(\omega)$ and electron density ρ_e with a great accuracy for the function $\phi(\beta)$ we have: $\phi(np_F/\mathcal{E}_F) \approx 2/3$ and the ultimate expression for the probability of the $e^- + e^+ \to \gamma$ process is rather simplified.

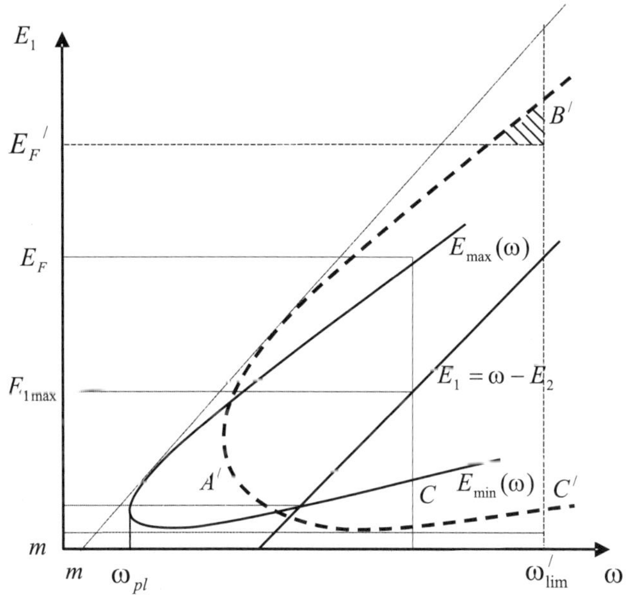

Fig. 6.1. Curves of $\mathcal{E}_{\min}(\omega)$, $\mathcal{E}_{\max}(\omega)$ and the lines corresponding to frequencies $\omega = \omega_{\lim} = (\pi/3)^{1/3} p_F$ and $\omega = \omega_{\max} = \mathcal{E}_F + \mathcal{E}_2$. The energies of the particles and γ-quantum can vary within the region $ABCA$, and the limits of integration with respect to the electron energy $\mathcal{E}_{1\,\min}$ and $\mathcal{E}_{1\,\max}$ are determined by the points at which the $\mathcal{E}_1 = \omega - \mathcal{E}_2$ line cuts the boundaries of this region.

The points of intersection of the line $\mathcal{E}_1 = \omega - \mathcal{E}_2$ and the boundaries of the region $ABCA$ then correspond to

$$\omega_1 = \frac{\omega_p}{2m^2}\left[\omega_p \mathcal{E}_2 - p_2\left(\omega_p^2 - 4m^2\right)^{1/2}\right],$$

$$\omega_2 = \begin{cases} \frac{\omega_p}{2m^2}\left[\omega_p \mathcal{E}_2 + p_2\left(\omega_p^2 - 4m^2\right)^{1/2}\right], & \mathcal{E}_2 \leq \mathcal{E}_{\min}\left(\omega = \omega_{\lim}\right), \\ \omega_{\lim}, & \mathcal{E}_{\min}\left(\omega = \omega_{\lim}\right) < \mathcal{E}_2 < \mathcal{E}_{\max}\left(\omega = \omega_{\lim}\right). \end{cases} \tag{6.100}$$

Finally, the total probability of the annihilation process is

$$W_\gamma = \frac{e^2}{4\pi p_2 \mathcal{E}_2}\left[\left(m^2 + \frac{\omega_p^2}{2}\right)(\omega_2 - \omega_1) + \frac{1}{2}\omega_p\left(\mathcal{E}_2^2 + \frac{\omega_p^2}{4}\right)\right.$$
$$\left.\times \ln\frac{(\omega_2 - \omega_p)(\omega_1 + \omega_p)}{(\omega_2 + \omega_p)(\omega_1 - \omega_p)} - \frac{\mathcal{E}_2 \omega_p^2}{2}\ln\frac{(\omega_2 - \omega_p)(\omega_2 + \omega_p)}{(\omega_1 - \omega_p)(\omega_1 + \omega_p)}\right]. \tag{6.101}$$

The lower limit for the density of the medium, above which pair annihilation is possible, can be defined from the reaction threshold condition (6.64) and the dispersion law (6.96). Thus, we obtain $\omega_p > 2m$, which is equivalent to $\mathcal{E}_F > \sqrt{3\pi}m/e \approx 36m$. The electron density of the plasma corresponding to this value of $\mathcal{E}_F$ is $\rho_e > p_F^3/3\pi^2 \approx 3 \cdot 10^{34}\mathrm{cm}^{-3}$.

For a nonrelativistic positron annihilation in an electron plasma we have a simple formula for the total probability:

$$W_\gamma = \frac{e^2 \omega_p^3}{8\pi m^3}\left(\omega_p^2 - 4m^2\right)^{1/2}, \qquad p_2 << m. \tag{6.102}$$

Let us now analyze the results for the electron–positron pair production in a superdense relativistic degenerate plasma with the dispersion law (6.96). The Pauli principle in this case demands the satisfaction of the condition $\mathcal{E}_1 > \mathcal{E}_F$ which together with conditions (6.64) and (6.65) substantially reduces the range of parameter values for this process to proceed even in the required superdense plasma. The range of integration with respect to $\mathcal{E}_1$ in Eq. (6.85) shrinks to a point and the probability of the process $\gamma \to e^- + e^+$ tends practically to zero. With the increase of the electron density when $\mathcal{E}_F \gtrsim 150m$ ($\mathcal{E}_{\max}(\omega_{\lim}) > \mathcal{E}_F$, see Fig. 6.1), a narrow region appears and Eqs. (6.65), (6.100) show that the creation of a pair by a γ-quantum with energy $\omega_1(\mathcal{E}_2 = \mathcal{E}_F) < \omega < \omega_{\lim}$ becomes possible in this region. As a result, the lower bound of the energy of a created electron instead of $\mathcal{E}_{\min}(\omega)$ should be $\mathcal{E}_F$ and from Eq. (6.85) we obtain

$$W = \frac{e^2 \left(\mathcal{E}_{\max}(\omega) - \mathcal{E}_F\right)}{4\pi a_\omega \omega^2 n^5(\omega)} \left\{ \frac{1-n^2(\omega)}{3} \left(\mathcal{E}^2_{\max}(\omega) + \mathcal{E}_F \mathcal{E}_{\max}(\omega) + \mathcal{E}_F^2\right) \right.$$

$$\left. - \frac{1-n^2(\omega)}{2} \omega \left(\mathcal{E}_{\max}(\omega) + \mathcal{E}_F\right) + n^2(\omega) m^2 + \frac{1-n^4(\omega)}{4} \omega^2 \right\}. \qquad (6.103)$$

However, it is important to recall that this region $\omega \simeq \omega_{\lim}$ lies at the limit of validity of the macroscopic concept for a refractive index of a medium (one particle within the length $\lambda/2$).

6.6 Electron–Positron Pair Production by Strong EM Wave in Nonstationary Medium

As the probability of the single-quantum production of an electron–positron pair in a stationary plasma, as a macroscopic dispersive medium, practically equals zero (even at the required superdensities of electrons) it is reasonable to consider an exclusive possibility for a single-photon pair production in a nonstationary medium of ordinary densities by strong light fields. Namely, we assume the abrupt temporal change of the dielectric permittivity of a medium which may be described by the stepwise function ε (6.1).

In order to describe pair production in the field (6.6), (6.7) we shall employ the Dirac model (all negative-energy states of the vacuum are filled with electrons). The Dirac equation in the field (6.6), (6.7) has the form ($\hbar = c = 1$)

$$i\frac{\partial \Psi}{\partial t} = \left[\widehat{\alpha}(\mathbf{p} - e\mathbf{A}) + \widehat{\beta} m\right] \Psi, \qquad (6.104)$$

where

$$\mathbf{A}(\mathbf{r}, t) = \begin{cases} i\frac{\mathbf{E}_0}{\omega_0} e^{i(\omega_0 t - \mathbf{k}_0 \mathbf{r})} + \text{c.c.}, & t < 0 \\ i\frac{\mathbf{E}_1}{\omega_1} e^{i\omega_1 t - \mathbf{k}_0 \mathbf{r}} - i\frac{\mathbf{E}_2}{\omega_1} e^{-i\omega_1 t - \mathbf{k}_0 \mathbf{r}} + \text{c.c.}, & t \geqslant 0 \end{cases} \qquad (6.105)$$

is the vector potential of the EM field and $\widehat{\alpha}$, $\widehat{\beta}$ are the Dirac matrices in the standard representation (3.2).

We solve Eq. (6.104) by perturbing in the field of the wave. This method is valid if

$$\left[1 + \left(\frac{\varepsilon_1}{\varepsilon_2}\right)^{1/2}\right] \xi_0 << 1, \qquad \xi_0 = \frac{eE_0}{m\omega_0}. \qquad (6.106)$$

We expand the perturbed first-order wave function $\Psi_1(\mathbf{r}, t)$ in a complete set of orthonormalized wave functions of the electrons (positrons) with momenta $\mathbf{p} - \mathbf{k}_0$ and $\mathbf{p} + \mathbf{k}_0$:

$$\Psi_1(\mathbf{r},t) = \Psi_1^{(-)}(t)e^{i(\mathbf{p}-\mathbf{k}_0)\mathbf{r}} + \Psi_1^{(+)}(t)e^{i(\mathbf{p}+\mathbf{k}_0)\mathbf{r}},$$

$$\Psi_1^{(-)}(t) = \sum_{l=1}^{4} a_l(t)u_l\left(\mathbf{p}-\mathbf{k}_0,t\right), \tag{6.107}$$

$$\Psi_1^{(+)}(t) = \sum_{j=1}^{4} b_j(t)u_j\left(\mathbf{p}+\mathbf{k}_0,t\right).$$

Here $a_l(t)$ and $b_j(t)$ are unknown functions and $u_i\left(\mathbf{p}',t\right)$ are orthonormalized bispinor functions which describe the particle states with energies $\pm\mathcal{E}' = \pm\sqrt{p'^2+m^2}$:

$$u_{1,2}\left(\mathbf{p}',t\right) = \left(\frac{\mathcal{E}'+m}{2\mathcal{E}'}\right)^{1/2}\begin{pmatrix}\varphi_{1,2}\\ \frac{\sigma\mathbf{p}'}{\mathcal{E}'+m}\varphi_{1,2}\end{pmatrix}\exp\left(-i\mathcal{E}'t\right), \tag{6.108}$$

$$u_{3,4}\left(\mathbf{p}',t\right) = \left(\frac{\mathcal{E}'+m}{2\mathcal{E}'}\right)^{1/2}\begin{pmatrix}\frac{-\sigma\mathbf{p}'}{\mathcal{E}'+m}\chi_{3,4}\\ \chi_{3,4}\end{pmatrix}\exp\left(i\mathcal{E}'t\right). \tag{6.109}$$

These functions are normalized to one particle per unit volume: $u_i^+u_j = \delta_{ij}$; the constant spinors $\varphi_{1,2}$ and $\chi_{3,4}$ are

$$\varphi_1 = \chi_3 = \begin{pmatrix}1\\0\end{pmatrix}, \qquad \varphi_2 = \chi_4 = \begin{pmatrix}0\\1\end{pmatrix}.$$

Under the transformations (6.107)–(6.109) the Dirac equation for the perturbed wave function $\Psi = \Psi_0 + \Psi_1 + \cdots$, $(|\Psi_1| << |\Psi_0|)$:

$$\left(i\frac{\partial}{\partial t} - \widehat{\alpha}\mathbf{p} - \widehat{\beta}m\right)\Psi_1 = -e\widehat{\alpha}\mathbf{A}\Psi_0 \tag{6.110}$$

transforms into a system of 16 equations for the unknown functions $a_l(t)$ and $b_j(t)$:

$$\left(i\frac{\partial}{\partial t} - \widehat{\alpha}\mathbf{p} - \widehat{\beta}m\right)\left[\sum_{l=1}^{4} a_l(t)u_l\left(\mathbf{p}-\mathbf{k}_0,t\right)e^{i(\mathbf{p}-\mathbf{k}_0)\mathbf{r}}\right.$$

$$\left.+\sum_{j=1}^{4} b_j(t)u_j\left(\mathbf{p}+\mathbf{k}_0,t\right)e^{i(\mathbf{p}+\mathbf{k}_0)\mathbf{r}}\right]$$

$$= -e\widehat{\alpha}\left[\mathbf{A}_{(-)}(t)e^{-i\mathbf{k}_0\mathbf{r}} + \mathbf{A}_{(+)}(t)e^{i\mathbf{k}_0\mathbf{r}}\right] u_s\left(\mathbf{p}, t\right) e^{i\mathbf{p}\mathbf{r}}, \tag{6.111}$$

where $s = 3, 4$ and

$$\mathbf{A}_{(-)}(t) = \begin{cases} i\frac{\mathbf{E}_0}{\omega_0}e^{i\omega_0 t}, & t < 0, \\ i\frac{\mathbf{E}_1}{\omega_1}e^{i\omega_1 t} - i\frac{\mathbf{E}_2}{\omega_1}e^{-i\omega_1 t}, & t \geqslant 0, \end{cases} \qquad \mathbf{A}_{(+)}(t) = \mathbf{A}^*_{(-)}(t). \tag{6.112}$$

The bispinor functions $u_s\left(\mathbf{p}, t\right)$ in Eq. (6.111) correspond to the unperturbed states of the Dirac vacuum (they are determined by Eq. (6.109) with $s = 3$ and $s = 4$, where $\mathbf{p}' = \mathbf{p}$ and $\mathcal{E}' = \mathcal{E}$ are the momenta and energies of the free vacuum electrons). According to this model, a pair is produced because of the interaction of the external field with the electrons of negative energies of the Dirac vacuum. In the first-order perturbation theory in the field this leads to electron states in the region of positive energies with the values

$$\mathcal{E}_{(-)} = \sqrt{\left(\mathbf{p} - \mathbf{k}_0\right)^2 + m^2}, \qquad \mathcal{E}_{(+)} = \sqrt{\left(\mathbf{p} + \mathbf{k}_0\right)^2 + m^2}.$$

The probabilities of these transitions are determined by the amplitudes $a_{1,2}$ and $b_{1,2}$, respectively (the indices 1 and 2 correspond to two different spin states). Therefore the problem reduces to determining the functions $a_{1,2}(t)$ and $b_{1,2}(t)$ by integrating the set of Eqs. (6.111). From the latter we obtain the following set of equations:

$$\sum_{l=1}^{4} i\frac{da_l}{dt}u_l\left(\mathbf{p} - \mathbf{k}_0, t\right) = -e\widehat{\alpha}\mathbf{A}_{(-)}(t)u_s\left(\mathbf{p}, t\right), \tag{6.113}$$

$$\sum_{j=1}^{4} i\frac{db_j}{dt}u_j\left(\mathbf{p} + \mathbf{k}_0, t\right) = -e\widehat{\alpha}\mathbf{A}_{(+)}(t)u_s\left(\mathbf{p}, t\right). \tag{6.114}$$

Multiplying Eq. (6.113) on the left by $u_l^\dagger\left(\mathbf{p} - \mathbf{k}_0, t\right)$ and Eq. (6.114) by $u_j^\dagger\left(\mathbf{p} + \mathbf{k}_0, t\right)$ and taking into account that the bispinors are orthonormal ($u_l^\dagger u_m = \delta_{lm}$) we obtain eight equations for the transitions amplitudes $a_l(t)$ and $b_j(t)$ for a given spinor state s of a vacuum electron ($s = 3$ or $s = 4$) :

$$\frac{da_l(t)}{dt} = ieu_l^\dagger\left(\mathbf{p} - \mathbf{k}_0, t\right)\widehat{\alpha}\mathbf{A}_{(-)}(t)u_s\left(\mathbf{p}, t\right), \qquad l = 1, ..., 4, \tag{6.115}$$

$$\frac{db_j(t)}{dt} = ieu_j^\dagger\left(\mathbf{p} + \mathbf{k}_0, t\right)\widehat{\alpha}\mathbf{A}_{(+)}(t)u_s\left(\mathbf{p}, t\right), \qquad j = 1, ..., 4. \tag{6.116}$$

Orienting the z axis parallel to the electric field $\mathbf{E}_0$ of the wave and the x axis parallel to the wave vector $\mathbf{k}_0$, we obtain for the amplitudes $a_{1,2}$ and $b_{1,2}$

$$a_{1,2}(t) = ieu^{\dagger}_{1,2}\left(\mathbf{p}-\mathbf{k}_0\right)\alpha_z u_s\left(\mathbf{p}\right)\int_{-\infty}^{t} A_{(-)}(t')e^{i\left(\mathcal{E}+\mathcal{E}_{(-)}\right)t'}dt', \qquad (6.117)$$

$$b_{1,2}(t) = ieu^{\dagger}_{1,2}\left(\mathbf{p}+\mathbf{k}_0\right)\alpha_z u_s\left(\mathbf{p}\right)\int_{-\infty}^{t} A_{(+)}(t')e^{i\left(\mathcal{E}+\mathcal{E}_{(+)}\right)t'}dt', \qquad (6.118)$$

where $u^{\dagger}_{1,2}\left(\mathbf{p}\mp\mathbf{k}_0\right)$ and $u_s\left(\mathbf{p}\right)$ are constant bispinors determined by Eqs. (6.108) and (6.109) (preexponential factors in Eqs. (6.108), (6.109)) .

The probability of electron production from a definite vacuum state $\mathbf{p}$, s is determined by the quantity $|a_1(t)|^2+|a_2(t)|^2+|b_1(t)|^2+|b_2(t)|^2$ (the probability of the production of a positron with a momentum $\mathbf{p}$ in a definite spinor state s). The differential probability of pair production, summed over the initial spin states of the Dirac vacuum, in an element of the phase volume $d\mathbf{p}/(2\pi)^3$ (the spatial normalization volume $V=1$), is

$$dW = 2\left[|a_1(t)|^2+|a_2(t)|^2+|b_1(t)|^2+|b_2(t)|^2\right]|_{t\to+\infty}\,\frac{d\mathbf{p}}{\left(2\pi\right)^3}. \qquad (6.119)$$

Integrating Eqs. (6.117), (6.118) over time with Eq. (6.112) and assuming that the EM wave is switched on and switched off adiabatically: $\mathbf{E}_0(t=-\infty)=\mathbf{E}_1(t=+\infty)=\mathbf{E}_2(t=+\infty)=0$ (the amplitudes of the incident, transmitted, and reflected waves are assumed to be slowly varying functions of time), we obtain the following expressions for the amplitudes $a_{1,2}$ and $b_{1,2}$ after the wave interaction with the Dirac vacuum:

$$a_{1,2}(t=+\infty) = \frac{ieE_0\left(\varepsilon_1-\varepsilon_2\right)\left(\mathcal{E}+\mathcal{E}_{(-)}\right)}{\varepsilon_2\left(\mathcal{E}+\mathcal{E}_{(-)}+\omega_0\right)\left(\left(\mathcal{E}+\mathcal{E}_{(-)}\right)^2-\omega_0^2\frac{\varepsilon_1}{\varepsilon_2}\right)}$$

$$\times\left[u^{\dagger}_{1,2}\left(\mathbf{p}-\mathbf{k}_0\right)\alpha_z u_s\left(\mathbf{p}\right)\right], \qquad (6.120)$$

$$b_{1,2}(t=+\infty) = \frac{ieE_0\left(\varepsilon_1-\varepsilon_2\right)\left(\mathcal{E}+\mathcal{E}_{(+)}\right)}{\varepsilon_2\left(\mathcal{E}+\mathcal{E}_{(+)}-\omega_0\right)\left(\left(\mathcal{E}+\mathcal{E}_{(+)}\right)^2-\omega_0^2\frac{\varepsilon_1}{\varepsilon_2}\right)}$$

$$\times\left[u^{\dagger}_{1,2}\left(\mathbf{p}+\mathbf{k}_0\right)\alpha_z u_s\left(\mathbf{p}\right)\right]. \qquad (6.121)$$

Evaluating the transition matrix elements in Eqs. (6.120), (6.121), we obtain with the help of Eq. (6.119) the differential probability of pair production by a strong EM wave in a nonstationary medium:

$$dW = \frac{e^2}{(2\pi)^3} \frac{E_0^2}{\mathcal{E}} \left(\frac{\varepsilon_1}{\varepsilon_2} - 1 \right)^2$$

$$\times \left\{ \frac{\left(\mathcal{E} + \mathcal{E}_{(-)}\right)^2 \left[\mathcal{E}\mathcal{E}_{(-)} + m^2 + p_x(p_x - k_0) + p_y^2 - p_z^2\right]}{\mathcal{E}_{(-)} \left(\mathcal{E} + \mathcal{E}_{(-)} + \omega_0\right)^2 \left[\left(\mathcal{E} + \mathcal{E}_{(-)}\right)^2 - \omega_0^2 \frac{\varepsilon_1}{\varepsilon_2}\right]^2} \right.$$

$$\left. + \frac{\left(\mathcal{E} + \mathcal{E}_{(+)}\right)^2 \left[\mathcal{E}\mathcal{E}_{(+)} + m^2 + p_x(p_x + k_0) + p_y^2 - p_z^2\right]}{\mathcal{E}_{(+)} \left(\mathcal{E} + \mathcal{E}_{(+)} - \omega_0\right)^2 \left[\left(\mathcal{E} + \mathcal{E}_{(+)}\right)^2 - \omega_0^2 \frac{\varepsilon_1}{\varepsilon_2}\right]^2} \right\} d\mathbf{p}. \qquad (6.122)$$

As one can see from Eq. (6.122), the process exhibits azimuthal asymmetry with respect to the direction of propagation of the wave. Orienting the polar axis in this direction ($d\mathbf{p} = p\mathcal{E}d\mathcal{E}\sin\theta d\theta d\varphi$, where θ is the angle between the vectors $\mathbf{p}$ and $\mathbf{k}_0$ and φ is the azimuthal angle relative to the direction of polarization of the wave) and integrating over the energy, we obtain the angular distribution of the produced electrons (positrons). As the case of physical interest is an EM wave of frequencies $\omega << m$, Eq. (6.122) simplifies greatly and takes the form

$$dW = \frac{e^2 E_0^2}{2\pi^3} \left(\frac{\varepsilon_1}{\varepsilon_2} - 1 \right)^2 \frac{\sqrt{\mathcal{E}^2 - m^2}}{\mathcal{E}}$$

$$\times \frac{m^2 \sin^2\theta \cos^2\varphi + \mathcal{E}^2 \left(1 - \sin^2\theta\cos^2\varphi\right)}{\left(4\mathcal{E}^2 - \omega_0^2 \frac{\varepsilon_1}{\varepsilon_2}\right)^2} \sin\theta d\theta d\varphi d\mathcal{E}. \qquad (6.123)$$

Integrating Eq. (6.123) over the energy we obtain the number of pairs produced in the element of solid angle $do = \sin\theta d\theta d\varphi$:

$$dW(\theta, \varphi) = \frac{e^2 E_0^2}{128\pi^2 m} \left(\frac{\varepsilon_1}{\varepsilon_2} - 1 \right)^2 \left[F\left(2; \frac{1}{2}; 2; \frac{\omega_0^2 \varepsilon_1}{4m^2 \varepsilon_2}\right) \right.$$

$$\left. \times \left(1 - \sin^2\theta\cos^2\varphi\right) + \frac{1}{4} F\left(2; \frac{3}{2}; 3; \frac{\omega_0^2 \varepsilon_1}{4m^2\varepsilon_2}\right) \sin^2\theta\cos^2\varphi \right] do, \qquad (6.124)$$

where $F(\nu; \mu; \lambda; z)$ is the hypergeometric function.

For the energy distribution of the produced electrons (positrons) we have

$$dW(\mathcal{E}) = \frac{2e^2E_0^2}{3\pi^2}\left(\frac{\varepsilon_1}{\varepsilon_2}-1\right)^2 \frac{\sqrt{\mathcal{E}^2-m^2}\,(2\mathcal{E}^2+m^2)}{\left(4\mathcal{E}^2-\omega_0^2\frac{\varepsilon_1}{\varepsilon_2}\right)^2}d\mathcal{E}. \tag{6.125}$$

Integrating Eq. (6.124) over the angles θ and φ (or Eq. (6.125) over the energy) we obtain the total number of electron–positron pairs produced by a strong EM wave in a nonstationary medium:

$$W = \frac{2e^2E_0^2}{48\pi m}\left(\frac{\varepsilon_1}{\varepsilon_2}-1\right)^2\left[F\left(2;\frac{1}{2};2;\frac{\omega_0^2\varepsilon_1}{4m^2\varepsilon_2}\right)\right.$$

$$\left.+\frac{1}{8}F\left(2;\frac{3}{2};3;\frac{\omega_0^2\varepsilon_1}{4m^2\varepsilon_2}\right)\right]. \tag{6.126}$$

Note that in Eqs. (6.123) and (6.125) the denominators become zero for $\omega_0\sqrt{\varepsilon_1/\varepsilon_2} = 2\mathcal{E}$. This is the conservation law for the single-photon pair production by a wave of the frequency $\omega_1 = \omega_0\sqrt{\varepsilon_1/\varepsilon_2}$ (by the transmitted and reflected waves) in a medium with the index of refraction $n_2 = \sqrt{\varepsilon_2} < 1$ (plasma). Since Eqs. (6.123)–(6.126) correspond to the case $\omega << m$, the pole in Eq. (6.123) can be reached, i.e., the conservation laws of energy and momentum for the process $\gamma \to e^- + e^+$ can be satisfied only if $\varepsilon_1/\varepsilon_2 >> 1$. Actually this is possible if $\varepsilon_2 << 1$, in agreement with the fact that pair production by a photon field requires a plasmalike medium. It is obvious from Eq. (6.126) that the total probability of the process diverges when $\omega_0^2\varepsilon_1/4m^2\varepsilon_2 = 1$. The latter is associated with the fact that these probabilities were determined for an infinitely long interaction time. In perturbation theory probabilities are proportional to the interaction time (under stationary conditions) and diverge as $t \to \infty$. Thus, this divergence is not associated with the process studied here, which is governed by the time dependence of the medium, and it can be eliminated by assuming $\omega_0^2\varepsilon_1/\varepsilon_2 < 4m^2$. Moreover, for laser frequencies and realistic values of the dielectric permittivities $\omega_0\sqrt{\varepsilon_1/\varepsilon_2} << 2\mathcal{E}$ and from Eq. (6.126) we obtain the following expression for the total number of e^-, e^+ pairs produced in the volume V due only to the medium nonstationary properties:

$$W = \frac{3e^2E_0^2V}{128\pi m}\left(1-\frac{\varepsilon_1}{\varepsilon_2}\right)^2. \tag{6.127}$$

In the general case, for arbitrary frequency of EM wave and temporal variation of the dielectric permittivity of the medium $\varepsilon_1/\varepsilon_2$ from Eq. (6.122) the following formula for the pair's probability distribution over the total energy $\mathcal{E}_t = \mathcal{E}_{e^-} + \mathcal{E}_{e^+}$ of the produced particles can be derived:

$$\frac{dW}{d\mathcal{E}_t} = \frac{e^2 E_0^2}{6\pi^2}\left(1 - \frac{\varepsilon_1}{\varepsilon_2}\right)^2 \left(1 - \frac{4m^2}{\mathcal{E}_t^2 - k_0^2}\right)^{1/2}$$

$$\times \frac{\mathcal{E}_t^2 \left(\mathcal{E}_t^2 + \omega_0^2\right)\left(\mathcal{E}_t^2 + 2m^2 - k_0^2\right)}{\left(\mathcal{E}_t^2 - \omega_0^2\right)\left(\mathcal{E}_t^2 - \omega_0^2 \frac{\varepsilon_1}{\varepsilon_2}\right)^2}. \tag{6.128}$$

Bibliography

G.S. Sahakyan, Zh. Éksp. Teor. Fiz. **38**, 843 (1960)
G.S. Sahakyan, Zh. Éksp. Teor. Fiz. **38**, 1593 (1960)
V.L. Ginzburg, Izv. VUZov, Radiofizika **16**, 512 (1973) [in Russian]
V.L. Ginzburg, V.N. Tsitovich, Zh. Éksp. Teor. Fiz. **65**, 132 (1973)
H.K. Avetissian, A.K. Avetissian, R.G. Petrossian, Zh. Éksp. Teor. Fiz. **75**, 382 (1978)
V.L. Ginzburg: Theoretical Physics and Astrophysics (Pergamon Press, Oxford 1979)
H.K. Avetissian, A.K. Avetissian, Kh.V. Sedrakian, Zh. Éksp. Teor. Fiz. **94**, 21 (1988)
H.K. Avetissian, A.K. Avetissian, Kh.V. Sedrakian, Zh. Éksp. Teor. Fiz. **100**, 82 (1991)

7 Induced Channeling Process in a Crystal

It is known that due to the relativistic motion of a charged particle in a crystal an exotic situation takes place when the effective potential of the crystal planes or axes becomes a potential well for the particle in the transversal direction with respect to its initial motion, and so-called channeling of the particle occurs accompanied by spontaneous channeling radiation.

The channeling radiation of ultrarelativistic electrons and positrons in a crystal is of great interest for two major reasons: the radiation is in the short-wave region (X-ray and γ-ray domains), and its spectral intensity considerably exceeds that of other types of radiation in this range of frequencies.

Induced channeling radiation in the presence of an external coherent radiation field becomes important as a potential source for short-wave coherent radiation, which may be considered as a version of a free electron laser.

As a periodic system with high coherency and owing to the similar periodic character of particle motion, the crystal channel may be compared with an undulator — it is a "micro-undulator" with the space period much smaller than that of an undulator.

On the other hand, the particle–external coherent EM wave interaction process in the channel of a crystal proceeds with the inverse stimulated effect reducing the particle acceleration and other classical and quantum coherent effects.

Hence, this chapter will consider the induced channeling process with regard to general aspects of coherent interaction of relativistic electrons and positrons with a plane transversal EM wave in a crystal.

7.1 Positron–Strong Wave Interaction at the Planar Channeling in a Crystal

If a charged particle with relativistic velocity enters a crystal at the angle with respect to a crystal plane or crystallographic axis smaller than some specified angle (Lindhard angle)

$$\theta_\alpha = \sqrt{\frac{2U_0}{\mathcal{E}}}, \tag{7.1}$$

then the effective electrostatic field of the crystal becomes a transversal potential well related to the particle motion and the latter moves in the crystal channel — the channeling of the particle occurs. Here U_0 is the depth of the potential well and $\mathcal{E}$ is the particle energy. In the most interesting case of ultrarelativistic energies for channeling phenomenon the transversal de Broglie wavelength of the particle

$$\lambda_D = \frac{\hbar c}{\sqrt{2U_0\mathcal{E}}} \tag{7.2}$$

is much smaller than the interplanar or interaxial distance d in a crystal (U_0 is of the order of the kinetic energy of the particle transversal motion) and consequently $d/\lambda_D >> 1$. On the other hand, the quantity d/λ_D with the coefficient coincides with the number of bound states l of the particle transversal motion in the crystal channel. Hence, in the most important region of energies $l >> 1$ and the particle motion at the channeling can be described classically.

We will study the induced interaction of a charged particle channeled in a crystal with the external coherent radiation field within the scope of the classical theory. In this section the case of the planar channeling will be considered.

As is known for a positron planar channeling the effective electrostatic potential of the crystal planes within the channel is well enough described by the parabolic law

$$U(x) = 4\frac{U_0}{d^2}x^2, \tag{7.3}$$

where d is the distance between the crystal planes, and the transversal coordinate x is evaluated from the median plane. The classical relativistic equation of motion for a positron in the fields (7.3) and an external plane monochromatic EM wave

$$\mathbf{E} = \mathbf{E}_0 \cos(\omega_0 t - \mathbf{k}_0\mathbf{r}); \qquad \mathbf{k}_0 = \nu\frac{n_0\omega_0}{c} \tag{7.4}$$

($n_0 = n_0(\omega_0)$ is the refractive index of the crystal on the wave frequency) is written as

$$\frac{d\mathbf{p}}{dt} = e\mathbf{E} + \frac{e}{c}[\mathbf{vH}] - \nabla U(x). \tag{7.5}$$

As for the permitted maximal values of the wave intensities in the dielectric media the characteristic interaction parameter $\xi_0 = eE_0/mc\omega_0 << 1$ (see Section 2.2), then for the ultrarelativistic energies of the channeled particles the interaction with the EM wave in a crystal with great accuracy can be described by the classical perturbation theory over the field (7.4). Consequently, in the zero order over the EM wave field from Eq. (7.5) we have the equations

$$\frac{dp_x}{dt} = -\frac{dU(x)}{dx}, \tag{7.6}$$

$$\frac{dp_y}{dt} = 0; \qquad \frac{dp_z}{dt} = 0. \tag{7.7}$$

Choosing the axis z along the initial motion of the particle from Eqs. (7.6) and (7.7) for the particle energy and momentum we obtain respectively

$$\mathcal{E} = \frac{mc^2}{\sqrt{1 - (\mathrm{v}_x^2 + \mathrm{v}_z^2)/c^2}} + U(x), \tag{7.8}$$

$$p_y = 0; \qquad p_z = \frac{m\mathrm{v}_z}{\sqrt{1 - (\mathrm{v}_x^2 + \mathrm{v}_z^2)/c^2}}. \tag{7.9}$$

For the transversal velocity of the particle from Eqs. (7.8) and (7.9) we have

$$\mathrm{v}_x^2 = c^2 \frac{[\mathcal{E} - U(x)]^2 - \mathcal{E}_{\shortparallel}^2}{[\mathcal{E} - U(x)]^2}, \tag{7.10}$$

where

$$\mathcal{E}_{\shortparallel} = c\sqrt{p_{\shortparallel}^2 + m^2c^2} \tag{7.11}$$

is the energy of the longitudinal motion. Equation (7.10) is the exact equation for the particle transversal motion. One can make some simplification of this equation taking into account the smallness of the potential energy related to the energy of the ultrarelativistic particle:

$$U_{\max}(x) \ll \mathcal{E}.$$

Representing the particle energy in the form

$$\mathcal{E} = \mathcal{E}_{\shortparallel} + \mathcal{E}_{\perp},$$

where $\mathcal{E}_{\perp}$ is the energy of the transversal motion, and taking into account that for the channeled particles

$$\mathcal{E}_{\perp} \lesssim U_{\max}(x) \ll \mathcal{E}_{\shortparallel},$$

then the equation for the particle transversal motion (7.10) with the accuracy of the small quantity $\mathcal{E}_{\perp}/\mathcal{E}_{\shortparallel} << 1$ will take the form

$$\mathrm{v}_x^2 = \frac{2c^2}{\mathcal{E}_{\shortparallel}} \left[\mathcal{E}_{\perp} - U(x)\right] . \qquad (7.12)$$

Formally Eq. (7.10) has a nonrelativistic character where instead of particle rest mass, the relativistic mass $m_{rel} \simeq \mathcal{E}_{\shortparallel}/mc^2$ stands.

The longitudinal velocity of the particle is determined from Eq. (7.9) and has the form

$$\mathrm{v}_z(t) \simeq c\left\{1 - \frac{1}{2}\left[\frac{\mathrm{v}_x^2}{c^2} + \left(\frac{mc^2}{\mathcal{E}_{\shortparallel}}\right)^2\right]\right\} . \qquad (7.13)$$

In the case of planar channeling of a positron when the effective electrostatic potential of the crystal may be approximated by Eq. (7.3), the integration of Eq. (7.12) gives the following law for the transversal motion:

$$x(t) = x_m \sin\left[\Omega(t - t_0) + \varphi\right] . \qquad (7.14)$$

Here

$$\Omega = \frac{2c}{d}\sqrt{\frac{2U_0}{\mathcal{E}_{\shortparallel}}} \qquad (7.15)$$

is the frequency of the positron transversal oscillations in the potential well of the crystal channel,

$$x_m = \frac{d}{2}\sqrt{\frac{\mathcal{E}_{\perp}}{U_0}} \qquad (7.16)$$

is the amplitude and φ is the phase of the transversal oscillations at the moment t_0 when the positron enters into the crystal. Corresponding to Eq. (7.14) the transversal velocity of the positron is

$$\mathrm{v}_x(t) = \mathrm{v}_{xm} \cos\left[\Omega(t - t_0) + \varphi\right] , \qquad (7.17)$$

where

$$\mathrm{v}_{xm} = \frac{d\Omega}{2}\sqrt{\frac{\mathcal{E}_{\perp}}{U_0}} \qquad (7.18)$$

is the maximal velocity of the transversal motion of the positron in the crystal channel. Then using Eq. (7.17) after the integration of Eq. (7.13) we will have

$$z(t) = \overline{\mathrm{v}}_z t - z_m \sin\left[2\Omega(t - t_0) + 2\varphi\right] + z_m \sin 2\varphi , \qquad (7.19)$$

where

$$\overline{\mathrm{v}}_z = c\left\{1 - \frac{1}{2}\left[\left(\frac{mc^2}{\mathcal{E}_{\parallel}}\right)^2 + \frac{\mathcal{E}_{\perp}}{\mathcal{E}_{\parallel}}\right]\right\} \tag{7.20}$$

is the mean longitudinal velocity of the positron, and the amplitude of the longitudinal oscillations z_m is

$$z_m = \frac{c\mathcal{E}_{\perp}}{4\Omega\mathcal{E}_{\parallel}}. \tag{7.21}$$

Now we can evaluate the induced channeling effect in the field of an external EM wave, by the classical perturbation theory in the first order over the field (7.4). The energy change of the channeled positron at the interaction with the plane transverse EM wave is given by

$$\Delta\mathcal{E} = e\int_{t_1}^{t_2} \mathbf{E}(t - \nu\mathbf{r}n_0/c)\mathbf{v}(t)dt, \tag{7.22}$$

where the law of motion $\mathbf{r} = \mathbf{r}(t)$ and velocity $\mathbf{v}(t)$ of the positron in the crystal channel are determined by Eqs. (7.14), (7.19) and Eqs. (7.13), (7.17), respectively. The induced interaction time $\Delta t = t_2 - t_1$ actually will be determined by the length of the channel ($t_1$and t_2 are correspondingly the moments of the wave entrance in the crystal and exit from the channel).

For the concreteness and evaluation of the energy change (7.22) we introduce a new Cartesian coordinate system x', y', z' and assume that a quasi-monochromatic EM wave linearly polarized along the axis x' propagates along the axis z', at a small angle with respect to a crystal plane (see Eq. (7.1)). The coordinate system x', y', z' is related to the system x, y, z via Eulerian angles α, β, γ as follows:

$$\begin{pmatrix} x' \\ y' \\ z' \end{pmatrix} = \begin{pmatrix} \cos\gamma & \sin\gamma & 0 \\ \sin\gamma & \cos\gamma & 0 \\ 0 & 0 & 1 \end{pmatrix} \begin{pmatrix} \cos\beta & 0 & -\sin\beta \\ 0 & 1 & 0 \\ \sin\beta & 0 & \cos\beta \end{pmatrix}$$

$$\times \begin{pmatrix} 1 & 0 & 0 \\ 0 & \cos\alpha & \sin\alpha \\ 0 & -\sin\alpha & \cos\alpha \end{pmatrix} \begin{pmatrix} x \\ y \\ z \end{pmatrix}. \tag{7.23}$$

At the motion of the positron in the crystal channel by the trajectory (7.14), (7.19), the wave phase in Eq. (7.22) corresponding to induced interaction is

$$\phi = \omega_0 t - \mathbf{k}_0\mathbf{r} = \omega t - \varkappa_1 \sin\left[\Omega\left(t - t_0\right) + \varphi\right]$$

$$+\varkappa_2 \sin 2\left[\Omega\left(t-t_0\right)+\varphi\right]+\psi, \tag{7.24}$$

where

$$\omega=\omega_0\left(1-\frac{n_0\overline{\mathrm{v}}_z}{c}\cos\alpha\cos\beta\right) \tag{7.25}$$

is the Doppler-shifted wave frequency, and the parameters $\varkappa_1$, $\varkappa_2$, ψ are

$$\varkappa_1=n_0\omega_0\frac{x_m}{c}\sin\beta; \qquad \varkappa_2=n_0\omega_0\frac{z_m}{c}\cos\alpha\cos\beta,$$

$$\psi=-n_0\frac{\omega_0}{c}\cos\alpha\cos\beta\left(z_m\sin 2\varphi-\overline{\mathrm{v}}_z t_0\right). \tag{7.26}$$

Substituting Eq. (7.24) as well as Eqs. (7.13) and (7.17) in Eq. (7.22) for the energy change of the positron due to the induced channeling effect, in the first order by the wave field we will have

$$\Delta\mathcal{E}=\sum_{s=-\infty}^{\infty}\frac{e}{\omega-s\Omega}\left\{E_{0x}\mathrm{v}_{xm}A_1\left(s,\varkappa_1,\varkappa_2\right)+E_{0z}\left(\overline{\mathrm{v}}_z+\mathrm{v}_{zm}\right)A_0\left(s,\varkappa_1,\varkappa_2\right)\right.$$

$$\left.-2E_{0z}\mathrm{v}_{zm}A_2\left(s,\varkappa_1,\varkappa_2\right)\right\}\left\{\sin\left[\left(\omega-s\Omega\right)t_2+s\Omega t_0-s\varphi+\psi\right]\right.$$

$$\left.-\sin\left[\left(\omega-s\Omega\right)t_1+s\Omega t_0-s\varphi+\psi\right]\right\}, \tag{7.27}$$

where

$$A_n\left(s,\alpha,\beta\right)=\frac{1}{2\pi}\int_{-\pi}^{\pi}\cos^n\varphi' e^{i\left(\alpha\sin\varphi'-\beta\sin 2\varphi'-s\varphi'\right)}d\varphi'$$

is the generalized Bessel function with the definitions

$$A_0\left(s,\alpha,\beta\right)=\sum_{k=-\infty}^{\infty}J_{s+2k}\left(\alpha\right)J_k\left(\beta\right),$$

$$A_1\left(s,\alpha,\beta\right)=\frac{1}{2}\left[A_0\left(s-1,\alpha,\beta\right)+A_0\left(s+1,\alpha,\beta\right)\right],$$

$$A_2\left(s,\alpha,\beta\right)=\frac{1}{4}\left[A_0\left(s-2,\alpha,\beta\right)+2A_0\left(s,\alpha,\beta\right)+A_0\left(s+2,\alpha,\beta\right)\right],$$

and

$$\mathrm{v}_{zm} = \frac{c\mathcal{E}_{\perp}}{2\mathcal{E}_{\parallel}} \tag{7.28}$$

is the amplitude of the positron longitudinal velocity oscillations.

Equation (7.27) shows that the energy change of the positron after the interaction differs from zero (will have nonoscillating character in the time) if the condition

$$\omega_0 \left(1 - n_0 \frac{\overline{\mathrm{v}}_z}{c} \cos\alpha \cos\beta\right) = s\Omega; \qquad s = 0, \pm1, \pm2, ... \tag{7.29}$$

is satisfied for a specified s. The latter is the condition of the resonance between the transversal oscillations of the positron in the potential well of the crystal channel and EM wave. Only at the fulfillment of this condition does the coherent energy exchange of the channeled positron with the monochromatic wave become real. Then for the energy change of the positron after the interaction we have

$$\Delta\mathcal{E} = eE_0\Delta t \left\{\mathrm{v}_{xm} \cos\beta \cos\gamma A_1 (s, \varkappa_1, \varkappa_2) + (\sin\alpha \sin\gamma - \cos\alpha \sin\beta \cos\gamma)\right.$$

$$\left. \times \left[(\overline{\mathrm{v}}_z + \mathrm{v}_{zm}) A_0 (s, \varkappa_1, \varkappa_2) - 2\mathrm{v}_{zm} A_2 (s, \varkappa_1, \varkappa_2)\right]\right\}$$

$$\times \cos\left[s\Omega t_0 - s\varphi + n_0 \frac{\omega_0}{c} \cos\alpha \cos\beta \left(\overline{\mathrm{v}}_z t_0 - z_m \sin 2\varphi\right)\right]. \tag{7.30}$$

Expressing the functions $A_{0,1,2}(s, \varkappa_1, \varkappa_2)$ via the ordinary Bessel functions, Eq. (7.30) can be presented in the form

$$\Delta\mathcal{E} = eE_0\Delta t \sum_{k=-\infty}^{\infty} \left\{\frac{1}{2}\mathrm{v}_{xm} \cos\beta \cos\gamma \left[J_{s-1+2k}(\varkappa_1) + J_{s+1+2k}(\varkappa_1)\right]\right.$$

$$+ \overline{\mathrm{v}}_z (\sin\alpha \sin\gamma - \cos\alpha \sin\beta \cos\gamma) J_{s+2k}(\varkappa_1)$$

$$\left. - \mathrm{v}_{zm} (\sin\alpha \sin\gamma - \cos\alpha \sin\beta \cos\gamma) \left[J_{s-2+2k}(\varkappa_1) + J_{s+2+2k}(\varkappa_1)\right]\right\} J_k(\varkappa_2)$$

$$\times \cos\left[s\Omega t_0 - s\varphi + n_0 \frac{\omega_0}{c} \left(\overline{\mathrm{v}}_z t_0 - z_m \sin 2\varphi\right) \cos\alpha \cos\beta\right]. \tag{7.31}$$

For the X-ray and γ-ray frequencies when $n_0(\omega_0) \lesssim 1$ the resonance condition (7.29) corresponds to the normal Doppler effect at which the energy absorption from the EM wave is accompanied by enhancement of the transversal oscillations of the positron (in these cases $s > 0$ in Eq. (7.31)). For the optical frequencies when $n_0(\omega_0) > 1$ the anomalous Doppler effect is possible as well:

$$1 - n_0 \frac{\overline{\mathrm{v}}_z}{c} \cos\alpha \cos\beta < 0, \tag{7.32}$$

which corresponds to enhancement of transversal oscillations of the positron at the induced radiation (in Eq. (7.31) in this case $s < 0$). Under the condition

$$1 - n_0 \frac{\overline{\mathrm{v}}_z}{c} \cos\alpha \cos\beta = 0, \tag{7.33}$$

that is, the Cherenkov condition in the crystal channel corresponding to $s = 0$, Eq. (7.29) expresses the real energy exchange at the positron–wave induced Cherenkov interaction.

Equation (7.31) for the general geometry of the positron planar channeling at the arbitrary propagation and polarization directions of the wave is very bulky. It can be simplified in the case of a particular geometry of the induced interaction — if the EM wave propagates along the direction of the positron motion in the channel (axis z) with the electric field directed along the axis x — and the positron energy $\mathcal{E}_{\parallel} \lesssim m^2c^4/\mathcal{E}_{\perp}$. Then, for the number of harmonic s we have: $s = 0, \pm 1$ (for the coherent accumulation of energy exchange), and for the frequencies satisfying the resonance condition (7.29) one can suppose $n_0(\omega_0) \simeq 1$. The latter excepts the possibility of the induced Cherenkov effect ($s = 0$) and the anomalous Doppler effect ($s = -1$) as well. Thus, for the induced energy exchange we have a simple formula

$$\Delta\mathcal{E} = \frac{eE_0 \mathrm{v}_{xm}}{2} \Delta t \cos\left[\left(\Omega + \omega_0 \frac{\overline{\mathrm{v}}_z}{c}\right) t_0 - \varphi\right]. \tag{7.34}$$

As is seen from Eqs. (7.31) and (7.34) depending on the initial conditions-a moment t_0 when the positron enters into the crystal and a phase φ of the transversal oscillations — either the direct or the inverse induced channeling effect occurs, i.e., positron deceleration or acceleration, respectively. Hence, at the interaction of the channeled positron beam with the monochromatic EM wave the diverse particles entering into a crystal at the different moments and in the different oscillation phases will acquire or lose different energies. As a result, the modulation of the particles' velocities will take place leading to beam bunching if the longitudinal size of the latter $l_z > \pi\overline{\mathrm{v}}_z/\omega_0$.

7.2 Induced Interaction of Electrons with Strong EM Wave at the Axial Channeling

As is known, for an electron axial channeling the effective electrostatic potential of the atomic chain along the crystal axis is well enough described by the two-dimensional Coulomb potential

$$U(\rho) = -\frac{\alpha_c}{\rho}, \tag{7.35}$$

where α_c is a constant depending on the type of crystal and the particular geometry, and ρ is the distance from the crystal axis. The transversal motion of the electron in the field (7.35) with a nonzero momentum occurs by the Keplerian elliptic trajectory. If one directs the coordinate axes OX and OY correspondingly along the major and minor semiaxes of the ellipse and the axis OZ along the crystal axis, and if at the moment $t = t_0$ the electron is situated in the perihelion of the orbit of the transversal motion with the coordinate $z = z_0$, then the electron trajectory may be presented in the known parametric form

$$x = a(\cos\zeta - \epsilon); \qquad y = (-1)^{s'} b\sin\zeta,$$

$$z = \overline{\mathrm{v}}_z (t - t_0) - a^2 \frac{\epsilon\Omega}{c} \sin\zeta + z_0, \tag{7.36}$$

$$t = \frac{\zeta - \epsilon\sin\zeta}{\Omega} + t_0,$$

where for a full rotation of the electron by the elliptic orbit the parameter ζ varies from zero to 2π. Here the parameters

$$a = \frac{\alpha_c}{2|\mathcal{E}_\perp|}; \qquad b = a\sqrt{1-\epsilon^2} \tag{7.37}$$

are the major and minor semiaxes of the ellipse,

$$\epsilon = \sqrt{1 - \frac{2|\mathcal{E}_\perp| M_z^2 c^2}{\mathcal{E}_{\shortparallel}\alpha_c^2}} \tag{7.38}$$

is the eccentricity (M_z is the z-component of the orbital moment),

$$\Omega = c\frac{(2|\mathcal{E}_\perp|)^{\frac{3}{2}}}{\alpha_c\sqrt{\mathcal{E}_{\shortparallel}}} \tag{7.39}$$

is the rotation frequency, and

$$\overline{\mathrm{v}}_z = c\left(1 - \frac{m^2c^4}{2\mathcal{E}_{\parallel}^2}\right) - \frac{c\left|\mathcal{E}_{\perp}\right|}{\mathcal{E}_{\parallel}} \tag{7.40}$$

is the mean longitudinal velocity of the electron. The parameter s' in Eq. (7.36) determines the right-hand or left-hand rotation of the electron by the elliptic orbit:

$$s' = \begin{cases} 0, & \frac{M_z}{|M_z|} > 0, \\ 1, & \frac{M_z}{|M_z|} < 0. \end{cases} \tag{7.41}$$

As the electron trajectory at the axial channeling is of helical type from the point of view of the symmetry in this issue we will suppose that an EM wave has a circular polarization:

$$E_{x'} = E_0 \cos\left(\omega_0 t - \mathbf{k}_0 \mathbf{r}\right); \quad E_{y'} = E_0 \left(-1\right)^{s''} \sin\left(\omega_0 t - \mathbf{k}_0 \mathbf{r}\right) \tag{7.42}$$

correspondingly with the left-hand and right-hand rotations:

$$s'' = \begin{cases} 0, & \text{left-hand,} \\ 1, & \text{right-hand.} \end{cases}$$

The coordinate system $x'y'z'$ relates to the xyz one in accordance with Eq. (7.23) and in the case of the wave circular polarization one can assume that the Eulerian angle $\gamma = 0$.

We will evaluate the induced effect at the axial channeling by Eq. (7.22) again in the first order by the EM wave field. As far as the particle velocity and law of motion in the channel in this case are determined in parametric form (Eq. (7.36)) it is necessary to pass in Eq. (7.22) from the variable t to ζ. Then the induced energy exchange between the channeled electron and EM wave will be written in the form

$$\Delta\mathcal{E} = e \int\limits_{\zeta(t_1)}^{\zeta(t_2)} \mathbf{E}(\phi\left(\zeta\right)) \frac{d\mathbf{r}(\zeta)}{d\zeta} d\zeta, \tag{7.43}$$

where $\Delta t = t_2 - t_1$ is the duration of electron–wave coherent interaction at the axial channeling. In the first-order approximation for the wave phase in the integral (7.43) with the help of Eqs. (7.36)–(7.41) we have

$$\phi\left(\zeta\right) = \omega_0 t - \mathbf{k}_0 \mathbf{r} = \frac{\omega_0 - k_{0z}\overline{\mathrm{v}}_z}{\Omega}\zeta - \varkappa_1 \sin\zeta - \varkappa_2 \cos\zeta + \psi, \tag{7.44}$$

where

$$\mathbf{k}_0 = n_0 \frac{\omega_0}{c} \left(\sin\beta, -\sin\alpha\cos\beta, \cos\alpha\cos\beta \right),$$

and the parameters $\varkappa_1$, $\varkappa_2$, ψ in this case are

$$\varkappa_1 = \frac{\epsilon}{\Omega} \left(\omega_0 - k_{0z}\overline{\mathrm{v}}_z \right) + (-1)^{s'} k_{0y} b - k_{0z} a^2 \epsilon \frac{\Omega}{c}; \qquad \varkappa_2 = a k_{0x},$$

$$\psi = \omega_0 t_0 + k_{0x} a\epsilon - k_{0z} z_0.$$

Performing integration in Eq. (7.43) with the help of Eqs. (7.36) and (7.44) we obtain the following ultimate equation for the coherent energy exchange between the electron and external strong EM wave at the axial channeling:

$$\Delta\mathcal{E} = -eE_0\Omega\Delta t \Bigg\{ J_s(\varkappa) \left[(-1)^{s''} \sin\alpha\sin\varphi - \cos\alpha\sin\beta\cos\varphi \right] \frac{\overline{\mathrm{v}}_z}{\Omega}$$

$$+ \frac{s}{\varkappa} J_s(\varkappa) \Bigg[a\cos\beta\sin\varphi_1\cos\varphi + (-1)^{s'} b\sin\alpha\sin\beta\cos\varphi\cos\varphi_1$$

$$+ (-1)^{s'+s''} b\cos\alpha\sin\varphi\cos\varphi_1 + \left(1 + \frac{2c\left|\mathcal{E}_\perp\right|}{\overline{\mathrm{v}}_z\mathcal{E}_\parallel} \right) \frac{\epsilon\overline{\mathrm{v}}_z}{\Omega}$$

$$\left(\cos\alpha\sin\beta\cos\varphi\cos\varphi_1 - (-1)^{s''} \sin\alpha\sin\varphi\cos\varphi_1 \right) \Bigg]$$

$$+ J_s'(\varkappa) \Bigg[a\cos\beta\sin\varphi\cos\varphi_1 + (-1)^{s'} b\sin\alpha\sin\beta\sin\varphi\sin\varphi_1$$

$$+ (-1)^{s'+s''} b\cos\alpha\cos\varphi\sin\varphi_1 - \left(1 + \frac{2c\left|\mathcal{E}_\perp\right|}{\overline{\mathrm{v}}_z\mathcal{E}_\parallel} \right) \frac{\epsilon\overline{\mathrm{v}}_z}{\Omega}$$

$$\times \left(\cos\alpha\sin\beta\sin\varphi\sin\varphi_1 + (-1)^{s''} \sin\alpha\sin\varphi_1\cos\varphi \right) \Bigg] \Bigg\}, \qquad (7.45)$$

where the parameters $\varkappa$, φ_1, and φ are

$$\varkappa = \sqrt{\varkappa_1^2 + \varkappa_2^2},$$

$$\varphi_1 = \frac{\varkappa_1}{|\varkappa_1|} \arcsin \frac{\varkappa_2}{\varkappa}, \tag{7.46}$$

$$\varphi = \omega_0 t_0 - n_0 \frac{\omega_0}{c} z_0 \cos\alpha \cos\beta + a\epsilon n_0 \frac{\omega_0}{c} \sin\beta - s\varphi_1 .$$

The physical analysis of Eq. (7.45) is the same as was made for the positron planar channeling. So, we will not repeat the analogous analysis, noting only that the condition of resonance at the axial channeling for coherent energy exchange (7.45) is given by Eq. (7.29), where the frequency of transversal oscillations Ω of the electron is determined by Eq. (7.39).

Equation (7.46) corresponding to general geometry of the electron axial channeling in the arbitrary propagation and polarization directions of the wave is very bulky. It is rather simplified if the wave propagates along the direction of the electron motion in the channel (axis z) with the components of the electric field strength directed along the axes x and y, as well as the electron energy should not exceed the value $m^2c^4/\mathcal{E}_\perp$. For the induced energy exchange we have the following ultimate equation:

$$\Delta\mathcal{E} = -eE_0\Omega\Delta t \left\{ aJ_s'(\varkappa) + b(-1)^{s'+s''} \frac{s}{\varkappa} J_s(\varkappa) \right\}$$

$$\times \sin\left(\omega_0 t_0 - n_0 \frac{\omega_0}{c} z_0\right) . \tag{7.47}$$

The existence of diverse harmonics in Eq. (7.47) is related to the anharmonic character of the electron transversal oscillations in the field (7.35) (in contrast to Eq. (7.34) for the planar channeling, at which the positron is a harmonic oscillator in the channel).

In addition, note that Eqs. (7.45) and (7.47) due to their coherent dependence on the interaction phase lead to the electron beam classical modulation and bunching after the interaction with the stimulating wave at the axial channeling analogously to the positron beam bunching at the planar channeling.

7.3 Quantum Description of the Induced Planar Channeling Effect

Consider the interaction of the particles channeled in a crystal and a plane monochromatic EM wave in the scope of the quantum theory. First we will study the case of a weak wave when the one-photon absorption and emission processes dominate and the induced channeling effect may be described within the quantum perturbation theory by the particle wave function in the linear over the field approximation with respect to the initial state in the potential field of the crystal channel. It means that the latter should be described exactly.

We will start from the Dirac equation which in the case of the planar channeling of a positron in the field of an external EM wave is written as

$$i\hbar\frac{\partial\Psi}{\partial t} = \left(\widehat{H}_0 + \widehat{V}\right)\Psi, \tag{7.48}$$

$$\widehat{H}_0 = c\widehat{\alpha}\widehat{\mathbf{p}} + \widehat{\beta}mc^2 + U(x); \quad \widehat{V} = -e\widehat{\alpha}\mathbf{A}, \tag{7.49}$$

where $\widehat{\alpha}$, $\widehat{\beta}$ are the Dirac matrices in the standard representation (3.2). According to perturbation theory we seek the solution of Eq. (7.49) in the form

$$\Psi = \Psi_0 + \Psi_1 + \cdots; \quad |\Psi_1| \ll |\Psi_0|, ...,$$

where Ψ_0 satisfies the following equation for the positron in the electrostatic field of the crystal channel:

$$i\hbar\frac{\partial\Psi_0}{\partial t} = \left[c\widehat{\alpha}\widehat{\mathbf{p}} + \widehat{\beta}mc^2 + U(x)\right]\Psi_0 \tag{7.50}$$

with the effective potential $U(x)$ (7.3). The particular solution of Eq. (7.50) may be presented in the form

$$\Psi_0(\mathbf{r}, t) = b\begin{pmatrix}\varphi \\ \chi\end{pmatrix} e^{-\frac{i}{\hbar}\mathcal{E}t}, \tag{7.51}$$

where φ and χ are spinor functions, $\mathcal{E}$ is the total energy of the positron in the potential field of the channel, and b is the normalization coefficient. From Eq. (7.50) for the spinor functions φ and χ we obtain the following set of equations:

$$\mathcal{E}\varphi = c(\sigma\widehat{\mathbf{p}})\chi + mc^2\varphi + U(x)\varphi,$$

$$\mathcal{E}\chi = c(\sigma\widehat{\mathbf{p}})\varphi - mc^2\chi + U(x)\chi, \tag{7.52}$$

where $\sigma = (\sigma_x, \sigma_y, \sigma_z)$ are the Pauli matrices (1.79). Eliminating χ from the first equation (7.52):

$$\chi = \frac{c\sigma\widehat{\mathbf{p}}}{\mathcal{E} + mc^2 - U(x)}\varphi, \tag{7.53}$$

for the spinor function φ we obtain a differential equation of the second order:

$$\Delta\varphi + \frac{1}{\hbar^2 c^2}\left(\left[\mathcal{E} - U(x)\right]^2 - m^2 c^4\right)\varphi + \frac{\sigma\nabla U(x)}{\mathcal{E} + mc^2 - U(x)}(\sigma\nabla)\varphi = 0. \quad (7.54)$$

The solution of Eq. (7.54) is sought in the form

$$\varphi = w\psi(x)\, e^{\frac{i}{\hbar}\mathbf{p}_{\shortparallel}\mathbf{r}} = \begin{pmatrix} w_1 \\ w_2 \end{pmatrix} \psi(x)\, e^{\frac{i}{\hbar}\mathbf{p}_{\shortparallel}\mathbf{r}}, \quad (7.55)$$

where $\psi(x)$ is the positron wave function corresponding to the transversal motion in the potential well of the channel, and w is a constant spinor which should be defined from the wave function normalization condition

$$w^{\dagger} w = w_1^* w_1 + w_2^* w_2 = 1.$$

Neglecting the small terms of the order $U_{\max}/\mathcal{E} << 1$ (or $\mathcal{E}_{\perp}/\mathcal{E} << 1$) in Eq. (7.54), for the positron wave function describing the transversal motion in the crystal channel we obtain a one-dimensional Schrödinger equation in the potential field $U(x)$

$$\frac{d^2\psi(x)}{dx^2} + \frac{2m_{eff}}{\hbar^2}\left[\mathcal{E}_{\perp} - U(x)\right]\psi(x) = 0, \quad (7.56)$$

with the effective mass m_{eff} corresponding to the energy $\mathcal{E}_{\shortparallel}$ of relativistic longitudinal motion

$$m_{eff} = \frac{\mathcal{E}_{\shortparallel}}{c^2} = \sqrt{\frac{\mathbf{p}_{\shortparallel}^2}{c^2} + m^2}. \quad (7.57)$$

In Eq. (7.56) $\mathcal{E}_{\perp} = \mathcal{E} - \mathcal{E}_{\shortparallel}$ is the energy of transversal motion, which parametrically depends on the energy of longitudinal motion $\mathcal{E}_{\perp} = \mathcal{E}_{\perp}(\mathcal{E}_{\shortparallel})$. In the case of planar channeling of positrons with the harmonic potential (7.3), Eq. (7.56) describes the quantum harmonic oscillator the solution of which is given by

$$\psi_n(x) = \left(\frac{\mathcal{E}_{\shortparallel}\Omega}{\pi\hbar c^2}\right)^{\frac{1}{4}} \frac{1}{\sqrt{2^n n!}} e^{-\frac{\mathcal{E}_{\shortparallel}\Omega}{2\hbar c^2}x^2} \mathcal{H}_n\left(\sqrt{\frac{\mathcal{E}_{\shortparallel}\Omega}{\hbar c^2}}x\right), \quad (7.58)$$

where

$$\mathcal{H}_n(\xi) = (-1)^n e^{\xi^2}\frac{d^n e^{-\xi^2}}{d\xi^n} \quad (7.59)$$

are the Hermit polynomials, and the quantization law for the positron transversal energy is

$$\mathcal{E}_{\perp}\left(n, \mathcal{E}_{\shortparallel}\right) = \left(n + \frac{1}{2}\right)\hbar\Omega, \tag{7.60}$$

where Ω is given by Eq. (7.15).

Finally, with the help of Eqs. (7.55) and (7.51) the solution of Eq. (7.48) for the positron wave function with the longitudinal momentum $\mathbf{p}_{\shortparallel}$ in the n-th bound state of the transversal motion and spin state σ can be written as

$$\Psi_{\mathbf{p}_{\shortparallel},n,\sigma}\left(\mathbf{r},t\right) = \sqrt{\frac{\mathcal{E}_{\shortparallel} + mc^2}{2\mathcal{E}_{\shortparallel}}}\begin{pmatrix} \varphi_\sigma \\ \frac{c\sigma\widehat{\mathbf{p}}}{\mathcal{E}+mc^2-U(x)}\varphi_\sigma \end{pmatrix}\psi_n\left(x\right)e^{\frac{i}{\hbar}\left(\mathbf{p}_{\shortparallel}\mathbf{r}-\mathcal{E}t\right)}, \tag{7.61}$$

where φ_σ are the spinors (3.11), and the total energy $\mathcal{E}$ is given by the relation

$$\mathcal{E}\left(\mathbf{p}_{\shortparallel}, n\right) = \sqrt{c^2\mathbf{p}_{\shortparallel}^2 + m^2c^4} + \left(n + \frac{1}{2}\right)\hbar\Omega. \tag{7.62}$$

Now we can evaluate the wave function of the channeled positron at the induced interaction with an external EM wave in the first approximation of perturbation theory (Ψ_1) on the basis of Eqs. (7.61), (7.62) for unperturbed (by the wave) state in the crystal channel (Ψ_0).

Before the interaction with a plane monochromatic EM wave assume that a positron with an initial longitudinal momentum $\mathbf{p}_{\shortparallel} = (0, p_y, p_z)$ is situated in the bound state of the crystal channel characterized by the quantum numbers n, σ, that is, the initial state is described by the wave function

$$\Psi_0\left(\mathbf{r},t\right) = \Psi_{\mathbf{p}_{\shortparallel},n,\sigma}\left(\mathbf{r},t\right). \tag{7.63}$$

The positron wave function Ψ_1 perturbed by the EM wave will be expanded in terms of the full basis of the eigenstates (7.63) with Eqs. (7.61), (7.62):

$$\Psi_1\left(\mathbf{r},t\right) = \sum_{\mathbf{p}'_{\shortparallel},n',\sigma'} a_{\mathbf{p}'_{\shortparallel},n',\sigma'}\left(t\right)\Psi_{\mathbf{p}'_{\shortparallel},n',\sigma'}\left(\mathbf{r},t\right), \tag{7.64}$$

where $a_{\mathbf{p}'_{\shortparallel},n',\sigma'}(t)$ are unknown functions, and the summation is made over all possible states of the positron transversal motion in the potential well corresponding to planar channeling. Substituting the wave function $\Psi = \Psi_0 + \Psi_1$ with Eqs. (7.63) and (7.64) in the Dirac equation (7.48) and neglecting the small terms of the second order by the quantity $\sim e\widehat{\alpha}\mathbf{A}\Psi_1$ (in accordance with the perturbation theory) we obtain the following differential equation for the expansion coefficients $a_{\mathbf{p}'_{\shortparallel},n',\sigma'}$:

$$\sum_{\mathbf{p}'_{\shortparallel},n',\sigma''} \hbar \frac{\partial a_{\mathbf{p}'_{\shortparallel},n',\sigma'}}{\partial t} \Psi_{\mathbf{p}'_{\shortparallel},n',\sigma'}(\mathbf{r},t) = ie\widehat{\alpha}\mathbf{A}(\mathbf{r},t)\,\Psi_{\mathbf{p}_{\shortparallel},n,\sigma}(\mathbf{r},t)\,. \tag{7.65}$$

Multiplying Eq. (7.65) on the left-hand side by $\Psi^{\dagger}_{\mathbf{p}'_{\shortparallel},n',\sigma'}(\mathbf{r},t)$ and integrating over $d\mathbf{r}dt$ one can present the solution of Eq. (7.65) in the form

$$a_{\mathbf{p}'_{\shortparallel},n',\sigma'} = i\frac{eA_0}{4}\sqrt{\frac{2\hbar\Omega}{\mathcal{E}_{\shortparallel}}}\delta_{\sigma'\sigma}\left[\sqrt{n}\delta_{n'+1,n} - \sqrt{n+1}\delta_{n'-1,n}\right]$$

$$\times\left[\delta_{\mathbf{p}'_{\shortparallel},\mathbf{p}_{\shortparallel}+\hbar\mathbf{k}_0}\frac{e^{-\frac{i}{\hbar}\left(\mathcal{E}(\mathbf{p}_{\shortparallel},n)-\mathcal{E}\left(\mathbf{p}'_{\shortparallel},n'\right)+\hbar\omega_0\right)t}}{\mathcal{E}(\mathbf{p}_{\shortparallel},n)-\mathcal{E}\left(\mathbf{p}'_{\shortparallel},n'\right)+\hbar\omega_0}\right.$$

$$\left.+\delta_{\mathbf{p}'_{\shortparallel},\mathbf{p}_{\shortparallel}-\hbar\mathbf{k}_0}\frac{e^{-\frac{i}{\hbar}\left(\mathcal{E}(\mathbf{p}_{\shortparallel},n)-\mathcal{E}\left(\mathbf{p}'_{\shortparallel},n'\right)-\hbar\omega_0\right)t}}{\mathcal{E}(\mathbf{p}_{\shortparallel},n)-\mathcal{E}\left(\mathbf{p}'_{\shortparallel},n'\right)-\hbar\omega_0}\right]. \tag{7.66}$$

In Eq. (7.66) it was assumed that the wave propagates in the plane yz with the vector potential directed along the axis x:

$$A_x = A_0\cos\left(\omega_0 t - \mathbf{k}_0\mathbf{r}\right),$$

and was taken into account that for actual cases $\hbar\omega_0/\mathcal{E}_{\shortparallel} << 1$ and the positron energies $\mathcal{E} < m^2c^4/U_0$ as well.

As is seen from Eq. (7.66) only the following expansion coefficients differ from zero

$$a_{\mathbf{p}_{\shortparallel}+\hbar\mathbf{k}_0,n-1,\sigma}(t) = \mathcal{D}\sqrt{n}\frac{e^{-i(\omega+\Omega)t}}{\omega+\Omega},$$

$$a_{\mathbf{p}_{\shortparallel}+\hbar\mathbf{k}_0,n+1,\sigma}(t) = -\mathcal{D}\sqrt{n+1}\frac{e^{-i(\omega-\Omega)t}}{\omega-\Omega},$$

$$a_{\mathbf{p}_{\shortparallel}-\hbar\mathbf{k}_0,n-1,\sigma}(t) = -\mathcal{D}\sqrt{n}\frac{e^{i(\omega-\Omega)t}}{\omega-\Omega}, \tag{7.67}$$

$$a_{\mathbf{p}_{\shortparallel}-\hbar\mathbf{k}_0,n+1,\sigma}(t) = \mathcal{D}\sqrt{n+1}\frac{e^{i(\omega+\Omega)t}}{\omega+\Omega},$$

where the quantity $\mathcal{D}$ is

$$\mathcal{D} = i\frac{eA_0}{2\hbar}\sqrt{\frac{\hbar\Omega}{2\mathcal{E}_{\shortparallel}}}, \tag{7.68}$$

and the Doppler-shifted wave frequency ω is

$$\omega = \omega_0 - \mathbf{k}_0 \mathbf{v}_{\shortparallel}; \quad \mathbf{v}_{\shortparallel} = \frac{c^2 \mathbf{p}_{\shortparallel}}{\mathcal{E}_{\shortparallel}}. \tag{7.69}$$

The expressions in Eq. (7.67) show that the second and third coefficients have a resonance character due to which the induced channeling effect occurs — resonance absorption of the wave photons by a channeled particle and coherent emission of the photons into the wave. Hence, neglecting in Eq. (7.64) the small terms with nonresonant expansion coefficients (first and fourth ones in Eq. (7.67)) of the perturbed wave function for the probability density of the positron at the planar channeling we will have

$$W(\mathbf{r}, t) = \varphi_n^2(x) + \frac{eA_0}{\hbar(\omega - \Omega)} \sqrt{\frac{\hbar\Omega}{2\mathcal{E}}} \varphi_n(x)$$

$$\times \left[\sqrt{n+1}\varphi_{n+1}(x) - \sqrt{n}\varphi_{n-1}(x)\right] \sin(\mathbf{k}_0 \mathbf{r} - \omega_0 t). \tag{7.70}$$

In the case of the exact resonance ($\omega = \Omega$) Eq. (7.70) is not applicable. In this case the solution of Eq. (7.65) for the probability density of the positron gives

$$W(\mathbf{r}, t) = \varphi_n^2(x) + \frac{eA_0}{\hbar} \sqrt{\frac{\hbar\Omega}{2\mathcal{E}}} \varphi_n(x)$$

$$\times \left[\sqrt{n}\varphi_{n-1}(x) - \sqrt{n+1}\varphi_{n+1}(x)\right] \Delta t \cos(\mathbf{k}_0 \mathbf{r} - \omega_0 t), \tag{7.71}$$

where Δt is the period of channeled positron interaction with EM wave.

As is seen from the Eqs. (7.70) and (7.71) the probability density of the positron due to the induced channeling effect is modulated at the stimulating wave frequency (in the one-photon approximation; in the next orders of perturbation theory we will obtain modulation at the harmonics of the wave fundamental frequency).

The condition of validity of the perturbation theory at which the obtained formulas are applicable we can obtain from Eq. (7.71):

$$\frac{eE_0 \mathrm{v}_{xm} \Delta t}{\hbar\omega_0} \ll 1, \tag{7.72}$$

where v_{xm} is the maximal velocity of transversal motion of the positron in the channel of the crystal (see Eq. (7.18)):

$$\mathrm{v}_{xm} = c\sqrt{\frac{2n\hbar\Omega}{\mathcal{E}_{\shortparallel}}} = c\sqrt{\frac{2\mathcal{E}_{\perp}}{\mathcal{E}_{\shortparallel}}}. \tag{7.73}$$

7.4 Quantum Description of the Induced Axial Channeling Effect

At the axial channeling the state of the electron is characterized by the projection of the momentum p_z on the crystal axis z, and due to the axial symmetry of the effective electrostatic potential of an atomic chain within the channel the projection of the orbital moment of the electron on the same axis is conserved.

The Dirac equation for an electron at the axial channeling is written in the form (7.48) with the Hamiltonian

$$\widehat{H}_0 = c\widehat{\alpha}\widehat{\mathbf{p}} + \widehat{\beta}mc^2 + U(\rho), \tag{7.74}$$

where $U(\rho)$ is given by Eq. (7.35). The interaction of the electron with the external EM wave will again be taken into account by perturbation theory (in the one-photon approximation):

$$\Psi = \Psi_0 + \Psi_1; \quad |\Psi_1| \ll |\Psi_0|,$$

where Ψ_0 is the electron wave function in a crystal at the axial channeling, which satisfies the equation

$$i\hbar\frac{\partial\Psi_0}{\partial t} = \left[c\widehat{\alpha}\widehat{\mathbf{p}} + \widehat{\beta}mc^2 + U(\rho)\right]\Psi_0. \tag{7.75}$$

The solution of Eq. (7.75) may be presented in the form

$$\Psi_0(\mathbf{r}, t) = b\begin{pmatrix}\Phi \\ \chi\end{pmatrix} e^{\frac{i}{\hbar}(p_z z - \mathcal{E}t)}, \tag{7.76}$$

where $\mathcal{E}$ is the total energy of the electron and b is the normalization coefficient. The bispinors Φ and χ are connected by the relation

$$\chi = \frac{cp_z\sigma_z + c\widehat{\mathbf{p}}\sigma}{\mathcal{E} + mc^2 - U(\rho)}\Phi. \tag{7.77}$$

From Eq. (7.75) for the wave function of the electron transversal motion in the channel with the accuracy of a small term $\sim U_0/\mathcal{E}$ we obtain the equation

$$\Delta_{\rho,\varphi}\Phi(\rho, \varphi) + \frac{2\mathcal{E}_{\parallel}}{\hbar^2 c^2}\left[\mathcal{E}_{\perp} - U(\rho)\right]\Phi(\rho, \varphi) = 0, \tag{7.78}$$

where

$$\Delta_{\rho,\varphi} = \frac{1}{\rho}\frac{\partial}{\partial\rho}\left(\rho\frac{\partial}{\partial\rho}\right) + \frac{1}{\rho^2}\frac{\partial^2}{\partial\varphi^2}$$

is the two-dimensional Laplacian,

$$\mathcal{E}_{\shortparallel} = \sqrt{c^2 p_z^2 + m^2 c^4}$$

is the energy of the electron longitudinal motion, and $\mathcal{E}_{\perp} = \mathcal{E} - \mathcal{E}_{\shortparallel}$ is the transversal one.

As is seen from Eq. (7.78) for wave function $\Phi(\rho, \varphi)$ the variables are separated and the eigenvalue of the operator

$$\widehat{L}_z = -i\hbar\frac{\partial}{\partial\varphi}$$

— the projection of the orbital moment of the electron on the z axis is conserved. Then the wave function $\Phi(\rho, \varphi)$ can be represented in the form

$$\Phi(\rho, \varphi) = \Phi(\rho)\, e^{i\mathfrak{m}\varphi}; \quad \mathfrak{m} = 0, \pm 1, \pm 2, ..., \tag{7.79}$$

where $\mathfrak{m}$ is the azimuthal quantum number, and from Eq. (7.78) for the function

$$R(\rho) = \frac{\Phi(\rho)}{\sqrt{\rho}} \tag{7.80}$$

we obtain the equation

$$R'' + \frac{2}{\rho}R' + \left[\frac{2\mathcal{E}_{\shortparallel}}{\hbar^2 c^2}\left(\mathcal{E}_{\perp} + \frac{\alpha_c}{\rho}\right) - \frac{\mathfrak{m}^2 - 1/4}{\rho^2}\right] R = 0. \tag{7.81}$$

For the solution of Eq. (7.81) we pass from ρ to a new variable

$$r = \frac{2}{\hbar c}\sqrt{2\mathcal{E}_{\shortparallel}\left|\mathcal{E}_{\perp}\right|}\rho, \tag{7.82}$$

and making a notation

$$n = \frac{\alpha_c}{\hbar c}\sqrt{\frac{\mathcal{E}_{\shortparallel}}{2\left|\mathcal{E}_{\perp}\right|}}, \tag{7.83}$$

then introducing the function $R(r)$ in the form

$$R(r) = r^{|\mathfrak{m}| - 1/2} e^{-r/2} w(r), \tag{7.84}$$

for the new function $w(r)$ we obtain the equation

$$rw'' + \left[2\left(|\mathfrak{m}| - \frac{1}{2}\right) + 2 - r\right] w' + \left(n - |\mathfrak{m}| - \frac{1}{2}\right) w = 0. \qquad (7.85)$$

The solution of Eq. (7.85) should not diverge at infinity more quickly than a limited power r and must be confined at $r = 0$. The function satisfying the second condition is the degenerated hypergeometric function

$$w(r) = F\left(-n + |\mathfrak{m}| + \frac{1}{2}, 2|\mathfrak{m}| + 1, r\right), \qquad (7.86)$$

and the solution satisfying the first condition at infinity will be obtained only at the integer negative (or equal to zero) values of the argument $-n+|\mathfrak{m}|+1/2$ when the function (7.86) turns to polynomial with the power $n - |\mathfrak{m}| - 1/2$. Otherwise it diverges at infinity as e^r. Hence, the number n must be a positive half-integer, and at the specified number $\mathfrak{m}$ it is necessary that

$$n \geq |\mathfrak{m}| + \frac{1}{2}; \quad n = |\mathfrak{m}| + \frac{1}{2} + n_\rho; \quad n_\rho = 0, 1, 2, \ldots. \qquad (7.87)$$

These conditions determine the quantization law of the electron transversal motion in the potential well of the crystal at the axial channeling. Thus, from Eq. (7.83) for the spectrum of the transversal energy eigenvalues of the electron bound states in the potential field (7.35) we obtain

$$\mathcal{E}_\perp = -\frac{\alpha_c^2 \mathcal{E}_{\shortparallel}}{2\hbar^2 c^2 n^2}. \qquad (7.88)$$

With the help of Eqs. (7.77), (7.79), (7.84) and (7.86) for the wave function of the channeled electron (7.76), normalized for one particle per unit volume, we will have the equation

$$\Psi_0(\mathbf{r}, t) = \Psi_{p_z, n, \mathfrak{m}, \sigma}(\mathbf{r}, t) = \sqrt{\frac{\mathcal{E}_{\shortparallel} + mc^2}{2\mathcal{E}_{\shortparallel}}} \begin{pmatrix} \varphi_\sigma \\ \frac{c\sigma \mathbf{p}}{\mathcal{E} + mc^2 - U(\rho)} \varphi_\sigma \end{pmatrix}$$

$$\times \sqrt{\frac{\rho}{2\pi}} R_{n,|\mathfrak{m}|-1/2}(\rho)\, e^{i\mathfrak{m}\varphi} e^{\frac{i}{\hbar}(p_z z - \mathcal{E}t)}, \qquad (7.89)$$

where φ_σ is a constant spinor determined in Eq. (7.61), and the function $R_{n,|\mathfrak{m}|-1/2}(\rho)$ is

$$R_{n,|\mathfrak{m}|-1/2}(\rho) = \left(\frac{\mathcal{E}_{\shortparallel}\alpha_c}{\hbar^2 c^2}\right)^{3/2} \frac{4}{n^{|\mathfrak{m}|+3/2}} \sqrt{\frac{2(n + |\mathfrak{m}| - 1/2)!}{(n - |\mathfrak{m}| - 1/2)!}} \left(\frac{4\mathcal{E}_{\shortparallel}\alpha_c \rho}{\hbar^2 c^2}\right)^{|\mathfrak{m}|-1/2}$$

$$\times \exp\left\{-\frac{2\mathcal{E}_{\shortparallel}\alpha_c}{n\hbar^2c^2}\rho\right\} F\left(-n+|\mathrm{m}|+1/2, 2|\mathrm{m}|+1, \frac{4\mathcal{E}_{\shortparallel}\alpha_c}{n\hbar^2c^2}\rho\right). \qquad (7.90)$$

The total energy $\mathcal{E}$ in Eq. (7.89) is given by the relation

$$\mathcal{E}(p_z, n) = \sqrt{c^2p_z^2 + m^2c^4} - \frac{2\alpha_c^2\mathcal{E}_{\shortparallel}}{\hbar^2c^2n^2}. \qquad (7.91)$$

To determine the electron wave function Ψ_1 perturbed by the EM wave in the next approximation of perturbation theory one needs the concrete form of the wave vector potential. Let it have the form

$$A_x = A_0 \cos(\omega_0 t - k_0 z),$$

$$A_y = A_0 \sin(\omega_0 t - k_0 z). \qquad (7.92)$$

Expanding Ψ_1 in terms of the full basis of the eigenstates (7.89)

$$\Psi_1(\mathbf{r}, t) = \sum_{p_z', n', m', \sigma'} c_{p_z', n', \mathrm{m}', \sigma'}(t)\, \Psi_{p_z', n', \mathrm{m}'\sigma'}(\mathbf{r}, t), \qquad (7.93)$$

and substituting the wave function in the first approximation of perturbation theory $\Psi_0 + \Psi_1$ into Eq. (7.48) with Eqs. (7.89)–(7.92), then after the solution of the obtained equation for unknown expansion coefficients $c_{p_z', n', \mathrm{m}'}(t)$ we will have

$$c_{p_z', n', \mathrm{m}', \sigma'} = -i\frac{eA_0}{2c}\Omega_{n'n}\mathcal{D}_{n'n}^{\mathrm{m}'\mathrm{m}}\delta_{\sigma\sigma'}\left\{\frac{e^{-\frac{i}{\hbar}\left(\mathcal{E}(p_z,n)-\mathcal{E}(p_z',n')+\hbar\omega_0\right)t}}{\mathcal{E}(p_z, n) - \mathcal{E}(p_z', n') + \hbar\omega_0}\delta_{\mathrm{m}',\mathrm{m}+1}\right.$$

$$\left.\times\delta_{p_z', p_z+\hbar k_0} + \frac{e^{-\frac{i}{\hbar}\left(\mathcal{E}(p_z,n)-\mathcal{E}(p_z',n')-\hbar\omega_0\right)t}}{\mathcal{E}(p_z, n) - \mathcal{E}(p_z', n') - \hbar\omega_0}\delta_{\mathrm{m}',\mathrm{m}-1}\delta_{p_z, p_z-\hbar k_0}\right\}, \qquad (7.94)$$

where

$$\mathcal{D}_{n'n}^{\mathrm{m}'\mathrm{m}} = \int_0^\infty \rho^3 R_{n', |\mathrm{m}'|-1/2}(\rho)\, R_{n, |\mathrm{m}|-1/2}(\rho)\, d\rho, \qquad (7.95)$$

and

$$\Omega_{n'n} = \frac{\mathcal{E}_{\perp n'} - \mathcal{E}_{\perp n}}{\hbar} = -\frac{2\mathcal{E}_{\shortparallel}\alpha_c^2}{\hbar^3c^2n'^2n^2}(n'+n)(n'-n) \qquad (7.96)$$

is the transition frequency between the initial and excited states of the transversal motion of the electron in the crystal channel.

Equations (7.93) and (7.94) determine the wave function of the one-photon induced axial channeling effect. With the help of the latter the probability density $(\Psi^+\Psi)$ of the electron after the interaction can be presented in the form

$$W = \frac{\rho}{2\pi} R^2_{n,|\mathrm{m}|-1/2}(\rho) + \frac{eA_0\rho}{2\pi\hbar} R_{n,|\mathrm{m}|-1/2}(\rho)$$

$$\times \left\{ \sum_{n' \geqslant |\mathrm{m}+1|+1/2} \Omega_{n'n} \frac{R_{n,|\mathrm{m}+1|-1/2}(\rho)}{\omega - \Omega_{n'n}} \mathcal{D}^{\mathrm{m}+1\mathrm{m}}_{n'n} \right.$$

$$\left. + \sum_{n' \geqslant |\mathrm{m}-1|+1/2} \Omega_{n'n} \frac{R_{n',|\mathrm{m}-1|-1/2}(\rho)}{\omega + \Omega_{n'n}} \mathcal{D}^{\mathrm{m}-1\mathrm{m}}_{n'n} \right\} \sin(k_0 z - \omega_0 t + \varphi), \quad (7.97)$$

where the Doppler-shifted wave frequency ω is

$$\omega = \omega_0 \left(1 - n_0 \frac{cp_z}{\mathcal{E}_{\shortparallel}}\right). \quad (7.98)$$

As in the case of the planar channeling the electron probability density is modulated at the wave frequency. Consequently, the electric current density in the case of an electron beam will be modulated at the stimulating wave frequency and its harmonics (corresponding equations for the modulation at the harmonics can be found in the next approximation of perturbation theory). Equation (7.97) is complicated enough for general forms of the functions $R_{n,\mathrm{m}}(\rho)$ and $\mathcal{D}^{\mathrm{m}'\mathrm{m}}_{n'n}$. It is rather simplified for resonant transitions of the electron from the initial bound state of transversal motion to the neighbor ones. Thus, from Eqs. (7.88), (7.95), and (7.96) we obtain that in the expression of the modulation depth quantity $\Omega_{n'n}\mathcal{D}^{\mathrm{m}'\mathrm{m}}_{n'n} \sim \sqrt{\mathcal{E}_\perp/\mathcal{E}_{\shortparallel}}$. The latter is the amplitude of the velocity of the electron transversal motion in the channel $\mathrm{v}_{\perp m}$. Besides, the resonant denominators in Eq. (7.97) define the period of coherent interaction of the electron with the EM wave in the channel: $(\omega - \Omega_{n'n})^{-1} \to \Delta t$. Hence, the modulation depth $\sim eE_0 \mathrm{v}_{\perp m} \Delta t/\omega << 1$ in accordance with the perturbation theory.

Note that in general the function $\mathcal{D}^{\mathrm{m}'\mathrm{m}}_{n'n}$ determined by Eq. (7.95) may be presented in the form

$$\mathcal{D}^{\mathrm{m}'\mathrm{m}}_{n'n} = \frac{\hbar^2 c^2}{\mathcal{E}_{\shortparallel}\alpha_c} \frac{2^{|\mathrm{m}|+|\mathrm{m}'|}}{n^{|\mathrm{m}|+3/2} n'^{|\mathrm{m}'|+3/2} (2|\mathrm{m}|)! (2|\mathrm{m}'|)!}$$

$$\times \sqrt{\frac{(n+|\mathrm{m}|-1/2)!(n'+|\mathrm{m}'|-1/2)!}{(n-|\mathrm{m}|-1/2)!(n'-|\mathrm{m}'|-1/2)!}} \int_0^\infty z^{|\mathrm{m}|+|\mathrm{m}'|+2} e^{-(1/n'+1/n)z} \quad (7.99)$$

$$\times F\left(-n+|\mathfrak{m}|+\frac{1}{2},2|\mathfrak{m}|+1,\frac{2z}{n}\right)F\left(-n'+|\mathfrak{m}'|+\frac{1}{2},2|\mathfrak{m}'|+1,\frac{2z}{n'}\right)dz.$$

In Eq. (7.95) integral is known as a function

$$\mathcal{J}_{\gamma}^{sp}(\alpha,\alpha')=\int_{0}^{\infty}e^{-\frac{\varkappa+\varkappa'}{2}z}z^{\gamma-1+s}F(\alpha,\gamma,\varkappa z)\,F(\alpha',\gamma-p,\varkappa' z)\,dz,$$

which is expressed via $\mathcal{J}_{\gamma}^{00}(\alpha,\alpha')$ by the recurrent relations.

7.5 Multiphoton Induced Channeling Effect

In the quantum description of the induced channeling effect in the previous two sections the wave field was weak enough so that the interaction process had mainly one-photon character. The coherent (resonant) interaction of the channeled particles with a strong EM wave from the quantum point of view has multiphoton character. Here we will consider the induced channeling effect in the strong wave fields in the scope of quantum theory, that is, we will solve the quantum equations of motion for channeled electrons or positrons in the strong plane EM wave field.

We will assume that the wave propagates in the yz plane of a crystal and is polarized in the xy plane with the vector potential

$$\mathbf{A}=\left\{A_x\left(t-n_0\frac{z}{c}\right),A_y\left(t-n_0\frac{z}{c}\right),0\right\},\tag{7.100}$$

where $n_0\equiv n(\omega_0)$ is the refractive index of the medium at the carrier frequency of the wave. We will consider the case when averaged potential of the crystal for a plane channeled particle is satisfactorily described by the harmonic potential

$$U(x)=\kappa\frac{x^2}{2}.\tag{7.101}$$

For the positron at the planar channeling

$$\kappa=\frac{8U_0}{d^2}\tag{7.102}$$

(see the potential (7.3)), while for the electrons the approximate potential of the channel is actually not harmonic and described by the potential

$$U(x)=-\frac{U_0}{\cosh^2\left(\frac{x}{b}\right)}.\tag{7.103}$$

Nevertheless, for the high energies it can be approximated by the harmonic potential (7.101). As we saw in previous sections, for the channeled particles the depth of the potential hole $U_0 << \mathcal{E}$, where $\mathcal{E}$ is the particle energy. The spin interaction, which is $\sim \bigtriangledown U(x)$, is again less than $\mathcal{E}$. For this reason the transverse motion of the channeled particle is described by the Schrödinger equation (7.56) with the effective mass $m_{eff} = \mathcal{E}_{\shortparallel}/c^2$. On the other hand, the spin interaction can play a role in the particle–wave interaction process at the energy of the photon comparable with the particle one: $\hbar\omega_0 \sim \mathcal{E}$. If the particle energy is not high enough, i.e., $\mathcal{E} << m^2c^4/\mathcal{E}_{\perp}$ (optimal cases for the channeling), then the resonant interaction of the channeled particles with an external EM wave takes place at $\hbar\omega_0 << \mathcal{E}$ and the spin effects are not essential. Hence, one may ignore the spin interaction and instead of the Dirac equation solve the Klein–Gordon equation

$$\left[i\hbar\frac{\partial}{\partial t} - U(x)\right]^2 \Psi = \left[c^2\left(\widehat{\mathbf{p}} - \frac{e}{c}\mathbf{A}\left(t - n_0\frac{z}{c}\right)\right)^2 + m^2c^4\right]\Psi. \tag{7.104}$$

As we saw in Section 7.3 the channeled particle initial motion (before the interaction with EM wave) is separated into longitudinal (y, z) and transversal (x) degrees of freedom. For the longitudinal motion we assume an initial state with a momentum $\mathbf{p}_{\shortparallel} = \{0, p_y, p_z\}$, while for the transversal motion we assume a quantum state $\{n\}$, where by n we indicate the energy levels in the harmonic potential (7.101). As the plane wave field depends only on the retarding coordinate $\tau = t - n_0 z/c$, then using the problem symmetry the wave function of a channeled particle can be sought in the form

$$\Psi(\mathbf{r}, t) = f(x, \tau)e^{\frac{i}{\hbar}(\mathbf{p}_{\shortparallel}\mathbf{r} - \mathcal{E}t)}. \tag{7.105}$$

The multiphoton interaction of the charged particles with a strong EM wave, in general, as was shown in diverse processes is well enough described by the eikonal-type wave function corresponding to a slowly varying function $f(x, \tau)$ on the wave coordinate τ. Hence, neglecting the second derivatives of this function compared with the first-order ones in accordance with the conditions (3.92) for the function $f(x, \tau)$ we will obtain the equation

$$\left[\hbar^2\frac{\partial^2}{\partial x^2} + \frac{2\mathcal{E}_{\shortparallel}}{c^2}(\mathcal{E}_{\perp} - U(x)) + 2i\frac{\widetilde{p}\hbar}{c}\frac{\partial}{\partial \tau} - 2i\frac{e\hbar}{c}A_x(\tau)\frac{\partial}{\partial x}\right.$$

$$\left. + 2\frac{e}{c}p_y A_y(\tau) - \frac{e^2}{c^2}\mathbf{A}^2(\tau)\right] f(x, \tau) = 0, \tag{7.106}$$

where

$$\widetilde{p} = \frac{1}{c}\left(\mathcal{E}_{\shortparallel} - n_0 c p_z\right). \tag{7.107}$$

In Eq. (7.106) the transversal and longitudinal motions are not separated. But after the definite unitarian transformation for the transformed function the variables are separated. The corresponding unitarian transformation operator is

$$\widehat{S} = e^{\frac{i}{\hbar}\{g_1(\tau)x - g_2(\tau)\widehat{p}_x\}}, \tag{7.108}$$

where the functions $g_1(\tau)$, $g_2(\tau)$ will be chosen to separate the transversal and longitudinal motions and to satisfy the initial condition. Taking into account Eq. (4.54) for transformed function

$$\Phi(x,\tau) = \widehat{S}f(x,\tau) \tag{7.109}$$

we obtain the equation

$$\left[\hbar^2\frac{\partial^2}{\partial x^2} + \frac{2\mathcal{E}_{\shortparallel}}{c^2}(\mathcal{E}_\perp - U(x)) + 2i\hbar\left(\frac{\widetilde{p}}{c}\frac{dg_2(\tau)}{d\tau} - g_1(\tau) - \frac{e}{c}A_x(\tau)\right)\frac{\partial}{\partial x}\right.$$

$$\left. + \frac{2}{c}\left(\widetilde{p}\frac{dg_1(\tau)}{d\tau} + \frac{\mathcal{E}_{\shortparallel}\kappa}{c}g_2(\tau)\right)x + \frac{2i\widetilde{p}\hbar}{c}\frac{\partial}{\partial\tau} + Q(\tau)\right]\Phi(x,\tau) = 0, \tag{7.110}$$

where

$$Q(\tau) = \frac{\widetilde{p}}{c}\left(\frac{dg_2(\tau)}{d\tau}g_1(\tau) - \frac{dg_1(\tau)}{d\tau}g_2(\tau)\right) - g_1^2(\tau) - \frac{\mathcal{E}_{\shortparallel}\kappa}{c^2}g_2^2(\tau)$$

$$-\frac{2e}{c}A_x(\tau)g_1(\tau) + \frac{2e}{c}p_yA_y(\tau) - \frac{e^2}{c^2}\mathbf{A}^2(\tau). \tag{7.111}$$

Let us choose $g_1(\tau)$ and $g_2(\tau)$ in such a form that the coefficients of x and $\partial/\partial x$ in Eq. (7.110) become zero. Then for the functions $g_1(\tau)$ and $g_2(\tau)$ we will obtain a classical equation of motion describing stimulated oscillations in the harmonic potential:

$$\frac{dg_1(\tau)}{d\tau} = -\frac{\mathcal{E}_{\shortparallel}\kappa}{c\widetilde{p}}g_2(\tau), \tag{7.112}$$

$$\frac{dg_2(\tau)}{d\tau} = \frac{c}{\widetilde{p}}g_1(\tau) + \frac{e}{\widetilde{p}}A_x(\tau). \tag{7.113}$$

The solutions of Eqs. (7.112) and (7.113) can be written as

$$g_1(\tau) = \frac{e\Omega'}{c}\mathrm{Im}\left[e^{-i\Omega'\tau}\int_{-\infty}^{\tau} A_x(\tau)\, e^{i\Omega'\tau'}d\tau'\right], \tag{7.114}$$

$$g_2(\tau) = \frac{e}{\widetilde{p}}\mathrm{Re}\left[e^{-i\Omega'\tau}\int_{-\infty}^{\tau} A_x(\tau)\, e^{i\Omega'\tau'}d\tau'\right], \tag{7.115}$$

where

$$\Omega' = \frac{\Omega}{1-n_0\frac{\mathrm{v}_z}{c}}; \quad \Omega = c\sqrt{\kappa/\mathcal{E}_{\shortparallel}}. \tag{7.116}$$

In Eqs. (7.114) and (7.115) we have taken into account the initial condition

$$g_1(-\infty) = g_2(-\infty) = 0.$$

After the unitarian transformation (7.109) for the function $\Phi(x,\tau)$ the following equation is obtained:

$$\left[\hbar^2\frac{\partial^2}{\partial x^2} + \frac{2\mathcal{E}_{\shortparallel}}{c^2}(\mathcal{E}_{\perp} - U(x)) + \frac{2i\widetilde{p}\hbar}{c}\frac{\partial}{\partial\tau} + Q(\tau)\right]\Phi(x,\tau) = 0. \tag{7.117}$$

Now in Eq. (7.117) the variables are separated and the solution can be written as follows:

$$\Phi(x,\tau) = N\varphi_n(x)\exp\left\{i\frac{c}{2\hbar\widetilde{p}}\int_{-\infty}^{\tau} Q(\tau)\, d\tau'\right\}, \tag{7.118}$$

where $\varphi_n(x)$ coincides with the harmonic oscillator wave function (7.58) and $N = 1/\sqrt{L_y L_z}$is the normalization constant (L_y and L_z are the quantization lengths). By inverse transformation

$$f(x,\tau) = \widehat{S}^{\dagger}\Phi(x,\tau),$$

with the help of Eq. (4.66) we obtain the solution of the initial equation (7.104) (taking into account Eq.(7.105)):

$$\Psi(\mathbf{r},t) = N\exp\left\{\frac{i}{\hbar}(\mathbf{p}_{\shortparallel}\mathbf{r} - \mathcal{E}t)\right\}\varphi_n(x + g_2(\tau))$$

$$\times\exp\left\{\frac{i}{\hbar}\left[\frac{c}{2\widetilde{p}}\int_{-\infty}^{\tau} Q(\tau)\, d\tau' - \frac{1}{2}g_1(\tau)g_2(\tau) - g_1(\tau)x\right]\right\}, \tag{7.119}$$

where the function $Q(\tau)$ can be represented in the form

$$Q(\tau) = \frac{2e}{c} p_y A_y(\tau) - \frac{e}{c} A_x(\tau) g_1(\tau) - \frac{e^2}{c^2} \mathbf{A}^2(\tau). \tag{7.120}$$

This wave function describes the multiphoton interaction of the channeled particle with the strong EM radiation field. Thus, for a monochromatic wave

$$\mathbf{A} = \{A_0 \cos(\omega_0 t - k_0 z), 0, 0\},$$

from Eqs. (7.114) and (7.115) for the functions $g_1(\tau)$ and $g_2(\tau)$ we obtain

$$g_1(\tau) = \frac{e}{c} A_0 \frac{\Omega'^2}{\Delta} \cos\omega_0\tau,$$

$$g_2(\tau) = \frac{eA_0}{\widetilde{p}} \frac{\omega_0}{\Delta} \sin\omega_0\tau, \tag{7.121}$$

and we will have the following wave function for the particle in the field of a strong EM wave at the planar channeling:

$$\Psi(\mathbf{r}, t) = N \exp\left\{\frac{i}{\hbar}\left(\mathbf{p}_{\shortparallel}\mathbf{r} - \mathcal{E}t - \frac{e^2 A_0^2 \omega_0^2}{4c\widetilde{p}\Delta}\tau\right)\right\} \varphi_n\left(x + \frac{eA_0\omega_0}{\widetilde{p}\Delta}\sin\omega_0\tau\right)$$

$$\times \exp\left\{-\frac{i}{\hbar}\left[\frac{eA_0\Omega'^2}{c\Delta} x\cos\omega_0\tau + \frac{e^2 A_0^2 \omega_0\left(\omega_0^2 + \Omega'^2\right)}{8c\widetilde{p}\Delta^2}\sin(2\omega_0\tau)\right]\right\}, \tag{7.122}$$

where

$$\Delta = \omega_0^2 - \Omega'^2$$

is the resonance detuning.

On the basis of the obtained wave function (7.119) consider the possibility of multiphoton excitation of transversal levels by the strong EM wave at the resonance

$$\omega_0 \simeq \frac{\Omega}{\left|1 - n_0 \frac{\mathrm{v}_z}{c}\right|}. \tag{7.123}$$

The Doppler factor $1 - n_0 \mathrm{v}_z / c$ may be positive as well as negative — anomalous Doppler effect at $n_0 > 1$. We will consider the actual case of a quasi-monochromatic EM wave with a slowly varying amplitude $A_0(\tau)$. After the interaction with the wave ($t \to +\infty$) from Eqs. (7.114) and (7.115) at the resonance condition (7.123) we have

$$g_1(\tau) = \frac{e\overline{A}_0 T \Omega'}{2c} \sin \omega_0 \tau, \tag{7.124}$$

$$g_2(\tau) = \frac{e\overline{A}_0 T}{2\widetilde{p}} \cos \omega_0 \tau, \tag{7.125}$$

where T is the coherent interaction time (for actual laser radiation T is the pulse duration) and $\overline{A}_0$ is the average value of the slowly varied envelope. Substituting Eqs. (7.124) and (7.125) into the expression for the wave function (7.119) and expanding the latter in terms of the full basis of the particle eigenstates

$$\Psi(\mathbf{r}, t) = \sum_{\mathbf{p}'_{\shortparallel}, n'} a_{\mathbf{p}'_{\shortparallel}, n'}(t) \, \Psi_{\mathbf{p}'_{\shortparallel}, n'}(\mathbf{r}, t), \tag{7.126}$$

we find the probabilities of the multiphoton induced transitions between the transversal levels. To calculate the expansion coefficients

$$a_{\mathbf{p}'_{\shortparallel}, n'}(t) = \int \Psi^*_{\mathbf{p}'_{\shortparallel}, n'}(\mathbf{r}, t) \, \Psi(\mathbf{r}, t) \, d\mathbf{r}, \tag{7.127}$$

we will take into account the result of the integration (4.73). Taking into account Eqs.(7.124), (7.125), (7.119), and (7.127) we get the following expansion coefficients:

$$a_{\mathbf{p}'_{\shortparallel}, n'}(t) = I_{n,n'}(\alpha) \, \delta_{p'_y, p_y} \delta_{p'_z, p_z + \mu\hbar k_0 (n'-n)}$$

$$\times \exp\left\{ \frac{i}{\hbar} (\mathcal{E}(\mathbf{p}'_{\shortparallel}, n') - \mathcal{E}(\mathbf{p}_{\shortparallel}, n) - \mu\hbar\omega_0 (n'-n)) t + i\phi \right\}, \tag{7.128}$$

where

$$\mu = \frac{1 - n_0 \frac{\mathrm{v}_z}{c}}{\left|1 - n_0 \frac{\mathrm{v}_z}{c}\right|},$$

and

$$\phi \equiv \frac{c}{2\hbar\widetilde{p}} \int_{-\infty}^{\infty} Q(\tau) \, d\tau'$$

is the constant phase. Here the argument of the Lagger function $I_{n,n'}(\alpha)$ is

$$\alpha = \frac{e^2 \overline{A}_0^2 T^2}{8\hbar} \frac{\Omega'}{c\widetilde{p}}. \tag{7.129}$$

According to Eq. (7.128) the transition of the particle from an initial state $\{p_y, p_z, n\}$ to a state $\{p'_y, p'_z, n'\}$ is accompanied by the emission or absorption of $|n - n'|$ number of photons. Consequently, substituting Eq. (7.128) into Eq. (7.126) we can rewrite the particle wave function in the form

$$\Psi(\mathbf{r}, t) = N \sum_{n'=0}^{\infty} I_{n,n'}(\alpha) \exp\left\{\frac{i}{\hbar}(p_y y + (p_z + \mu\hbar k_0(n' - n))z)\right\}$$

$$\times \exp\left\{-\frac{i}{\hbar}(\mathcal{E}(\mathbf{p}_{\shortparallel}, n) + \mu\hbar\omega_0(n' - n))t + i\phi\right\}\varphi_{n'}(x). \qquad (7.130)$$

Hence, the probability of the induced transitions $n \to n'$ between the energy levels of the particle transversal motion in the channel finally is defined from Eq. (7.130):

$$W_{n,n'} = I^2_{n,n'}\left(\frac{e^2\overline{A}_0^2 T^2 \Omega'}{8\hbar c\widetilde{p}}\right). \qquad (7.131)$$

Equation (7.130) shows that in the field of a strong EM wave the transversal levels are excited at the absorption of the wave quanta if $1 - n_0 \mathrm{v}_z/c > 0$ and $\mu = 1$, corresponding to the normal Doppler effect, while in the case $1 - n_0 \mathrm{v}_z/c < 0$ and $\mu = -1$ the transversal levels are excited at the emission of coherent quanta due to the anomalous Doppler effect.

Let us now estimate the average number of emitted (absorbed) photons by the particle at the resonance for the high excited levels ($n >> 1$) and for the strong EM wave. In this case the most probable number of photons in the strong wave field corresponds to the quasiclassical limit ($|n - n'| >> 1$) when multiphoton processes dominate and the nature of the interaction process is very close to the classical one. In this case the argument of the Lagger function can be represented as

$$\alpha = \frac{1}{4n}\left(\frac{\triangle\mathcal{E}_{cl}}{\hbar\omega_0}\right)^2, \qquad (7.132)$$

where

$$\triangle\mathcal{E}_{cl} = \frac{eE_0 T}{2}\frac{\overline{\mathrm{v}}_\perp}{\left|1 - n_0\frac{\mathrm{v}_z}{c}\right|}$$

is the maximal energy change of the particle according to classical perturbation theory (E_0 is the amplitude of the electric field strength of the EM wave, $\overline{\mathrm{v}}_\perp \simeq c\sqrt{2n\hbar\Omega/\mathcal{E}_{\shortparallel}}$ is the particle mean transversal velocity). Note that according to conditions (3.92) of the considered eikonal approximation $\Delta\mathcal{E} << \mathcal{E}$.

The Lagger function is maximal at $\alpha \to \alpha_0 = \left(\sqrt{n'} - \sqrt{n}\right)^2$, exponentially falling beyond α_0. Hence, for the transition $n \to n'$ and when $|n - n'| << n$ we have

$$\alpha_0 \simeq \frac{(n' - n)^2}{4n}.$$

The comparison of this expression with Eq. (7.132) shows that the most probable transitions are

$$|n - n'| \simeq \frac{\triangle \mathcal{E}_{cl}}{\hbar \omega_0},$$

in accordance with the correspondence principle.

Bibliography

I. Lindhard, Usp. Fiz. Nauk **99**, 249 (1969)
M.A. Kumakhov, Phys. Lett. A **57**, 17 (1976)
M.A. Kumakhov, Phys. Status Solidi B **84**, 41 (1977)
V.A. Bazylev, N.K. Zhevago, Zh. Éksp. Teor. Fiz. **73**, 1697 (1977)
M.A. Kumakhov, Zh. Éksp. Teor. Fiz. **72**, 1489 (1977)
M.A. Kumakhov, R. Wedell, Phys. Status Solidi B **84**, 581 (1977)
V.V. Beloshitski, M.A. Kumakhov, Zh. Éksp. Teor. Fiz. **74**, 1244 (1978)
N.K. Zhevago, Zh. Éksp. Teor. Fiz. **75**, 1389 (1978)
V.A. Bazylev, N.K. Zhevago, Usp. Fiz. Nauk **127**, 529 (1979)
A.V. Andreev et al., Zh. Éksp. Teor. Fiz. **84**, 798 (1979)
V.A. Bazylev, N.K. Zhevago, Phys. Status Solidi B **97**, 63 (1980)
V.A. Bazylev, I.V. Globov, N.K. Zhevago, Zh. Éksp. Teor. Fiz. **78**, 62 (1980)
A.V. Tulupov, Pis'ma Zh. Tekh. Fiz. **7**, 460 (1981) [in Russian]
A.V. Tulupov, Zh. Éksp. Teor. Fiz. **81**, 1639 (1981)
V.A. Bazylev et al., Zh. Éksp. Teor. Fiz. **80**, 608 (1981)
V.A. Bazylev, N.K. Zhevago, Usp. Fiz. Nauk **137**, 605 (1982)
R.H. Pantell, M.J.Alguard, J. Appl. Phys. **50**, 5433 (1982)
A.V. Tulupov, Zh. Éksp. Teor. Fiz. **86**, 1365 (1984)
I.M. Ternov, V.R. Khalilov, B.V. Kholomay, Zh. Éksp. Teor. Fiz. **88**, 329 (1985)
V.N. Bayer, A.I. Milstein, Nucl. Instrum. Methods Phys. Res. **17**, 25 (1986)
M.A. Kumakhov: Emission By Channeled Particles In Crystals (Energoatomizdat, Moscow 1986) [in Russian]
S.A. Bogacz, J.B. Kefterson, G.K. Wong, Nucl. Instrum. Methods Phys. Res. **250**, 328 (1986)
H.K. Avetissian et al., Dokl. Acad. Nauk Arm. SSR **85**, 164 (1987) [in Russian]
V.A. Bazylev and N.K. Zhevago: Emission by Fast Particles in Matter and

External Fields (Nauka, Moscow 1987) [in Russian]
G. Kurizki, Adv. Laser Sci. **3**, 56 (1988)
K.B. Oganesyan, A.M. Prokhorov, M.V. Fedorov, Zh. Éksp. Teor. Fiz. **94**, 80 (1988)
M.V. Fedorov et al., Appl. Phys. Lett. **53**, 353 (1988)
H.K. Avetissian, A.K. Avetissian, Kh.V. Sedrakian, Zh. Éksp. Teor. Fiz. **100**, 82 (1991)
H.K. Avetissian et al., Phys. Lett. A **206**, 141 (1995)
H.K. Avetissian et al., Zh. Éksp. Teor. Fiz. **109**, 1159 (1996)
A.K. Avetissian et al., Phys. Lett. A **299**, 331 (2002)

8 Nonlinear Mechanisms of Free Electron Laser

The problem of creation of short-wave coherent EM radiation sources in general aspects reduces to the implementation of free electron lasers (FEL). The principal advantage of a FEL with respect to traditional quantum generators operating on discrete transitions in atomic/molecular systems is that the radiation frequency is continuously Doppler upshifted due to high relativism of electron beams, providing rapid tunability over a broad range of frequencies up to γ-ray.

Among the diverse versions of FEL at present the undulator scheme is being actively developed. Although the amplifying frequencies are still far from X-ray, the main hopes for an efficient X-ray FEL remain associated with the undulator scheme based on the accumulation of coherent radiation of ultrarelativistic electron beams in the Self-Amplified Spontaneous Emission (SASE) regime, in which the initial shot noise on the electron beam is amplified over the course of propagation through a long wiggler. For that it is required that the lengths are on the order of several ten to hundred meters. The recent experimental success shows the feasibility of construction of such facilities.

Nevertheless, because there are no drivers or mirrors operable at X-ray wavelengths the problem reduces to amplification/generation of coherent radiation in the single-pass regime. It is clear that the latter can be achieved with more efficiency via the nonlinear schemes of FEL induced by strong pump EM fields. The latter will considerably abbreviate the amplification length as well and one can expect small setup FEL devices.

On the other hand, as the photon wavelength moves into the deep UV and X-ray regions the interaction becomes quantum mechanical, i.e., quantum recoil becomes comparable to or larger than the gain bandwidth and quantum effects play an essential role. The quantum effects are also essential if one considers the FEL versions where one or two degrees of freedom of the charged particles are quantized and the resonant enhancement of electron–photon interaction cross section holds. This takes place for the X-ray laser schemes based on the electron/positron beam channeling radiation in crystals.

The smallness of the electron–photon interaction cross section can also be compensated and the quality of the output X-ray radiation can be enhanced in the hybrid schemes of FEL and atomic laser. It can be achieved by means of fast high-density ion beam interaction with a strong counterpropagating pump laser field or with a crystal periodic electrostatic potential.

Investigation of the nonlinear schemes and quantum aspects of FEL on the basis of a self-consistent set of Maxwell and quantum kinetic equations is the subject of the present chapter.

8.1 Self-Consistent Maxwell and Relativistic Quantum Kinetic Equations for Compton FEL with Strong Pump Laser Field

In contrast to conventional laser devices in atomic systems, the FEL is usually regarded as a classical device that also exhibits non-Poissonian photon statistics. But this is not a universal property of FELs as in some cases quantum effects may play a significant role. In the quantum description the small-signal gain of the FEL is usually represented as a convolution integral of the electron beam momentum distribution with the difference between the probability distributions of emission and absorption per photon. Since the electron recoils in opposite directions depending on whether it emits or absorbs photons with the same wave vector $\mathbf{k}'$, the resonant momenta of an electron for emission p_e and absorption p_a are different. Hence, the probability distributions of emission and absorption are centered at p_e and p_a, and when these distributions are much narrower than the spread of the electron beam distributions $f(p)$, the small-signal gain is proportional to the so-called "population inversion" $f(p_e) - f(p_a)$. In the quasiclassical limit when photon energy $\hbar\omega'$ satisfies the condition

$$\hbar\omega' << \max\{\Delta\varepsilon_\gamma, \Delta\varepsilon_\vartheta, \Delta\varepsilon_L\} \tag{8.1}$$

($\Delta\varepsilon_\gamma$ and $\Delta\varepsilon_\vartheta$ are the resonance widths due to energetic and angular spreads, and $\Delta is\varepsilon_L$ the resonance width caused by the finite interaction length) the quantum expression for the gain coincides with its classical counterpart, being antisymmetric about the classical resonant momentum $p_c = (p_e + p_a)/2$ and proportional to the derivative of the momentum distribution $df(p)/dp$ at resonant value p_c. The result is that amplification takes place only if the initial momentum distribution is centered above p_c as the electrons whose momenta are above p_c contribute on average to the small-signal gain, and the electrons whose momenta are below p_c contribute on average to the corresponding loss. This severely limits the FEL gain performance at short wavelengths. In the more conventional undulator devices, to achieve the X-ray frequency domain one should increase the electron energies up to several gigaelectron volts, which in turn significantly reduces the small-signal gain ($\sim \gamma_L^{-3}$). To achieve the X-ray domain with moderate relativistic electron beams (energy of electrons $\leq$ 50 MeV), the frequency of electron self-oscillation should be high enough $\sim 10^{14} \div 10^{15}\text{s}^{-1}$ (in undulator 10^{10}s^{-1}). The latter can be realized, e.g., in the Compton backscattering scheme suggested over 40 years ago.

Another way to increase the efficiency of a FEL is to achieve the quantum regime of generation

$$\hbar\omega' \geq \max\left\{\varDelta\varepsilon_{\gamma}, \varDelta\varepsilon_{\vartheta}, \varDelta\varepsilon_{L}\right\}, \tag{8.2}$$

as in this case the absorption and emission line shapes are separated and the simultaneous absorption of a probe wave is excluded. From this point of view the scheme of an X-ray Compton laser has an advantage with respect to the conventional undulator devices connected with the satisfaction of condition (8.2) for the quantum regime of generation. To achieve this condition for current FEL devices operating in undulators is problematic as it presumes severe restrictions on the beam spread. Thus, the scheme of an X-ray Compton laser in the quantum regime of generation is preferable, since it requires considerably lower energies of the electron beam and moderate restrictions on the beam spreads.

Consider a scheme of X-ray coherent radiation generation in the nonlinear quantum regime by means of a mildly relativistic high-density electron beam and a strong pump laser field. This makes it possible to achieve the quantum regime of generation at X-ray frequencies as well, due to radiation of high harmonics of Doppler-shifted pump frequencies in the strong laser field. In addition, concerning the further process of X-ray radiation amplification it is necessary to realize a single-pass FEL, as long as the construction of resonators in the X-ray domain is problematic. In the linear regime this demands very long interaction lengths. Here the main emphasis is on the nonlinear regime of generation. The consideration is based on a self-consistent set of Maxwell and quantum kinetic equations. Because the energy-momentum levels are not equidistant, the probe wave resonantly couples only two Volkov states, and the coupled equations will be solved in the slowly varying envelope approximation.

We will consider given pump EM wave with four-wave vector $k \equiv (\omega/c, \mathbf{k})$ which is described by the four-vector potential

$$A^{\mu} = (0, \mathbf{A}), \tag{8.3}$$

where $\mathbf{A}$ is defined by Eq. (1.48). As we saw in Section 1.4 the Dirac equation allows the exact solution in the field of a plane EM wave (Volkov solution). Although the Volkov states are not stationary, as there are no real transitions in the monochromatic EM wave (due to violation of energy and momentum conservation laws) the state of a particle in an EM wave can be characterized by the quasimomentum $\mathbf{\Pi}$ and polarization σ and the particle state in the field (8.3) is given by the wave function (1.94).

We assume the probe EM wave to be linearly polarized with the carrier frequency ω' and four-vector potential

$$A_w = \frac{\epsilon}{2}\left\{A_e(t,\mathbf{r})e^{-ik'x} + \text{c.c.}\right\}, \tag{8.4}$$

where $A_e(t,\mathbf{r})$ is a slowly varying envelope, $k' = (\omega'/c, \mathbf{k}')$ is the four-wave vector and ϵ is the unit polarization four vector $\epsilon k' = 0$, and $x = (ct, \mathbf{r})$ is the four-component radius vector.

Cast in the second quantization formalism, the Hamiltonian is

$$\widehat{H} = \int \widehat{\Psi}^{+}\widehat{H}_0\widehat{\Psi}d\mathbf{r}+\widehat{H}_{int}, \tag{8.5}$$

where $\widehat{\Psi}$ is the fermionic field operator, $\widehat{H}_0$ is the one-particle Hamiltonian in the plane EM wave (8.3), and the interaction Hamiltonian is

$$\widehat{H}_{int} = \frac{1}{c}\int \widehat{j}A_w d\mathbf{r}, \tag{8.6}$$

with the current density operator

$$\widehat{j} = e\widehat{\Psi}^{+}\gamma_0\gamma\widehat{\Psi}. \tag{8.7}$$

We pass to the furry representation and write the Heisenberg field operator of the electron in the form of an expansion in the quasistationary Volkov states (1.97)

$$\widehat{\Psi}(\mathbf{r},t) = \sum_{\mathbf{\Pi},\sigma} \widehat{a}_{\mathbf{\Pi},\sigma}(t)\Psi_{\mathbf{\Pi}\sigma}(\mathbf{r},t), \tag{8.8}$$

where we have excluded the antiparticle operators, since contribution of particle–antiparticle intermediate states will lead only to small corrections to the processes considered. The creation and annihilation operators, $\widehat{a}^{+}_{\mathbf{\Pi},\sigma}(t)$ and $\widehat{a}_{\mathbf{\Pi},\sigma}(t)$, associated with positive energy solutions satisfy the anticommutation rules at equal times

$$\{\widehat{a}^{\dagger}_{\mathbf{\Pi},\sigma}(t), \widehat{a}_{\mathbf{\Pi}',\sigma'}(t')\}_{t=t'} = \delta_{\mathbf{\Pi},\mathbf{\Pi}'}\delta_{\sigma,\sigma'}\ , \tag{8.9}$$

$$\{\widehat{a}^{\dagger}_{\mathbf{\Pi},\sigma}(t), \widehat{a}^{\dagger}_{\mathbf{\Pi}',\sigma'}(t')\}_{t=t'} = \{\widehat{a}_{\mathbf{\Pi},\sigma}(t), \widehat{a}_{\mathbf{\Pi}',\sigma'}(t')\}_{t=t'} = 0. \tag{8.10}$$

Taking into account Eqs. (8.8), (8.7), (8.6), and (1.97), the second quantized interaction Hamiltonian can be expressed in the form

$$\widehat{H}_{int} = \sum_{s=-\infty}^{\infty} \sum_{\mathbf{\Pi},\sigma,\sigma'} \left\{ \frac{eA_e}{2c} M^{(-s)} \left(\mathbf{\Pi},\sigma; \mathbf{\Pi} - \hbar\mathbf{k}' + s\hbar\mathbf{k},\sigma' \right) e^{-i\Delta(s,\mathbf{\Pi})t} \right.$$

$$\times \widehat{a}^{\dagger}_{\mathbf{\Pi},\sigma}(t) \widehat{a}_{\mathbf{\Pi}-\hbar\mathbf{k}'+s\hbar\mathbf{k},\sigma'}(t) + \frac{eA_e^*}{2c} M^{(s)} \left(\mathbf{\Pi} - \hbar\mathbf{k}' + s\hbar\mathbf{k},\sigma'; \mathbf{\Pi},\sigma \right) e^{i\Delta(s,\mathbf{\Pi})t}$$

$$\left. \times \widehat{a}^{\dagger}_{\mathbf{\Pi}-\hbar\mathbf{k}'+s\hbar\mathbf{k},\sigma'}(t) \widehat{a}_{\mathbf{\Pi},\sigma}(t) \right\}. \tag{8.11}$$

Here

$$M^{(s)} \left(\mathbf{\Pi}',\sigma'; \mathbf{\Pi},\sigma \right) = \frac{1}{2\sqrt{\Pi_0' \Pi_0}} \overline{u}_{\sigma'}(p') \left\{ \frac{e^2 (k\epsilon) Q_{2s}(\alpha,\beta,\varphi)}{2c^2 (kp')(kp)} \widehat{k} \right.$$

$$\left. + \left(\frac{e\widehat{Q}_{1s}(\alpha,\beta,\varphi)\widehat{k}\widehat{\epsilon}}{2c(kp')} + \frac{e\widehat{\epsilon}\widehat{k}\widehat{Q}_{1s}(\alpha,\beta,\varphi)}{2c(kp)} \right) + \widehat{\epsilon} Q_{0s}(\alpha,\beta,\varphi) \right\} u_{\sigma}(p), \tag{8.12}$$

where the vector functions $Q^{\mu}_{1s} = (0, \mathbf{Q}_{1s})$ and scalar functions Q_{0s}, Q_{2s} are expressed via generalized Bessel functions $G_s(\alpha,\beta,\varphi)$:

$$Q_{0s} = G_s(\alpha,\beta,\varphi), \tag{8.13}$$

$$\mathbf{Q}_{1s} = \frac{A_0}{2} \left\{ \mathbf{e}_1 \left(G_{s-1}(\alpha,\beta,\varphi) + G_{s+1}(\alpha,\beta,\varphi) \right) \right.$$

$$\left. + i\mathbf{e}_2 g \left(G_{s-1}(\alpha,\beta,\varphi) - G_{s+1}(\alpha,\beta,\varphi) \right) \right\}, \tag{8.14}$$

$$Q_{2s} = A_0^2 \frac{(1+g^2)}{2} G_s(\alpha,\beta,\varphi)$$

$$+ A_0^2 \frac{(1-g^2)}{2} \left(G_{s-2}(\alpha,\beta,\varphi) + G_{s+2}(\alpha,\beta,\varphi) \right). \tag{8.15}$$

The definition of arguments α, β, φ are the same as in Eqs. (1.103)–(1.105). The resonance detuning in Eq.(8.11) is

$$\hbar\Delta(s,\mathbf{\Pi}) = \sqrt{c^2 \left(\mathbf{\Pi} - \hbar\mathbf{k}' + s\hbar\mathbf{k} \right)^2 + m^{*2}c^4} + \hbar\omega'$$

$$- \sqrt{c^2 \mathbf{\Pi}^2 + m^{*2}c^4} - s\hbar\omega. \tag{8.16}$$

We will use Heisenberg representation, where evolution of the operators are given by the equation

$$i\hbar\frac{\partial\widehat{L}}{\partial t}=\left[\widehat{L},\widehat{H}\right], \tag{8.17}$$

and expectation values are determined by the initial density matrix $\widehat{D}$

$$<\widehat{L}>=Sp\left(\widehat{D}\widehat{L}\right). \tag{8.18}$$

Equations (8.17) should be supplemented by the Maxwell equation for $\overline{A}_e$ which is reduced to

$$\frac{\partial A_e}{\partial t}+\frac{c^2\mathbf{k}'}{\omega'}\frac{\partial A_e}{\partial \mathbf{r}}=-i\frac{4\pi c}{\omega'}\overline{<\widehat{\epsilon j}>\exp(ik'x)}, \tag{8.19}$$

where the bar denotes averaging over time and space much larger than $(1/\omega'$, $1/k')$ and

$$<\widehat{\epsilon j}>=Sp\left(\widehat{\epsilon j}\widehat{D}\right). \tag{8.20}$$

Taking into account Eqs. (8.7) and (8.8) we obtain

$$\widehat{\epsilon j}\exp(ik'x)=e\sum_{s=-\infty}^{\infty}\sum_{\mathbf{\Pi}',\mathbf{\Pi},\sigma',\sigma}\left\{\widehat{a}^{+}_{\mathbf{\Pi}',\sigma'}(t)\widehat{a}_{\mathbf{\Pi},\sigma}(t)\right.$$

$$\left.\times M^{(s)}\left(\mathbf{\Pi}',\sigma';\mathbf{\Pi},\sigma\right)e^{\frac{1}{\hbar}\left(\Pi'-\Pi-s\hbar k+\hbar k'\right)x}\right\}. \tag{8.21}$$

As we are interested in amplification of the wave with a certain $\omega',\mathbf{k}'$, then we can keep only resonant terms in Eq. (8.21) with $\mathbf{\Pi}'=\mathbf{\Pi}-\hbar\mathbf{k}'+s\hbar\mathbf{k}$. In principle, because of the electron beam energy and angular spreads different harmonics may contribute to the process considered, but in the quantum regime (see below Eqs. (8.44), (8.45)) we can keep only one harmonic $s=s_0$. For the resonant current amplitude we will have the expression

$$-i\overline{(\widehat{\epsilon j})\exp(ik'x)}=\int\widehat{J}(\mathbf{\Pi},t)d\mathbf{\Pi}, \tag{8.22}$$

where

$$\widehat{J}(\mathbf{\Pi},t) = -\frac{ie}{(2\pi\hbar)^3}\sum_{\sigma',\sigma}\widehat{a}^{+}_{\mathbf{\Pi}_f,\sigma'}(t)\widehat{a}_{\mathbf{\Pi},\sigma}(t)M^{(s_0)}\left(\mathbf{\Pi}_f,\sigma';\mathbf{\Pi},\sigma\right)e^{i\Delta(s_0,\mathbf{\Pi})t} \tag{8.23}$$

and the summation over $\mathbf{\Pi}$ has been replaced by integration according to

$$\sum_{\mathbf{\Pi}}\to\frac{1}{(2\pi\hbar)^3}\int d\mathbf{\Pi}.$$

Here we have introduced the notation

$$\mathbf{\Pi}_f = \mathbf{\Pi} - \hbar\mathbf{k}' + s_0\hbar\mathbf{k}. \tag{8.24}$$

The physical meaning of Eq. (8.23) with Eq. (8.24) is obvious: it describes the process where a particle with quasimomentum $\mathbf{\Pi}$ is annihilated and is created in the state with quasimomentum $\mathbf{\Pi} - \hbar\mathbf{k}' + s_0\hbar\mathbf{k}$ with the emission of a photon with the frequency ω' and momentum $\mathbf{k}'$.

Taking into account Eqs.(8.11), (8.17), (8.9), and (8.10) for the operator $\widehat{J}(\mathbf{\Pi},t)$ we obtain the equation

$$\frac{\partial\widehat{J}(\mathbf{\Pi},t)}{\partial t} - i\Delta\left(s_0,\mathbf{\Pi}\right)\widehat{J}(\mathbf{\Pi},t) = \frac{e^2A_e}{2c\hbar(2\pi\hbar)^3}$$

$$\times\sum_{\sigma',\sigma,\sigma_1}\Big\{M^{(s_0)}\left(\mathbf{\Pi}_f,\sigma';\mathbf{\Pi},\sigma\right)M^{(-s_0)}\left(\mathbf{\Pi},\sigma_1;\mathbf{\Pi}_f,\sigma'\right)\widehat{a}^{\dagger}_{\mathbf{\Pi},\sigma_1}(t)\widehat{a}_{\mathbf{\Pi},\sigma}(t)$$

$$-M^{(s_0)}\left(\mathbf{\Pi}_f,\sigma';\mathbf{\Pi},\sigma\right)M^{(-s_0)}\left(\mathbf{\Pi},\sigma;\mathbf{\Pi}_f,\sigma_1\right)\widehat{a}^{+}_{\mathbf{\Pi}_f,\sigma'}(t)\widehat{a}_{\mathbf{\Pi}_f,\sigma_1}(t)\Big\}, \tag{8.25}$$

where we have kept only resonant terms. These terms are predominant in near-resonant emission/absorption, since their detuning is much smaller than that of nonresonant terms, which are detuned from resonance by $\omega >> \left|\Delta\left(s_0,\mathbf{\Pi}\right)\right|$.

We will assume that the electron beam is nonpolarized. This means that the initial single-particle density matrix in momentum space is

$$\rho_{\sigma_1\sigma_2}(\mathbf{\Pi}_1,\mathbf{\Pi}_2,0) =< \widehat{a}^{+}_{\mathbf{\Pi}_2,\sigma_2}(0)\widehat{a}_{\mathbf{\Pi}_1,\sigma_1}(0) >= \rho_0(\mathbf{\Pi}_1,\mathbf{\Pi}_2)\delta_{\sigma_1,\sigma_2}. \tag{8.26}$$

Here $\rho_0(\mathbf{\Pi},\mathbf{\Pi})$ is connected to the classical momentum distribution function $F(\mathbf{\Pi})$ by the equation

$$\rho_0(\mathbf{\Pi},\mathbf{\Pi}) = \frac{(2\pi\hbar)^3}{2}F_0(\mathbf{\Pi}). \tag{8.27}$$

For the expectation value of $\widehat{J}(\mathbf{\Pi},t)$ from Eq. (8.25) we have

$$\frac{\partial J(\mathbf{\Pi},t)}{\partial t} - i\Delta\left(s_0,\mathbf{\Pi}\right) J(\mathbf{\Pi},t) = \frac{e^2 M^2}{4\hbar c} A_e \left(F(\mathbf{\Pi},t) - F(\mathbf{\Pi}_f,t)\right), \quad (8.28)$$

where

$$F(\mathbf{\Pi}_1,t) = \frac{2}{(2\pi\hbar)^3} < \widehat{a}_{\mathbf{\Pi}_1,\sigma_1}(t)\widehat{a}_{\mathbf{\Pi}_1,\sigma_1}(t) >, \quad (8.29)$$

$$M^2 = \sum_{\sigma',\sigma} M^{(s_0)}\left(\mathbf{\Pi}_f,\sigma';\mathbf{\Pi},\sigma\right) M^{(-s_0)}\left(\mathbf{\Pi},\sigma;\mathbf{\Pi}_f,\sigma'\right). \quad (8.30)$$

The M^2 is reduced to the usual calculation of a trace (see Eq. (1.112), where summation over the photon polarizations should not be made), and in our notations we have

$$M^2 = \frac{2c^4}{\Pi_{f0}\Pi_0}\left\{\left|\left[(p\epsilon')\, Q_{0s} - \frac{e}{c}\left(Q_{1s}\epsilon'\right)\right]\right|^2\right.$$

$$\left. - \frac{e^2}{4c^2}\frac{(\hbar k'k)^2}{(kp')(kp)}\left[\left|Q_{1s}\right|^2 + Re\left(Q_{2s}Q_{0s}^*\right)\right]\right\}, \quad (8.31)$$

where

$$\epsilon' = \epsilon - k'\left(\frac{k\epsilon}{kk'}\right). \quad (8.32)$$

In Eq.(8.31) one can neglect the terms on the order of $(\hbar k'k/(kp))^2 << 1$ as for a FEL this condition is always satisfied. Taking into account Eqs. (8.11), (8.17), (8.9), (8.10), and (8.29) for $F(\mathbf{\Pi},t)$ and $F(\mathbf{\Pi}_f,t)$ we obtain

$$\frac{\partial F(\mathbf{\Pi},t)}{\partial t} = -\frac{1}{2\hbar c}\left(A_e^* J(\mathbf{\Pi},t) + A_e J^*(\mathbf{\Pi},t)\right), \quad (8.33)$$

$$\frac{\partial F(\mathbf{\Pi}_f,t)}{\partial t} = \frac{1}{2\hbar c}\left(A_e^* J(\mathbf{\Pi},t) + A_e J^*(\mathbf{\Pi},t)\right). \quad (8.34)$$

To take into account the pulse propagation effects we can replace the time derivatives by the following expression:

$$\frac{\partial}{\partial t} \to \frac{\partial}{\partial t} + \overline{\mathbf{v}}\frac{\partial}{\partial \mathbf{r}},$$

where $\overline{\mathbf{v}} = c^2\mathbf{\Pi}/\Pi_0$ is the mean velocity of the electron beam and the convectional part of the derivative expresses the pulse propagation effects. Introducing the new quantity

$$\delta F(\mathbf{\Pi}, t) = F(\mathbf{\Pi}, t) - F(\mathbf{\Pi}_f, t), \tag{8.35}$$

which physically expresses population inversion in momentum space, from Eqs.(8.19), (8.22), (8.28), (8.33), and (8.34) we obtain the self-consistent set of equations:

$$\frac{\partial J(\mathbf{\Pi})}{\partial t} + \overline{\mathbf{v}}\frac{\partial J(\mathbf{\Pi})}{\partial \mathbf{r}} - i\Delta(s_0, \mathbf{\Pi}) J(\mathbf{\Pi}) = \frac{e^2 M^2}{4\hbar c} A_e \delta F(\mathbf{\Pi}),$$

$$\frac{\partial \delta F(\mathbf{\Pi})}{\partial t} + \overline{\mathbf{v}}\frac{\partial \delta F(\mathbf{\Pi})}{\partial \mathbf{r}} = -\frac{1}{\hbar c}\left(A_e^* J(\mathbf{\Pi}) + A_e J^*(\mathbf{\Pi})\right), \tag{8.36}$$

$$\frac{\partial A_e}{\partial t} + \frac{c^2\mathbf{k}'}{\omega'}\frac{\partial A_e}{\partial \mathbf{r}} = \frac{4\pi c}{\omega'}\int J(\mathbf{\Pi})\, d\mathbf{\Pi}.$$

These equations yield the conservation laws for the energy of the system and particle number:

$$\frac{\partial |A_e|^2}{\partial t} + \frac{c^2\mathbf{k}'}{\omega'}\frac{\partial |A_e|^2}{\partial \mathbf{r}} = -\frac{4\pi\hbar c^2}{\omega'}\int \left(\frac{\partial}{\partial t} + \overline{\mathbf{v}}\frac{\partial}{\partial \mathbf{r}}\right)\delta F(\mathbf{\Pi})\, d\mathbf{\Pi}, \tag{8.37}$$

$$\left(\frac{\partial}{\partial t} + \overline{\mathbf{v}}\frac{\partial)}{\partial \mathbf{r}}\right)\left(\left(\delta F(\mathbf{\Pi})\right)^2 + \frac{8}{e^2 M^2}\left|J(\mathbf{\Pi})\right|^2\right) = 0. \tag{8.38}$$

Note that from the set of Eqs. (8.36) one can obtain a small signal gain passing into perturbation theory which in the quasiclassical limit will coincide with the classical one (the latter will be done for a wiggler).

8.2 Nonlinear Quantum Regime of X-Ray Compton Backscattering Laser

In the quantum regime the emission and absorption are characterized by the widths

$$\Delta_e = \Delta(s_0, \mathbf{\Pi}) = \omega'(1 - \frac{\overline{\mathrm{v}}}{c}\cos\theta)$$

$$-s_0\omega(1 - \frac{\overline{\mathrm{v}}}{c}\cos\vartheta_0) + \frac{s_0\hbar\omega\omega'}{\Pi_0}(1 - \cos\theta_r), \tag{8.39}$$

$$\Delta_a = \Delta\left(s_0, \mathbf{\Pi}+\hbar\mathbf{k}' - s_0\hbar\mathbf{k}\right) = \Delta_e - \frac{2s_0\hbar\omega\omega'}{\Pi_0}(1-\cos\theta_r), \tag{8.40}$$

where ϑ_0, ϑ are the incident and scattering angles of the pump and probe photons with respect to the direction of the particle mean velocity $\overline{\mathbf{v}}$, and ϑ_r is the angle between the propagation directions of the pump and probe photons.

The quantum regime assumes that

$$\Delta_e - \Delta_a = \frac{2s_0\hbar\omega\omega'}{\Pi_0}(1-\cos\theta_0)$$

$$> \max\left\{\left|\frac{\partial\Delta_e}{\partial\eta_i}\delta\eta_i + \frac{\partial^2\Delta_e}{\partial\eta_i^2}(\delta\eta_i)^2\right|, \frac{\omega}{N_\omega}\right\}, \tag{8.41}$$

where by η_i we denote the set of quantities characterizing the electron beam and pump field and by $\delta\eta_i$ their spreads. The second term in the curly brackets of Eq. (8.41) expresses the resonance width caused by the finite interaction length and N_ω is the number of periods of the pump field. In particular, for the energetic ($\Delta\mathcal{E}$) and angular ($\Delta\vartheta$) spreads from Eq. (8.41) (for $\theta_r = \theta_0 \simeq \pi$, $\theta << 1$) we will have

$$\Delta\mathcal{E} \prec \hbar\omega', \tag{8.42}$$

$$\left|\theta\Delta\vartheta + \frac{\Delta\vartheta^2}{2}\right| < \frac{4s_0\hbar\omega}{\mathcal{E}}. \tag{8.43}$$

The conditions for keeping only one harmonic $s = s_0$ in the resonant current are

$$\frac{\Delta\mathcal{E}}{\mathcal{E}} << \frac{1}{s_0}, \tag{8.44}$$

$$\left|\theta\Delta\vartheta + \frac{\Delta\vartheta^2}{2}\right| << \frac{\omega}{\omega'}. \tag{8.45}$$

As we see, for not very high harmonics the conditions (8.44) and (8.45) are weaker than the conditions in the quantum regime (8.42), (8.43) , or (8.2) and are well enough satisfied for current accelerator beams.

Our goal is to determine the conditions under which we will have nonlinear amplification. We assume steady-state operation, i.e., dropping of all partial time derivatives in Eqs. (8.36). The considered setup is either a single-pass amplifier for which an injected input signal is necessary, or self-amplified coherent spontaneous emission for which a modulated beam is necessary. In addition, we will consider the case of exact resonance neglecting detuning in

Eqs. (8.36) assuming that electron beam momentum distribution is centered at $\Delta_e = 0$, i.e.

$$J(\mathbf{r}, t, \mathbf{\Pi}) = \overline{J}(\mathbf{r}, t)\, \delta(\mathbf{\Pi} - \mathbf{\Pi}_e), \tag{8.46}$$

$$\delta F(\mathbf{r}, t, \mathbf{\Pi}) = F(\mathbf{r}, t)\, \delta(\mathbf{\Pi} - \mathbf{\Pi}_e), \tag{8.47}$$

where for $\mathbf{\Pi}_e$

$$\Delta_e = \Delta(s_0, \mathbf{\Pi}_e) = 0.$$

To achieve maximal Doppler shift and optimal conditions of amplification we will assume counterpropagating electron and pump photon beams (X axis, $\theta_r = \theta_0 = \pi$). In this case the optimal condition for the linearly polarized pump wave is $\theta = 0$, while for the circular wave $\theta \sim \xi/\gamma_L$ ($\theta << 1$). For the on axis radiation we have the following known formula for the radiation wavelengths

$$\lambda' = \frac{1}{4} \frac{\lambda}{s_0 \gamma_L^2} \left(1 + \frac{1 + g^2}{2} \xi_0^2 \right), \tag{8.48}$$

where λ is the wavelength of the pump wave. For both cases we will assume that the envelope of the probe wave depends only on x . Then the set of Eqs. (8.36) and conservation laws (8.37), (8.38) are reduced to

$$\frac{d\overline{J}}{dx} = \frac{e^2 M^2}{4\hbar c \overline{\mathrm{v}}} A_e F,$$

$$\frac{dF}{dx} = -\frac{2}{\hbar c \overline{\mathrm{v}}} A_e \overline{J}, \tag{8.49}$$

$$\frac{dA_e}{dx} = \frac{4\pi}{\omega'} \overline{J},$$

$$F^2 + \frac{8}{e^2 M^2} \left| \overline{J} \right|^2 = N_0^2,$$

$$W = W_0 + \frac{\hbar \omega' \overline{\mathrm{v}}}{2} (F_0 - F),$$

where N_0 is the electron beam density, W is the probe wave intensity, and W_0 is the initial one. From Eq.(8.49) we have the following expressions for $\overline{J}$ and F:

$$F = N_0 \cos \left\{ \frac{e\,|M|}{2^{1/2} \hbar c \overline{\mathrm{v}}} \int_0^x A_e dx' + \varphi_0 \right\}, \tag{8.50}$$

$$\overline{J} = \frac{e\,|M|}{2^{3/2}} N_0 \sin\left\{ \frac{e\,|M|}{2^{1/2}\hbar c\overline{v}} \int_0^x A_e dx' + \varphi_0 \right\}, \tag{8.51}$$

where φ_0 is determined by boundary conditions. Denoting

$$\varphi = \frac{e\,|M|}{2^{1/2}\hbar c\overline{v}} \int_0^x A_e dx' + \varphi_0, \tag{8.52}$$

we arrive at the nonlinear pendulum equation

$$\frac{d^2\varphi}{dx^2} = \chi^2 \sin\varphi, \tag{8.53}$$

where

$$\chi^2 = \frac{\pi e^2 M^2 N_0}{\hbar\omega' c\overline{v}} \tag{8.54}$$

is the main characteristic parameter of amplification: $L_c = 1/\chi$ is the characteristic length of amplification. For the linearly polarized pump wave from Eqs.(8.13), (8.14), (8.15), and (8.31) we have

$$\chi_L = \frac{\xi_0\,|\Lambda_1(0,\beta,s_0)|}{2\gamma_L^2} \sqrt{\alpha_0 \frac{c\lambda}{s_0\overline{v}} N_0 (1+\xi_0^2/2)}. \tag{8.55}$$

Here α_0 is the fine structure constant and the function $\Lambda_1(0,\beta,s)$ is expressed by the ordinary Bessel functions:

$$\Lambda_1(0,\beta,s_0) \simeq \frac{1}{2}\left\{ J_{\frac{s_0-1}{2}}\left(\frac{s_0\xi_0^2}{4+2\xi_0^2}\right) - J_{\frac{s_0+1}{2}}\left(\frac{s_0\xi_0^2}{4+2\xi_0^2}\right) \right\}. \tag{8.56}$$

In this case only odd harmonics are possible. For the circularly polarized pump wave we have

$$\chi_c = \frac{\xi_0}{2\gamma_L^2}\left(\frac{\theta\gamma_L}{\xi_0} + \frac{s_0}{\alpha}\right) |J_{s_0}(\alpha)| \sqrt{\alpha_0 \frac{c\lambda}{s_0\overline{v}} N_0 (1+\xi_0^2+\theta^2\gamma_L^2)}, \tag{8.57}$$

and the argument of the Bessel function is

$$\alpha \simeq \frac{2s_0\xi_0\theta\gamma_L}{1+\xi_0^2+\theta^2\gamma_L^2}. \tag{8.58}$$

We will consider two regimes of amplification which are determined by initial conditions. For the first regime the initial macroscopic transition current of the electron beam is zero and it is necessary to have a seeding electromagnetic wave. In this case the following boundary conditions are imposed:

$$F\mid_{x=0}= N_0; \quad \overline{J}\mid_{x=0}= 0; \quad W\mid_{x=0}= W_0. \tag{8.59}$$

The solution for the probe wave intensity in this case is written as

$$W(x) = W_0 dn^{-2}\left(\frac{\chi}{\kappa}x;\kappa\right), \tag{8.60}$$

$$\kappa = \left(1 + \frac{W_0}{N_0\hbar\omega'\overline{\mathrm{v}}}\right)^{-\frac{1}{2}}, \tag{8.61}$$

where $dn(x,\kappa)$ is the elliptic function of Jacobi and κ its module.

As is known $dn(x,\kappa)$ is the periodic function with the period $2K(\kappa)$, where $K(\kappa)$ is the complete elliptic integral of first order. At the distances $L = (2r+1)\kappa K(\kappa)/\chi$ $(r = 0,1,2,...)$ the wave intensity reaches its maximal value which equals

$$W_{\max} = W_0 + N_0\hbar\omega'\overline{\mathrm{v}}. \tag{8.62}$$

For the short interaction length $x \ll L_c$ from Eq.(8.60) we have

$$W(x) = W_0\left(1 + \chi^2 x^2\right),$$

and the wave gain is rather small. To extract maximal energy from the electron beam the interaction length should be at least on the order of half the spatial period of the wave envelope variation — $\kappa K(\kappa)/\chi$. Under this condition the intensity value $W_{\max} = W_0 + N\hbar\omega'\overline{\mathrm{v}}$ is achieved, because all electrons make a contribution in the radiation field. Taking into account that seed power is much smaller than $W_{\max}$ and if $1-\kappa << 1$

$$K(\kappa) \to \frac{1}{2}\ln\left[\frac{16}{1-\kappa^2}\right],$$

for amplification length we will have

$$L \simeq L_c \ln\left(4\frac{W_{\max}}{W_0}\right). \tag{8.63}$$

Let us now consider the other regime of wave amplification when the electron beam is modulated — "macroscopic transition current" J differs from zero. This regime can operate without any initial seeding power ($W_0 = 0$). Thus, we will consider the optimal case with the following initial conditions:

$$\overline{J}\mid_{x=0}= J_0; \qquad F\mid_{x=0}= \delta N_0; \qquad W\mid_{x=0}= 0. \tag{8.64}$$

Then the wave intensity is expressed by

$$W(x) = \frac{N_0\hbar\omega'\overline{\mathrm{v}}}{2}\left(1-\frac{\delta N_0}{N_0}\right)\left[\frac{1}{dn^2(\chi x;k)}-1\right], \tag{8.65}$$

and module κ of Jacobi elliptic function is determined by

$$\kappa = \frac{1}{2}(1+\frac{\delta N_0}{N_0}). \tag{8.66}$$

As is seen from Eq. (8.65) in this case the intensity varies periodically with the distances as well, with the maximal value of intensity

$$W'_{\max} = \frac{N_0\hbar\omega'\overline{\mathrm{v}}}{2}(1+\frac{\delta N_0}{N_0}). \tag{8.67}$$

The second regime is more interesting. It is the regime of amplification without initial seeding power and has superradiant nature. For the short interaction length $x \ll L_c$ according to Eq. (8.65)

$$W(x) = \frac{N_0\hbar\omega'\overline{\mathrm{v}}\chi^2x^2}{4}\left(1-\frac{\delta N_0}{N_0}\right). \tag{8.68}$$

The intensity is scaled as N_0^2 ($\chi^2 \sim N_0$) which means that we have a superradiation. The radiation intensity in this regime reaches a significant value even at $x \ll L_c$.

The coherent interaction time of electrons with probe radiation is confined by the several relaxation processes. To be more precise in the self-consistent set of Eqs. (8.36) we should add the terms describing spontaneous transitions and other relaxation processes. Since we have not taken into account the relaxation processes, this consideration is correct only for the distances $L \preceq c\tau_{\min}$, where $\tau_{\min}$ is the minimum of all relaxation times. Due to spontaneous radiation electrons will lose energy $\sim \hbar\omega'$ at the distances

$$L_s \simeq c\frac{\hbar\omega'}{W_s} = \frac{3}{2\pi}\frac{s_0\lambda}{\alpha_0(1+\xi_0^2/2)\xi_0^2}, \tag{8.69}$$

where W_s is the intensity of spontaneous radiation (for linearly polarized pump wave; for circularly polarized wave one should replace $\xi_0^2 \rightarrow 2\ \xi_0^2$). Although the cutoff harmonic increases with the increasing of ξ_0 ($s_c \sim \xi_0^3$), for the high laser intensities $\xi_0 \gtrsim 1$ the role of spontaneous radiation increases as $L_s \sim \xi_0^{-4}$ and the above mentioned regimes will be interrupted. Therefore, the obtained solutions are correct at the distances $\sim L_s$. At $\xi_0 \gtrsim 1$ for the high harmonics L_c decreases and simultaneously the quantum recoil $\hbar\omega'/\mathcal{E}$ increases, but $L_s \sim L_c$. The first regime will effectively work as a single-pass amplifier if $L_c \gtrsim 10L_s$ (see Eq. (8.63): $W_{\max} \simeq e^{L/L_c}W_0/4$).

The second regime may be more promising as it allows considerable output intensities even for the small interaction lengths (8.68). It is expected that the effects of energy and angular spreads will not have a significant influence on this regime as it is governed by the initial current and only Doppler dephasing and spontaneous lifetime may interrupt the superradiation process. Note that necessary for the second regime initially quantum modulation of the particle beam at the above optical frequencies can be obtained through multiphoton transitions in the laser field at the presence of a "third body". The possibilities of quantum modulation at hard X-ray frequencies in the induced Compton, undulator, and Cherenkov processes have been studied in Chapters 3 and 5.

8.3 Quantum Description of FEL Nonlinear Dynamics in a Wiggler

To evaluate the nonlinear gain of a FEL in a wiggler on the basis of quantum theory we need the relativistic wave function of an electron in a wiggler. We will consider linear (LW) as well as helical wigglers (HW). The magnetic field of a wiggler is described by the following vector potential:

$$\mathbf{A}_H = \{0, A_0\cos(\mathbf{k}_0\mathbf{r}), gA_0\sin(\mathbf{k}_0\mathbf{r})\}\,, \tag{8.70}$$

where

$$\mathbf{k}_0 \equiv \left\{\frac{2\pi}{\ell}, 0, 0\right\}, \tag{8.71}$$

with the wiggler step ℓ. In Eq. (8.70) $g = \pm 1$ correspond to HW, while $g = 0$ corresponds to LW.

The quantum dynamics of an electron in a wiggler will be described by the Dirac equation which in the quadratic form (see Eqs. (1.82), (1.83)), taking into account the specified field configuration (8.70), can be represented in the form

$$\left\{\hbar^2\frac{\partial^2}{\partial t^2} + c^2\hat{\mathbf{p}}^2 - 2ce\mathbf{A}_H\hat{\mathbf{p}} + e^2\mathbf{A}_H^2 + m^2c^4 - ech\widehat{\mathbf{\Sigma}}\mathbf{H}\right\}\Psi = 0, \tag{8.72}$$

where

$$\widehat{\mathbf{\Sigma}} = \begin{pmatrix} \sigma & 0 \\ 0 & \sigma \end{pmatrix} \tag{8.73}$$

is the spin operator with the $\widehat{\sigma}$ Pauli matrices and

$$\mathbf{H} = \mathrm{rot}\mathbf{A}_H \tag{8.74}$$

is the magnetic field of a wiggler.

As the magnetic field depends only on the $\phi = \mathbf{k}_0\mathbf{r}$, then raising from the symmetry, we seek a solution of Eq. (8.72) in the form

$$\Psi(\mathbf{r}, t) = F(\phi)e^{\frac{i}{\hbar}(\mathbf{pr}-\mathcal{E}t)}, \tag{8.75}$$

where $\mathcal{E}$ and $\mathbf{p}$ are the energy and momentum of a free electron.

To solve Eq. (8.72) we will consider $F(\phi)$ as a slowly varying bispinor function of ϕ (on the scale of $\mathbf{pk}_0/(\hbar k_0^2)$) and neglect the second derivative compared with the first order, which restricts the magnetic field strength by the condition

$$\xi_H \equiv \frac{eA_0}{mc^2} = \frac{eH_0\ell}{2\pi mc^2} << \gamma_L. \tag{8.76}$$

Here $\gamma_L = \mathcal{E}/mc^2$ is the Lorentz factor (ξ_H is the so-called wiggler parameter (5.26)).

Hence, from Eqs. (8.72) and (8.75) for $F(\phi)$ we will have the following equation:

$$\left\{2i\hbar\left(\mathbf{pk}_0\right)\frac{d}{d\phi} + 2\frac{e}{c}\mathbf{pA}_H - \frac{e^2}{c^2}\mathbf{A}_H^2 + \frac{e\hbar}{c}\widehat{\mathbf{\Sigma}}\mathbf{H}\right\} F(\phi) = 0. \tag{8.77}$$

The solution of Eq. (8.77) can be written in the operator form

$$F(\phi) = \exp\left\{\frac{i}{2\hbar\mathbf{pk}_0}\int\limits_{-\infty}^{\phi}\left(\frac{2e}{c}\mathbf{pA}_H - \frac{e^2}{c^2}\mathbf{A}_H^2\right)d\phi'\right\}$$

$$\times\exp\left\{\frac{ie}{2c\mathbf{pk}_0}\widehat{\mathbf{\Sigma}}\left[\mathbf{k}_0\mathbf{A}_H\right]\right\}\frac{u_\sigma}{\sqrt{2\mathcal{E}}}, \tag{8.78}$$

where u_σ is the bispinor amplitude of a free electron with polarization σ (it is assumed adiabatic entry of the electron into the wiggler — $\mathbf{H}(-\infty) = 0$).

Then taking into account the property of spin operator

$$\exp\left[\widehat{\mathbf{\Sigma}}\mathbf{a}\right] = \frac{1}{2}(\exp(a) + \exp(-a)) + \widehat{\mathbf{\Sigma}}\mathbf{a}\frac{1}{2a}(\exp(a) - \exp(-a)),$$

and taking into account the condition (8.76), which in this case restricts the parameter $a << 1$, for the wave function (8.75) we will have the expression

$$\Psi(\mathbf{r},t) = \left(1 + \frac{ie}{2c\,(\mathbf{pk}_0)} \widehat{\boldsymbol{\Sigma}}\left[\mathbf{k}_0 \mathbf{A}_H\right]\right) \frac{u_\sigma}{\sqrt{2\mathcal{E}}}$$

$$\times \exp\left\{ \frac{i}{\hbar}\left[\mathbf{pr} - \mathcal{E}t + \frac{1}{2\,(\mathbf{pk}_0)} \int\limits_{-\infty}^{\phi} \left(2\frac{e}{c}\mathbf{pA}_H - \frac{e^2}{c^2}\mathbf{A}_H^2\right) d\phi' \right]\right\}. \tag{8.79}$$

The wave function (8.77) is an analogy of the Volkov wave function (1.93). Therefore, it is reasonable to represent the wave function in the four-dimensional notation making analogy more evident. Introducing four-dimensional vector potential and "wave vector"

$$A_H = (0, \mathbf{A}_H); \quad k \equiv (0, -\frac{2\pi}{\ell}, 0, 0),$$

and taking into account that

$$\widehat{\boldsymbol{\Sigma}}\left[\mathbf{k}_0 \mathbf{A}_H\right] = i\widehat{k}\widehat{A}_H,$$

the wave function (8.77) will be written as

$$\Psi(\mathbf{r},t) = \left(1 + \frac{e}{2c\,(pk)} \widehat{k}\widehat{A}_H\right) \frac{u_\sigma}{\sqrt{2\mathcal{E}}}$$

$$\times \exp\left\{ -\frac{i}{\hbar}\left[px + \frac{1}{2\,(pk)} \int\limits_{-\infty}^{\phi} \left(2\frac{e}{c}pA_H - \frac{e^2}{c^2}A_H^2\right) d\phi' \right]\right\}. \tag{8.80}$$

Here $p = (\mathcal{E}/c, \mathbf{p})$ is the four-momentum of a free electron and $\widehat{a} = a^\mu \gamma_\mu$. As we see this wave function by the form coincides with the Volkov wave function (1.93). Hence, we will not repeat all calculations which have been done for the Compton effect and use the obtained results for spontaneous as well as for induced undulator radiation. The main difference in this case is that $k^2 \neq 0$ but taking into account Eq. (8.76) we can neglect the terms which come from $k^2 \neq 0$ (quantum recoil). This will be more evident in the Weizsäcker-Williams approach, when in the frame concerned with electrons the wiggler field is well enough described by a plane EM wave field.

Performing integration in Eq. (8.80), taking into account Eq. (8.70), for the electron wave function we will have

$$\Psi_{\mathbf{\Pi}\sigma} = \left[1 + \frac{e\widehat{k}\widehat{A}_H}{2c(kp)}\right] \frac{u_\sigma(p)}{\sqrt{2\Pi_0}} \exp\left[-\frac{i}{\hbar}\Pi x - \frac{i}{\hbar}\frac{e^2 A_0^2}{8c^2\,(pk)}\left(1 - g^2\right)\sin\left(2\mathbf{k}_0\mathbf{r}\right)\right]$$

$$\times \exp\left[\frac{i}{\hbar}\frac{eA_0}{c\,(pk)}\left(p_y \sin \mathbf{k}_0\mathbf{r} - p_z g \cos \mathbf{k}_0\mathbf{r}\right)\right], \tag{8.81}$$

where by further analogy with the Volkov states we have introduced four-quasimomentum

$$\Pi = p + k\frac{m^2c^2}{4kp}(1+g^2)\xi_H^2. \tag{8.82}$$

Hence, the state of an electron in the wiggler field (8.70) is characterized by the quasimomentum $\mathbf{\Pi}$ and polarization σ and the wave function (8.81) is normalized by the condition

$$\frac{1}{(2\pi\hbar)^3}\int \Psi^{\dagger}_{\mathbf{\Pi}'\sigma'}\Psi_{\mathbf{\Pi}\sigma}d\mathbf{r} = \delta(\mathbf{\Pi}-\mathbf{\Pi}')\delta_{\sigma,\sigma'}.$$

The FEL dynamics in the wiggler will be described by the same self-consistent set of Eqs. (8.36) with

$$M^2 = \frac{2c^4}{\mathcal{E}^2}\left|\left[(p\epsilon')\,Q_{0s}(\alpha,\beta,\varphi) - \frac{e}{c}\left(Q_{1s}(\alpha,\beta,\varphi)\epsilon'\right)\right]\right|^2, \tag{8.83}$$

and the parameters α, β, and φ are

$$\alpha = \frac{eA_0}{\hbar c}\left[\left(\frac{p_y}{\mathbf{pk}_0} - \frac{p'_y}{\mathbf{p}'\mathbf{k}_0}\right)^2 + g^2\left(\left(\frac{p_z}{\mathbf{pk}_0} - \frac{p'_z}{\mathbf{p}'\mathbf{k}_0}\right)^2\right)^2\right]^{1/2},$$

$$\beta = \frac{e^2A_0^2}{8c^2}(g^2-1)\frac{\mathbf{k}'\mathbf{k}_0}{(\mathbf{pk}_0)^2}, \tag{8.84}$$

$$\tan\varphi = \frac{g\left(\frac{p_z}{\mathbf{pk}_0} - \frac{p'_z}{\mathbf{p}'\mathbf{k}_0}\right)^2}{\left(\frac{p_y}{\mathbf{pk}_0} - \frac{p'_y}{\mathbf{p}'\mathbf{k}_0}\right)}.$$

The resonance detuning for the wiggler is

$$\hbar\Delta\left(s_0,\mathbf{\Pi}\right) = \sqrt{c^2\left(\mathbf{\Pi} - \hbar\mathbf{k}' - s_0\hbar\mathbf{k}_0\right)^2 + m^{*2}c^4}$$

$$-\sqrt{c^2\mathbf{\Pi}^2 + m^{*2}c^4} + \hbar\omega'. \tag{8.85}$$

The spectrum of emitted photons is determined from the conservation laws $\Delta\left(s_0,\mathbf{\Pi}\right) = 0$:

$$\omega' = \frac{s_0 \frac{2\pi}{\ell} \overline{\mathrm{v}} \cos \vartheta_0}{1 - \frac{\overline{\mathrm{v}}}{c} \cos \theta + \frac{2\pi c \hbar s_0}{\mathcal{E} \ell} \cos \theta_r}, \tag{8.86}$$

where θ and θ_r are the scattering angles of probe photons with respect to the electron beam direction of motion and undulator axis, respectively, and ϑ_0 is the angle of the electron beam direction of motion with respect to undulator axis. The last term in the denominator is the quantum recoil. Neglecting the latter for the on axis radiation $\theta = \vartheta_0$ we obtain the following known formula for the radiation wavelengths

$$\lambda' = \frac{1}{2} \frac{\ell}{s_0 \gamma_L^2} \left(1 + \frac{1 + g^2}{2} \xi_H^2 \right). \tag{8.87}$$

Note that the spectrum (8.87) coincides with the spectrum of Compton effect (8.48) with the factor 1/4 instead of 1/2. As has been mentioned the scheme of an X-ray Compton laser has an advantage with respect to the conventional undulator devices concerned with the satisfaction of condition (8.2) for the quantum regime of generation. To achieve this condition for current FEL devices operating in undulators is problematic as it presumes severe restrictions on the beam spreads.

8.4 High-Gain Regime of FEL

Now we will solve the self-consistent set of Eqs. (8.36) for FEL at an arbitrary detuning of resonance. As the most effective case the hydrodynamic instability of a cold electron beam will be considered and the criteria will be obtained showing that either High Gain or quantum regime of generation takes place depending on the beam parameters and amplifying photon energy.

We assume steady-state operation of FEL at which one can drop all partial time derivatives in Eqs. (8.36). To achieve maximal Doppler shift and optimal conditions of amplification we will assume that the electron beam propagates along the wiggler axis (OX) (or counterpropagating electron and pump photon beams). Consequently, the electron beam dynamics will be considered one dimensional.

Our goal is to determine the conditions under which we will have collective instability, which causes exponential growth of the probe wave. Hence, we will assume a small density perturbation for the electron beam and seek the solution of Eq. (8.36) in the form

$$\delta F = \delta F_0 \left(\Pi_x \right) + \delta F_1 \left(\Pi_x, x \right).$$

Then in the first order by the field we will obtain the following set of linear equations:

$$\overline{\mathrm{v}}\frac{dJ(x,\Pi_x)}{dx} - i\Delta\left(s_0,\Pi_x\right)J(x,\Pi_x) = \frac{e^2M^2}{4\hbar c}\delta F_0\left(\Pi_x\right)A_e\left(x\right), \tag{8.88}$$

$$\frac{dA_e\left(x\right)}{dx} = \frac{4\pi}{\omega'}\int J(x,\Pi_x)d\Pi_x, \tag{8.89}$$

where

$$\delta F_0\left(\Pi_x\right) = F_0\left(\Pi_x\right) - F_0\left(\Pi_x - \hbar k' - s_0\hbar k_0\right) \tag{8.90}$$

is defined via initial distribution function $F_0\left(\Pi_x\right)$.

Performing Laplace transformation

$$f\left(q\right) = \int_0^\infty f(x)e^{-qx}dx \tag{8.91}$$

for the functions $J(q,\Pi_x)$ and $A_e\left(q\right)$ we obtain

$$\left(\overline{\mathrm{v}}q - i\Delta\left(s_0,\Pi_x\right)\right)J(q,\Pi_x) = \frac{e^2M^2}{4\hbar c}\delta F_0\left(\Pi_x\right)A_e\left(q\right), \tag{8.92}$$

$$qA_e\left(q\right) = \frac{4\pi}{\omega'}\int J(q,\Pi_x)d\Pi_x. \tag{8.93}$$

From these equations we arrive at the following characteristic equation for variable q:

$$q = \frac{\pi e^2M^2}{\hbar\omega' c}\int\frac{\delta F_0\left(\Pi_x\right)}{\overline{\mathrm{v}}q - i\Delta\left(s_0,\Pi_x\right)}d\Pi_x. \tag{8.94}$$

For the initial cold electron beam with the distribution function

$$F_0\left(\Pi_x\right) = N_0\delta\left(\Pi_x - \Pi_{0x}\right) \tag{8.95}$$

from Eq. (8.94) one can obtain the equation

$$q = \chi^2\left[\frac{1}{q - i\frac{\Delta_e}{\overline{\mathrm{v}}}} - \frac{1}{q - i\frac{\Delta_a}{\overline{\mathrm{v}}}}\right], \tag{8.96}$$

where

$$\Delta_e = \Delta\left(s_0,\Pi_{0x}\right),$$

$$\Delta_a = \Delta\left(s_0,\Pi_{0x} + \hbar k' + s_0\hbar k_0\right)$$

are the resonance widths for the emission and absorption and χ is the main characteristic parameter of amplification in the quantum regime (see Eq. (8.54)). Equation (8.96) is the cubic equation known in the FEL theory, but it is more generalized and includes the quantum effects. We will solve the latter in the opposite limits, which characterize the quantum and classical high-gain regimes.

In the quantum regime when the electron beam momentum distribution is centered at $\Delta_e = 0$ and

$$|\chi| << \frac{|\Delta_a|}{\overline{\mathrm{v}}}, \tag{8.97}$$

the second term in the square brackets of Eq. (8.96) can be neglected and we obtain

$$q = \pm\chi,$$

whence the exponential growth rate in the quantum regime will be

$$G_q = \chi. \tag{8.98}$$

This result is predictable from the nonlinear solutions (8.54) and (8.65) for the short interaction lengths.

In the classical limit the quantum recoil can be neglected and since in this limit $\Delta_a = -\Delta_e$ (classical resonance), Eq.(8.96) under the condition

$$|q|^2 >> \frac{\Delta_e^2}{\overline{\mathrm{v}}^2} \tag{8.99}$$

can be rewritten as

$$q^3 = 2i\chi^2 \frac{\Delta_e}{\overline{\mathrm{v}}}, \tag{8.100}$$

whence the unstable root defines the classical result for exponential growth rate:

$$G_{cl} \equiv \frac{\sqrt{3}}{2}\left(2\chi^2 \frac{\Delta_e}{\overline{\mathrm{v}}}\right)^{1/3}. \tag{8.101}$$

For joint consideration of Compton and undulator FELs the resonance widths (8.39) and (8.85) at the classical resonance for the emission/absorption can be written as

$$\Delta_e = \epsilon \frac{\hbar\omega'}{\mathcal{E}} \frac{2\pi c s_0}{\lambda}, \tag{8.102}$$

where the factor $\epsilon = 2$ for Compton FEL and $\epsilon = 1$ for undulator FEL, and λ is the wavelength of the pump wave or wiggler step. Recalling the definition

(8.54) for the parameter χ and using Eq. (8.102) the classical exponential growth rate can be written as

$$G_{cl} \equiv \frac{\sqrt{3}}{2}\left(4\epsilon s_0 \frac{\pi^2 e^2 M^2 N_0}{\overline{\mathrm{v}}^2 \mathcal{E}\lambda}\right)^{1/3}. \tag{8.103}$$

In particular, at the linear polarization of the pump field for the on axis radiation from Eqs. (8.31) and (8.83) we have

$$M^2 = c^2 \frac{\xi_p^2}{2\gamma_L^2} \Lambda^2,$$

where

$$\Lambda = J_{\frac{s_0+1}{2}}\left(\frac{s_0 \xi_p^2}{4+2\xi_p^2}\right) - J_{\frac{s_0-1}{2}}\left(\frac{s_0 \xi_p^2}{4+2\xi_p^2}\right),$$

and $\xi_p = \xi_0$ and $\xi_p = \xi_H$ for Compton and undulator FELs, respectively. Then for the classical exponential growth rate (8.103) we obtain the known equation

$$G_{cl} \equiv \frac{\sqrt{3}}{2}\left(\frac{2\epsilon s_0 \pi^2 c^2 r_e N_0 \Lambda^2}{\overline{\mathrm{v}}^2 \lambda} \frac{\xi_p^2}{\gamma_L^3}\right)^{1/3}. \tag{8.104}$$

Finally we note that the condition (8.99) for the classical high-gain regime can be written as

$$\chi >> \frac{\Delta_e}{\overline{\mathrm{v}}}, \tag{8.105}$$

which is opposite the condition for the quantum regime (8.97).

8.5 Quantum SASE Regime of FEL

In the previous sections we have described the FEL dynamics by the universal self-consistent set of Eqs. (8.36) which were derived in detail to reveal the FEL dynamics in general. In particular, it has been solved in the steady-state regime neglecting the dependence on time. This is appropriate for the FEL when slippage due to the difference between the light and electron velocities is neglected. Here we describe the FEL dynamics in the Self-Amplified Spontaneous Emission (SASE) regime taking into account the propagation effects. Thus, we will not consider diffraction or saturation effects and the FEL dynamics will be considered to be one dimensional. Taking into account the mentioned fact and keeping the time derivatives in Eqs. (8.36), in a similar way as was done with respect to Eqs. (8.88) and (8.89) we will obtain the following set of linear equations:

$$\frac{\partial J(x,t,\Pi_x)}{\partial t}+\overline{\mathrm{v}}\frac{\partial J(x,t,\Pi_x)}{\partial x}-i\Delta\left(s_0,\Pi\right)J(x,t,\Pi_x)$$

$$=\frac{e^2M^2}{4\hbar c}\delta F_0\left(\Pi_x\right)A_e\left(x,t\right), \tag{8.106}$$

$$\frac{\partial A_e\left(x,t\right)}{\partial t}+c\frac{\partial A_e\left(x,t\right)}{\partial x}=\frac{4\pi c}{\omega'}\int J(x,t,\Pi_x)d\Pi_x, \tag{8.107}$$

where $\delta F_0\left(\Pi_x\right)$ is defined again by Eq. (8.90) via initial distribution function of the electron beam.

By Fourier transformation for slowly varying envelopes of the probe EM wave and electric current density

$$A_e(x,t)=\int_{-\infty}^{\infty}A_{\varpi}(x)e^{i\varpi t}d\varpi, \tag{8.108}$$

$$J(x,t,\Pi_x)=\int_{-\infty}^{\infty}J_{\varpi}(x,\Pi_x)e^{i\varpi t}d\varpi, \tag{8.109}$$

Eqs. (8.106) and (8.107) are reduced to the equations

$$\frac{\partial J_{\varpi}(x,\Pi_x)}{\partial x}-i\Theta_{\varpi}\left(\Pi_x\right)J_{\varpi}(x,\Pi_x)=\frac{e^2M^2}{4\hbar c\overline{\mathrm{v}}}A_{\varpi}(x)\delta F_0\left(\Pi_x\right), \tag{8.110}$$

$$\frac{\partial A_{\varpi}(x)}{\partial x}+i\frac{\varpi}{c}A_{\varpi}(x)=\frac{4\pi}{\omega'}\int J_{\varpi}(x,\Pi_x)d\Pi_x, \tag{8.111}$$

where

$$\Theta_{\varpi}\left(\Pi_x\right)=\frac{\Delta\left(s_0,\Pi_x\right)-\varpi}{\overline{\mathrm{v}}}. \tag{8.112}$$

The solution of Eq. (8.110) can be written as

$$J_{\varpi}(x,\Pi_x)=J_{\varpi}(0,\Pi_x)e^{i\Theta_{\varpi}(\Pi_x)x}$$

$$+\frac{e^2M^2}{4\hbar c\overline{\mathrm{v}}}\int_0^x e^{i\Theta_\varpi(\Pi_x)(x-x')}A_\varpi(x')\delta F_0(\Pi_x)\,dx'. \tag{8.113}$$

Here it is assumed that

$$J_\varpi(0,\Pi_x)=\overline{J}_\varpi\delta(\Pi_x-\Pi_{0x}), \tag{8.114}$$

where $\overline{J}_\varpi$ characterizes the shot noise in the electron beam or modulation depth for the initially modulated beam. Substituting Eq. (8.113) into Eq. (8.111) we obtain an integro-differential equation for the phase transformed amplitude $\widetilde{A}_\varpi(x)$ of the amplifying wave field:

$$\frac{\partial\widetilde{A}_\varpi(x)}{\partial x}+i\left(\frac{\varpi}{c}+\Theta_\varpi(\Pi_{0x})\right)\widetilde{A}_\varpi(x)=\frac{4\pi}{\omega'}\overline{J}_\varpi,$$

$$+\frac{\pi e^2M^2}{\hbar\omega' c\overline{\mathrm{v}}}\int\int_0^x e^{i(\Theta_\varpi(\Pi_x)-\Theta_\varpi(\Pi_{0x}))(x-x')}\widetilde{A}_\varpi(x')\delta F_0(\Pi_x)\,dx'd\Pi_x, \tag{8.115}$$

where

$$\widetilde{A}_\varpi(x)=A_\varpi(x)e^{-i\Theta_\varpi(\Pi_{0x})x}. \tag{8.116}$$

In the quantum regime, when condition (8.97) holds one can neglect the second term in Eq. (8.90) (which is equivalent to neglecting the absorption probability compared with the emission one) and put

$$\delta F_0(\Pi_x)\simeq N_0\delta(\Pi_x-\Pi_{0x});\quad \Delta(s_0,\Pi_{0x})=0 \tag{8.117}$$

in Eq. (8.115). Then we will obtain

$$\frac{\partial\widetilde{A}_\varpi(x)}{\partial x}-i\left(1-\frac{\overline{\mathrm{v}}}{c}\right)\frac{\varpi}{\overline{\mathrm{v}}}\widetilde{A}_\varpi(x)=\frac{4\pi}{\omega'}\overline{J}_\varpi+\chi^2\int_0^x\widetilde{A}_\varpi(x')dx'. \tag{8.118}$$

Performing Laplace transformation (8.91) on Eq. (8.118) we arrive at the following characteristic equation for variable q:

$$q^2-i\left(1-\frac{\overline{\mathrm{v}}}{c}\right)\frac{\varpi}{\overline{\mathrm{v}}}q-\chi^2=0, \tag{8.119}$$

and the solution of Eq. (8.118) can be written as

$$\widetilde{A}_{\varpi}(x) = \frac{1}{2i\pi} \oint \frac{q\widetilde{A}_{\varpi}(0) + \frac{4\pi}{\omega'}\overline{J}_{\varpi}}{(q - q_1)(q - q_2)} e^{qx} dq, \tag{8.120}$$

where $\widetilde{A}_{\varpi}(0)$ characterizes a coherent input signal. The contour integration in Eq. (8.120) is the result of the inverse Laplace transformation and encloses the poles which are the solutions of the characteristic equation (8.119):

$$q_1 = \frac{i}{2}\left(1 - \frac{\overline{\mathrm{v}}}{c}\right)\frac{\varpi}{\overline{\mathrm{v}}} + \chi\sqrt{1 - \frac{\left(1 - \frac{\overline{\mathrm{v}}}{c}\right)^2}{4\chi^2}\frac{\varpi^2}{\overline{\mathrm{v}}^2}}, \tag{8.121}$$

$$q_2 = \frac{i}{2}\left(1 - \frac{\overline{\mathrm{v}}}{c}\right)\frac{\varpi}{\overline{\mathrm{v}}} - \chi\sqrt{1 - \frac{\left(1 - \frac{\overline{\mathrm{v}}}{c}\right)^2}{4\chi^2}\frac{\varpi^2}{\overline{\mathrm{v}}^2}}. \tag{8.122}$$

In Eq. (8.120) the term proportional to $\widetilde{A}_{\varpi}(0)$ describes the amplification of the coherent input signal, while the second term proportional to $\overline{J}_{\varpi}$ describes either the amplification of the shot noise or coherent spontaneous emission (for the initially modulated electron beam). Since the main propose of this section is to study the amplification process without initial seed the first term will not be considered further. Hence, at $\widetilde{A}_{\varpi}(0) = 0$ Eq. (8.120) yields

$$\widetilde{A}_{\varpi}(x) = \frac{4\pi}{\omega'}\frac{\overline{J}_{\varpi}}{q_1 - q_2}e^{q_1 x} + \frac{4\pi}{\omega'}\frac{\overline{J}_{\varpi}}{q_2 - q_1}e^{q_2 x}. \tag{8.123}$$

The root q_1 has a positive real part that gives rise to an exponentially growing term in the radiation intensity. Keeping only this term and taking into account that $q_1 - q_2 \simeq 2\chi$, we have

$$\widetilde{A}_{\varpi}(x) = \frac{2\pi}{\omega'\chi}\overline{J}_{\varpi}e^{q_1 z}. \tag{8.124}$$

The spectral property of output radiation is defined by the dependence of q_1 on ϖ and from Eq. (8.121) we obtain

$$Req_1 \simeq \chi - \frac{\left(1 - \frac{\overline{\mathrm{v}}}{c}\right)^2}{8\chi}\frac{\varpi^2}{\overline{\mathrm{v}}^2}. \tag{8.125}$$

For the average spectral intensity

$$I_{\varpi}(x) = \frac{c}{8\pi}\left\langle \left|E_{\varpi}(x)\right|^2 \right\rangle = \frac{\omega'^2}{8\pi c}\left\langle \left|\widetilde{A}_{\varpi}(x)\right|^2 \right\rangle \tag{8.126}$$

with the help of Eqs. (8.124) and (8.125) we will have

$$I_{\omega}(x) = \frac{\pi}{2c\chi^2} \left\langle \left|\overline{J}_{\varpi}\right|^2 \right\rangle \exp\left[-\frac{(\omega - \omega')^2}{2\Delta_q^2(x)}\right] e^{2\chi x}, \tag{8.127}$$

where ω' is the resonant frequency ($\varpi \to \omega - \omega'$) and the spectral width in the quantum SASE regime is defined as follows:

$$\Delta_q(x) = \sqrt{\frac{2\chi}{x}\frac{\overline{\mathrm{v}}}{1 - \frac{\overline{\mathrm{v}}}{c}}}. \tag{8.128}$$

In the classical regime when condition (8.105) holds the electrons have almost the same probability of absorption or emission of a photon and the net gain factor is proportional to the derivative of the momentum distribution function $F_0(\Pi_x)$. Hence, from Eq. (8.90) one can put

$$\delta F_0(\Pi_x) \simeq \frac{\partial F_0(\Pi_x)}{\partial \Pi_x}\frac{\hbar\omega'}{c}. \tag{8.129}$$

For the initial cold electron beam (8.95) from Eq. (8.115) in this case we obtain

$$\frac{\partial \widetilde{A}_{\varpi}(x)}{\partial x} - i\left(1 - \frac{\overline{\mathrm{v}}}{c}\right)\frac{\varpi}{\overline{\mathrm{v}}}\widetilde{A}_{\varpi}(x) = \frac{4\pi}{\omega'}\overline{J}_{\varpi}$$

$$+iG_{cl}^3 \int_0^x (x - x')\,\widetilde{A}_{\varpi}(x')dx', \tag{8.130}$$

where G_{cl} is the classical exponential growth rate (8.101). Without initial seed the solution of Eq.(8.130) is given as

$$\widetilde{A}_{\varpi}(x) = -i\frac{2}{\omega'}\oint \frac{\overline{J}_{\varpi} q e^{qx}}{(q - q_1)(q - q_2)(q - q_3)}dq, \tag{8.131}$$

where $q_{1,2,3}$ are the solutions of the characteristic equation

$$q^3 - i\left(1 - \frac{\overline{\mathrm{v}}}{c}\right)\frac{\varpi}{\overline{\mathrm{v}}}q^2 - iG_{cl}^3 = 0. \tag{8.132}$$

The unstable solution (suppose $Req_1 > 1$) in this case is given as

$$\widetilde{A}_{\varpi}(x) = \frac{4\pi}{\omega'}\overline{J}_{\varpi}\frac{q_1}{(q_1 - q_2)(q_1 - q_3)}e^{q_1 x}, \tag{8.133}$$

where one can put

$$Req_1 \simeq G_{cl} - \frac{\left(1-\frac{\overline{\mathrm{v}}}{c}\right)^2}{12G_{cl}} \frac{\varpi^2}{\overline{\mathrm{v}}^2}, \tag{8.134}$$

$$\left|\frac{q_1}{(q_1-q_2)(q_1-q_3)}\right|^2 \simeq \frac{1}{12G_{cl}^2}. \tag{8.135}$$

Hence, for the average spectral intensity (8.126) we have

$$I_{\varpi}(x) = \frac{\pi}{6cG_{cl}^2} \left\langle \left|\overline{J}_{\varpi}\right|^2 \right\rangle \exp\left[-\frac{(\omega-\omega')^2}{2\Delta_{cl}^2(x)}\right] e^{2G_{cl}x}. \tag{8.136}$$

The spectral width in the classical SASE regime is defined as follows:

$$\Delta_{cl}(x) = \sqrt{\frac{3G_{cl}}{x}} \frac{\overline{\mathrm{v}}}{1-\frac{\overline{\mathrm{v}}}{c}}. \tag{8.137}$$

Comparing Eq. (8.136) with its quantum counterpart (8.127) one can see that for the same initial shot noise in the quantum regime the start-up intensity is enhanced by the factor $G_{cl}^2/\chi^2 >> 1$ (see conditions (8.97), (8.99)) and the spectrum of the SASE intensity is narrowed by the factor $\sqrt{2\chi/3G_{cl}} << 1$, while for the quantum SASE regime a longer amplification length is required.

8.6 High-Gain FEL on the Coherent Bremsstrahlung in a Crystal

To achieve the condition of coherency for generation of shortwave radiation by electron beams of considerably lower energies, in the problem of X-ray FEL it may be reasonable to consider other versions of stimulated radiation in the crystals, based on the coherent bremsstrahlung of charged particles on the periodic ionic lattice. It is clear that the coherent length in this scheme is confined by the multiple scattering of electrons in a crystal. The latter drastically increases the lasing threshold for the beam density. To compensate it we will consider the case when the electron beam current density is initially modulated.

Thus, we will investigate the lasing in the X-ray domain due to the coherent bremsstrahlung in a crystal, in the high-gain regime, when the electron beam moves close to the crystal lattice plane or axis. To avoid the channeling effect in a crystal we assume that the incident angle θ of an electron with respect to a crystalline plane or axis is larger than the Lindhard angle $\theta_L = \sqrt{2U_0/\mathcal{E}}$, where U_0 is the height of the barrier of a crystal plane (axis) potential, and $\mathcal{E}$ is the energy of an electron. In this case, when the radiation coherence length $l_c \sim \gamma^2 \mathrm{v}/\omega$ (γ being the Lorentz factor, v the electron velocity, and ω the radiation frequency) exceeds the crystal lattice periods: $l_c >> d_i$, the bremsstrahlung emitted from the various centers interfere with each other and the enhancement of radiation occurs, which is referred to as coherent bremsstrahlung. The trajectory of a particle can be considered as quasi-linear and the trajectory period will be determined by the space period of the crystal potential. In this respect the coherent bremsstrahlung is close to the undulator radiation, where the trajectory period is determined by the space period of the magnetic field. We will assume that

$$N_c Z_a e^2/\hbar \mathrm{v} >> 1; \quad l_c > R/\theta, \tag{8.138}$$

where Z_a is the nuclear charge number of the crystal atoms, R is the radius of screening, N_c is the number of atoms on the radiation coherence length l_c, and $\theta << 1$. In this case one can treat the particle motion by the classical theory (the first condition is contrary to the Born one) and approximate the interaction of the particle with the crystal by the continuous potential (second condition of (8.138)) of atomic planes or strings, i.e., the atomic potential is averaged over the given crystallographic plane or axis, which is oriented at a small angle to the incident beam. For the concreteness we will consider the case of the atomic plane, then the generalization for the crystal axis will be obvious. The potential of the atomic plane, which governs the particle motion, can be represented as a superposition of the potentials

$$U(x) = \sum_l U_p(x - ld_1),$$

$$U_p(x) = \frac{1}{d_2 d_3} \int u(\mathbf{r}) dy dz, \tag{8.139}$$

where $u(\mathbf{r})$ is the single atomic potential. Considering $U(x)$ as a perturbation, from the classical equations of motion one can obtain the perturbed velocity of the electron, which is responsible for the coherent bremsstrahlung. The latter can be expressed in the form

$$\mathrm{v}'_x \simeq \sum_n \frac{u_n}{m\mathrm{v}\theta\gamma} \exp\left[i\frac{2\pi}{d_1} n\mathrm{v}\theta t\right], \tag{8.140}$$

where

$$u_n = \frac{4\pi N_a Z_a e^2}{\left(\frac{2\pi n}{d_1}\right)^2 + \frac{1}{R^2}} \tag{8.141}$$

is the Fourier component of the potential $U_p(x)$ (8.139). Here N_a is the atomic concentration in the crystal and for the single atomic potential we have taken a screening Coulomb potential. We will consider the more reasonable case of amplification of forward radiation of the electrons. Ignoring space charge effects, the probe EM wave can be treated as transversal, propagating parallel to the electron beam. We assume the probe wave to be linearly polarized with the carrier frequency ω, wave vector k, and electric field strength

$$E = E_0(t, z_l)e^{i(kz_l - \omega t)} + \text{c.c.}, \tag{8.142}$$

where $E_0(t, z_l)$ is a slowly varying envelope and z_l is the coordinate along the electron beam propagation. Taking into account (8.140), the rate of energy exchange between the electrons and probe wave can be expressed in the form

$$\frac{d\mathcal{E}}{dz_l} \simeq \sum_n \frac{u_n e E_0(t, z_l)}{m^2 \text{v}\theta\gamma} \exp\left[i\Psi_n\right] + \text{c.c.}, \tag{8.143}$$

where

$$\Psi_n = kz_l - \omega t + \frac{2\pi}{d_1} n \text{v}\theta t. \tag{8.144}$$

Then, the coherence condition, at which the bremsstrahlung emitted from various crystal centers along the electron path interfere constructively, is the following

$$\frac{d\Psi_n}{dz_l} = 0; \quad \omega = \frac{2\pi n \text{v}\theta}{d_1(1 - \frac{\text{v}}{c})}, \tag{8.145}$$

which represents the general resonance condition for the forward radiation. Though the consideration can be easily generalized to higher harmonics, we will consider the fundamental resonance and keep only the resonant term ($n = 1$) in Eq. (8.143). For the formulation of the Maxwell–Vlasov equations it is convenient to change the independent variables from (z_l, t) to $(z_l, \Psi_1 \equiv \psi)$ and the conjugate variable to ψ will be

$$\chi = (\gamma - \gamma_0)/\gamma_0, \tag{8.146}$$

where $mc^2\gamma_0$ is the electron resonant energy defined from Eq. (8.145). From Eqs. (8.143) and (8.144) one can obtain the equations for (ψ, χ), generally known as the pendulum equations in the conventional undulator version of FEL:

$$\frac{d\psi}{dz_l} = \frac{4\pi}{d_1}\theta\chi, \tag{8.147}$$

$$\frac{d\chi}{dz_l} = \frac{e\xi_{cb}}{2mc\text{v}\gamma_0^2} E_0(\psi, z_l)e^{i\psi} + \text{c.c.}, \tag{8.148}$$

where by further analogy with the undulator or Compton FEL we have introduced the effective interaction parameter ξ_{cb} for coherent bremsstrahlung

$$\xi_{cb} = \frac{8\pi c N_a Z_a r_e R^2}{\mathrm{v}\theta}, \tag{8.149}$$

which has the same physical meaning as the usual ξ_H parameter for conventional undulators (r_e is the electron classical radius). Hence, taking into account Eqs. (8.147) and (8.148) the Vlasov equation for the phase-space distribution function $F(z_l, \psi, \chi)$ will be

$$\frac{\partial F}{\partial z_l} + \frac{4\pi\theta\chi}{d_1}\frac{\partial F}{\partial \psi} + \frac{e\xi_{cb}}{2mc\mathrm{v}\gamma_0^2}$$

$$\times \left(E_0(t, z_l)e^{i\psi} + \text{c.c.}\right)\frac{\partial F}{\partial \chi} = 0. \tag{8.150}$$

The Maxwell equation for the slowly varying envelope of the probe wave can be written as

$$\frac{\partial E_0}{\partial z_l} + \frac{2\pi\theta}{d_1}\frac{\partial E_0}{\partial \psi} + \mu E_0 = -\pi e\frac{\xi_{cb}}{\gamma_0}\overline{e^{-i\psi}\int F d\chi}, \tag{8.151}$$

where the bar denotes averaging over time and space much larger than $(1/\omega, 1/k)$, and to take into account the probe wave damping because of absorption and scattering in the crystal, we have introduced absorption coefficient μ. Equations (8.150) and (8.151) are the self-consistent set of equations for the considered scheme of FEL. The main impending factor in the coherent bremsstrahlung process, which we have not taken into account in Eq. (8.150), is the multiple scattering of electrons in a crystal. The latter will not violate the electron coupling with the radiation field and, consequently, will not have essential bearing on the amplification process, if the detuning of the phase ψ due to multiple scattering is less than π. For the forward radiation we have the condition $L\delta\vartheta_{ms}^2/2 < \lambda$ (where λ is the wavelength of the amplifying wave), which restricts the effective interaction length of the electrons in a crystal

$$L < L_{ms} = \left(8\pi r_e^2 Z_a^2 N_a d^{-1}\theta \ln 183 Z_a^{-1/3}\right)^{-1/2}, \tag{8.152}$$

where L_{ms} and ϑ_{ms} are the characteristic length and angle of multiple scattering.

We shall determine the conditions under which the collective instability develops in the coherent bremsstrahlung process causing the exponential growth of the probe wave. Correspondingly, we will assume steady-state operation and a small density perturbation for the electron beam and seek the solution of Eq.(8.150) in the form

$$F = F_0 + F_1 e^{i\psi} + \text{c.c.},$$

dropping all partial derivatives with respect to ψ in the equations for F_1 and E_0. For the initial cold electron beam at the exact resonance with distribution function

$$F_0(\chi) = N_0 \delta(\chi)$$

(N_0 is the mean density of the electron beam) from Eqs.(8.150) and (8.151) one can obtain the integro-differential equation for the slowly varying envelope E_0:

$$\frac{dE_0}{dz_l} + \mu E_0 = -\frac{\pi e \xi_{cb} \delta N_0}{\gamma_0} + i\alpha_g \int_0^z (z - z') E_0(z') dz'. \tag{8.153}$$

Here we introduced the gain parameter

$$\alpha_g = \frac{2\pi^2 c r_e \xi_{cb}^2 N_0 \theta}{\mathrm{v} \gamma_0^3 d_1}, \tag{8.154}$$

and for the initially modulated electron beam it was assumed that

$$F_1(z = 0, \chi) = \delta N_0 \delta(\chi),$$

where $\delta N_0 / N_0$ is the modulation depth. Performing Laplace transformation (8.91) on Eq. (8.153), we obtain the following characteristic equation:

$$q^3 + \mu q^2 - i\alpha_g = 0, \tag{8.155}$$

which for the values $\alpha_g > \mu$ gives the exponential growth rate for coherent bremsstrahlung

$$G = \frac{\sqrt{3}}{2} \left(\frac{2\pi^2 c r_e \xi_{cb}^2 N_0 \theta}{\mathrm{v} \gamma_0^3 d_1} \right)^{1/3}. \tag{8.156}$$

For the high-gain regime the growth rate (8.156) is required to be larger than the characteristic ones for the impending effects of radiation absorption and multiple scattering of electrons in the crystal: $G > \max\{\mu, L_{ms}^{-1}\}$.

For the electron beam low currents $G << \{\mu, L_{ms}^{-1}\}$ and for the initially modulated current densities ($\delta N_0 \neq 0$), with no input signal, the solution of Eq. (8.153) gives

$$E_0 \simeq \frac{\pi e \xi_{cb} \delta N_0}{\gamma_0 \mu} \left(e^{-\frac{\mu}{2} z} - 1 \right) . \tag{8.157}$$

In Eq. (8.157) the amplification length z is restricted by the length of the multiple scattering of electrons in the crystal L_{ms}. At the large absorption of amplifying radiation in the crystal, when $\mu >> 1/L_{ms}$, for the maximal power of output radiation, which has a superradiant nature, we have

$$I \simeq \frac{c}{2\pi} \left[\frac{\pi e \xi_{cb} \delta N_0}{\gamma_0 \mu} \right]^2 . \tag{8.158}$$

In the inverse case of small absorption $\mu << 1/L_{ms}$ from Eq. (8.157) we have

$$I \simeq \frac{c}{8\pi} \left[\frac{\pi e \xi_{cb} \delta N_0}{\gamma_0} L_{ms} \right]^2 . \tag{8.159}$$

Although the regimes (8.158) and (8.159) require low electron beam currents, they nevertheless may provide considerable output intensities for coherent X-ray. Hence, the considered setup of coherent bremsstrahlung in a crystal may serve as a powerful mechanism for prebunched electron beam superradiation, at moderate relativistic energies of electron beams.

8.7 Nonlinear Scheme of X-Ray FEL on the Channeling Particle Beam in a Crystal

As the channeling radiation of ultrarelativistic electrons and positrons lies in the X-ray and γ-ray domain, and its spectral intensity exceeds that of other radiation sources in this frequency range, hence, the stimulated channeling radiation of charged particles is of certain interest as a potential FEL in the short wavelength domain. As the absorption coefficients of X-rays and γ-rays in crystals are very high ($\sim 10^2 \div 10^3 \mathrm{cm}^{-1}$) and the construction of mirrors in this domain is very problematic, it is necessary to study the possibilities of realization of the single-pass nonlinear regimes of X-ray amplification.

To obtain coherent radiation in the crystal channel it is most appropriate to use electron beams with comparatively low energies ($\mathcal{E} \lesssim 50$ MeV for planar channeled electrons and $\mathcal{E} \lesssim 10$ MeV for axial ones). First, the states of channeled electrons are most stable in this energy region, i.e., the scattering of channeled particles on atomic electrons and nuclei of the lattice are suppressed. Then, at these energies a few discrete energy levels in the transverse potential well of the channeled electron are formed that are not equidistant. In this case by means of varying the angle of incidence of the electron beam to the crystal an inverted population of electron states in the transverse potential can be reached. In addition, at low energies it is possible to use electron beams with high densities and increase the population inversion. Because the energy levels are not equidistant the stimulating EM wave resonantly couples only two energy levels, and the physical processes in the above-mentioned case of the channeling are similar to those of a two-level atom (two-dimensional "atom" in the case of axial channeling, and one-dimensional in the case of planar channeling) moving with relativistic velocity.

The problem concerned with controlling the overpopulation of channeled particles can be overcome by two component laser-assisted schemes. In particular, the stimulated Compton scattering by channeled particles is of certain interest as the necessity of inverse population of transverse levels for lasing vanishes and the cross section of the considered process is resonantly enhanced by several orders with respect to the Compton process on free electrons.

For the description of a FEL operating in the crystal, where transverse degrees of freedom of the particles are fully quantized, we will begin from the second quantization formalism. The second quantized Hamiltonian is

$$\widehat{H} = \int \widehat{\Psi}^{+} \widehat{H}_0 \widehat{\Psi} d\mathbf{r} + \widehat{H}_{int}, \tag{8.160}$$

where $\widehat{\Psi}$ is the fermionic field operator, $\widehat{H}_0$ is the one-particle Hamiltonian in the channel of the crystal (along the axis OZ) with average electrostatic potential $U(\rho)$ ($\rho \equiv x$ in case of a planar channeling and $\rho \equiv \sqrt{x^2+y^2}$ for the axial one), and $\widehat{H}_{int}$ is the interaction Hamiltonian:

$$\widehat{H}_{int} = -\frac{1}{c} \int \widehat{\mathbf{j}} \left(\mathbf{A}_e + \mathbf{A} \right) d\mathbf{r}. \tag{8.161}$$

Here $\widehat{j} = e\widehat{\Psi}^{+} \widehat{\alpha} \widehat{\Psi}$ is the current density operator ($\widehat{\alpha}$ is the Dirac matrix) and $\mathbf{A}_e$, $\mathbf{A}$ are the vector potentials of the probe and pump EM waves, respectively. To achieve maximal Doppler shift and optimal conditions of amplification, we will assume a co-propagating probe EM wave and channeled particle beam and counterpropagating pump EM wave. We will consider a linearly polarized (along OX) pump EM wave with the frequency ω and wave vector k that is described by the vector potential

$$\mathbf{A} = \widehat{\mathbf{x}} \frac{A_0}{2} \left\{ e^{i(\omega t + kz)} + \text{c.c.} \right\}. \tag{8.162}$$

We assume the probe wave to be linearly polarized with the carrier frequency ω', wave vector k', and vector potential

$$\mathbf{A}_e = \widehat{\mathbf{x}} \frac{1}{2} \left\{ A_e(t,z) e^{i(\omega' t - k' z)} + \text{c.c.} \right\}, \tag{8.163}$$

where $A_e(t,z)$ is a slowly varying envelope.

As in Section 8.1 we write the Heisenberg field operator of the particles in the form of an expansion in the stationary states

$$\widehat{\Psi}(\mathbf{r},t) = \sum_{\mu,p_z} \widehat{a}_{\mu,p_z}(t) e^{-\frac{i}{\hbar}\mathcal{E}_\mu(p_z)t} \psi_{\mu,p_z}. \tag{8.164}$$

The creation and annihilation operators $\widehat{a}^+_{\mu,p_z}(t)$ and $\widehat{a}_{\mu,p_z}(t)$, associated with positive energy $\mathcal{E}_\mu(p_z)$ solutions of the Dirac equation, satisfy the usual anti-commutation rules at equal times (see Eqs. (8.9), (8.10)). Here μ, p_z are the complete set of quantum numbers $\mu = \{p_y, n, \sigma\}$ for the planar channeling and $\mu = \{\mathfrak{m}, n, \sigma\}$ for the axial one, n is the main quantum number and $\mathfrak{m}$ is the magnetic quantum number, σ characterizes spin polarization and p_y, p_z are the components of particle momentum; ψ_{μ,p_z} are the normalized eigenvectors of channeled particle corresponding to the given set of quantum numbers. We will assume that probe and pump waves resonantly couple only two transverse levels, which will be labeled (0) and (1). It is also assumed that the particle beam is nonpolarized and the probability of transitions with the spin flip is negligible (this imposes a restriction on the wave frequency $\hbar\omega' << \mathcal{E}_\mu(p_z)$). As a result, taking into account Eqs. (8.161)–(8.164) and keeping only the resonant terms (Rotating Frame Approximation) the Hamiltonian (8.160) can be reduced to the form

$$\widehat{H} = \sum_{p_z} \left[\mathcal{E}_0(p_z)\widehat{a}^+_{0,p_z}\widehat{a}_{0,p_z} + \mathcal{E}_1(p_z)\widehat{a}^+_{1,p_z}\widehat{a}_{1,p_z}\right] + \widehat{H}_{int} \tag{8.165}$$

with the interaction Hamiltonian:

$$\widehat{H}_{int} = \sum_{p_z} \left[\frac{\beta_\perp}{2c}\left\{ieA_0\widehat{a}^+_{0,p_z+\hbar k}\widehat{a}_{1,p_z}e^{i\Gamma(p_z+\hbar k,p_z,\omega)t}\right.\right.$$

$$\left.\left.+ieA_e\widehat{a}^+_{0,p_z-\hbar k'}\widehat{a}_{1,p_z}e^{i\Gamma\left(p_z-\hbar k',p_z,\omega'\right)t} + \text{h.c.}\right\}\right]. \tag{8.166}$$

Included in Eq. (8.166) the resonance detuning $\Gamma(p,p',\varpi)$ as a function of any three parameters has the following definition:

$$\Gamma(p,p',\varpi) = \frac{\mathcal{E}_0(p) - \mathcal{E}_1(p') + \hbar\varpi}{\hbar}, \tag{8.167}$$

and $\beta_\perp$ is the transition matrix element for the transverse velocity operator:

$$\beta_\perp = \Omega_{nn'} x_{\mu\mu'}, \tag{8.168}$$

where

$$\Omega_{nn'} = \frac{\mathcal{E}_{\perp n'} - \mathcal{E}_{\perp n}}{\hbar} \tag{8.169}$$

is the transition frequency between the initial and excited states of the transversal motion of the particle in the crystal channel. The resonant frequencies of the probe and pump waves for resonant coupling of the two transverse levels are defined from the conditions

$$\Gamma\left(p_z + \hbar k, p_z, \omega\right) = 0, \quad \Gamma\left(p_z - \hbar k', p_z, \omega'\right) = 0$$

and are written as

$$\omega = \frac{\Omega_{01}}{1 + n\left(\omega\right)\frac{\mathrm{v}_z}{c}}, \tag{8.170}$$

$$\omega' = \frac{\Omega_{01}}{1 - n\left(\omega'\right)\frac{\mathrm{v}_z}{c}}. \tag{8.171}$$

Here v_z is the electrons' mean longitudinal velocity in the beam and $n\left(\varpi\right)$ is the index of refraction of a crystal medium ($n\left(\omega'\right) \simeq 1$ for the frequency region under consideration).

The energy spectrum of the planar channeled electron in the potential well (7.103) has the form

$$\mathcal{E}_{\perp n} = -\frac{\hbar^2}{2b^2 m\gamma}\left[s - n\right]^2; \quad n = 0, 1, \ldots, [s], \tag{8.172}$$

where

$$s = -\frac{1}{2} + \sqrt{\frac{1}{4} + \frac{2b^2 m\gamma U_0}{\hbar^2}},$$

and for the axial channeled electron in the potential (7.35):

$$\varepsilon_{\perp n} = -\frac{m\gamma\alpha^2}{2\hbar^2}\frac{1}{\left(n + \frac{1}{2}\right)^2}; \quad n = 0, 1, 2, \ldots. \tag{8.173}$$

The selection rules for transitions are determined by the matrix element of dipole momentum and for the axial channeling are: $\Delta\mathfrak{m} = \pm 1$. For the planar channeling, $x_{\mu\mu'}$ differs from zero between the states having different parities. For the axial channeling there is degeneracy by the magnetic quantum number and in the case of the wave of linear polarization both of the states $\mathfrak{m} = \pm 1$ will have a contribution in the resonant interaction process. Because $\beta_\perp$ depends on $|\mathfrak{m}|$ for $\Delta\mathfrak{m} = \pm 1$ transitions, so the $\mathfrak{m} = \pm 1$ states are equally populated if the initial populations are also equal.

In the channeling potential (7.103) for the $\mu_0 = \{0,0\} \longrightarrow \mu = \{0,1\}$ transition we have

$$\beta_\perp = \frac{\hbar}{2bm\gamma}(2s-1)\left(\frac{s-1}{2}\right)^{\frac{1}{2}}\frac{\Gamma^2\left(s-\frac{1}{2}\right)}{\Gamma^2(s)}, \tag{8.174}$$

where $\Gamma(s)$ is the Euler gamma function. In the potential (7.35) for the transition $\mu_0 = \{0,0\} \longrightarrow \mu = \{\pm1,1\}$ we have

$$\beta_\perp = \sqrt{2}\frac{\alpha_c}{\hbar}\sqrt{\frac{3}{32}}, \tag{8.175}$$

where the factor $\sqrt{2}$ is related to the degeneracy for axial channeling.

For the determination of the self-consistent field we need the evolution equation for the single-particle density matrix $\rho_{ij}(p_z,p_z') =< \widehat{a}^+_{j,p_z'}\widehat{a}_{i,p_z} >$. From the Heisenberg equation (8.17) in the interaction representation we obtain the following equations for the populations of ground and excited states:

$$\begin{aligned}\frac{\partial\rho_{00}(p_z,p_z',t)}{\partial t} &= \frac{e}{2\hbar c}\beta_\perp\Big[A_0\rho_{01}(p_z,p_z'-\hbar k,t)e^{-i\Gamma(p_z',p_z'-\hbar k,\omega)t}\\ &+A_0\rho_{10}(p_z-\hbar k,p_z',t)e^{i\Gamma(p_z,p_z-\hbar k,\omega)t}\\ &+A_e^*\rho_{01}(p_z,p_z'+\hbar k',t)e^{-i\Gamma(p_z',p_z'+\hbar k',\omega')t}\\ &+A_e\rho_{10}(p_z+\hbar k',p_z',t)e^{i\Gamma(p_z,p_z+\hbar k',\omega')t}\Big],\end{aligned} \tag{8.176}$$

$$\begin{aligned}\frac{\partial\rho_{11}(p_z,p_z')}{\partial t} &= -\frac{e}{2\hbar c}\beta_\perp\Big[A_0\rho_{10}(p_z,p_z'+\hbar k,t)e^{i\Gamma(p_z'+\hbar k,p_z',\omega)t}\\ &+A_e\rho_{10}(p_z,p_z'-\hbar k',t)e^{i\Gamma(p_z'-\hbar k',p_z',\omega')t}\\ &+A_0\rho_{01}(p_z+\hbar k,p_z',t)e^{-i\Gamma(p_z+\hbar k,p_z,\omega)t}\\ &+A_e^*\rho_{01}(p_z-\hbar k',p_z',t)e^{-i\Gamma(p_z-\hbar k',p_z,\omega')t}\Big],\end{aligned} \tag{8.177}$$

and for the nondiagonal elements we have

$$\frac{\partial\rho_{01}(p_z,p_z')}{\partial t} = -\frac{e}{2\hbar c}\beta_\perp\Big[A_0\rho_{00}(p_z,p_z'+\hbar k)e^{i\Gamma(p_z'+\hbar k,p_z',\omega)t}$$

$$+A_e\rho_{00}(p_z, p_z' - \hbar k')e^{i\Gamma(p_z'-\hbar k', p_z', \omega')t}$$

$$-A_0\rho_{11}(p_z - \hbar k, p_z')e^{i\Gamma(p_z, p_z-\hbar k, \omega)t}$$

$$-A_e\rho_{11}(p_z + \hbar k', p_z')e^{i\Gamma(p_z, p_z+\hbar k', \omega')t}\Big], \tag{8.178}$$

$$\rho_{10}(p_z, p_z') = \rho_{01}^*(p_z', p_z). \tag{8.179}$$

This set of equations should be supplemented by the Maxwell equation, which is reduced to

$$\frac{\partial A_e}{\partial t} + c\frac{\partial A_e}{\partial z} = \frac{4\pi ce}{\omega'}\beta_\perp \sum_{p_z} \rho_{01}(p_z, p_z + \hbar k')e^{-i\Gamma(p_z, p_z+\hbar k', \omega')t}. \tag{8.180}$$

Equations (8.176)–(8.180) define the FEL dynamics in the crystal channel with the pump EM wave.

First we consider the case when there is no pump field ($A_0 = 0$). In this case for the X-ray generation process it is necessary to have an inverted population of the energy levels in transverse potential or one should have an initial macroscopic dipole momentum, i.e., the electrons should be in the coherent superposition state of transverse levels.

If $A_0 = 0$ from Eqs. (8.176)–(8.179) one can find the closed set of equations for the density matrix elements $\rho_{00}(p_z, p_z)$, $\rho_{11}(p_z + \hbar k', p_z + \hbar k')$, and $\rho_{01}(p_z, p_z + \hbar k', t)$:

$$\frac{\partial \rho_{00}(p_z, p_z, t)}{\partial t} = \frac{e}{2\hbar c}\beta_\perp \Big[A_e^*\rho_{01}(p_z, p_z + \hbar k', t)e^{-i\Gamma(p_z, p_z+\hbar k', \omega')t}$$

$$+A_e\rho_{01}^*(p_z, p_z + \hbar k', t)e^{i\Gamma(p_z, p_z+\hbar k', \omega')t}\Big], \tag{8.181}$$

$$\frac{\partial \rho_{11}(p_z + \hbar k', p_z + \hbar k', t)}{\partial t} = -\frac{\partial \rho_{00}(p_z, p_z, t)}{\partial t}, \tag{8.182}$$

and

$$\frac{\partial \rho_{01}(p_z, p_z + \hbar k', t)}{\partial t} = \frac{e}{2\hbar c}\beta_\perp A_e e^{i\Gamma(p_z, p_z+\hbar k', \omega')t}$$

$$\times \left[\rho_{11}(p_z + \hbar k', p_z + \hbar k', t) - \rho_{00}(p_z, p_z, t)\right]. \tag{8.183}$$

Introducing the new quantities

$$\rho_{11}(p_z + \hbar k', p_z + \hbar k', t) - \rho_{00}(p_z, p_z, t) = 2\pi\hbar\delta F(p_z),$$

$$J(p_z) = \frac{e\beta_\perp}{2\pi\hbar}\rho_{01}(p_z, p_z + \hbar k', t)e^{-i\Gamma(p_z, p_z + \hbar k', \omega')t}$$

and replacing the time derivatives $\partial/\partial t \to \partial/\partial t + \mathrm{v}_z \partial/\partial z$, we obtain

$$\frac{\partial \delta F(p_z)}{\partial t} + \mathrm{v}_z \frac{\partial \delta F(p_z)}{\partial z} = -\frac{e}{\hbar c}\left(A_e^* J(p_z) + A_e J^*(p_z)\right),$$

$$\frac{\partial J(p_z)}{\partial t} + \mathrm{v}_z \frac{\partial J(p_z)}{\partial z} + i\Gamma(p_z - \hbar k', p_z, \omega') J(p_z) = \frac{e^2\beta_\perp^2}{2\hbar c} A_e \delta F(p_z),$$

$$\frac{\partial A_e}{\partial t} + c\frac{\partial A_e(t,z)}{\partial z} = \frac{4\pi c}{\omega'}\int J(p_z)\, dp_z. \tag{8.184}$$

This set of equations is equivalent to the set (8.36) for the Compton and undulator FELs. One should make only the replacement in Eqs. (8.36)

$$M^2 \to 2\beta_\perp^2. \tag{8.185}$$

Hence, we will not repeat all calculations which have been done for Compton and undulator FELs and will use the obtained results. In particular, for steady-state regimes we have the same solutions (8.60), (8.65), where the main characteristic parameter of amplification (the characteristic length of amplification) will be

$$L_{ch} = \frac{1}{\chi_{ch}}; \quad \chi_{ch} = \sqrt{\frac{2\pi\beta_\perp^2 e^2 N_0}{\hbar\omega' c \mathrm{v}_z}}. \tag{8.186}$$

The coherent interaction time of channeled particles with EM radiation is confined by the lifetime of eigenstates of channeled particles and dechanneling effects. For the axial channeling of mildly relativistic electrons the eigenstate width is of order of 1 eV (at $\hbar\omega' \sim 1$ keV) which corresponds to relaxation length $L_r \sim 1$ μm. For planar channeling this length is a little large. To fulfill the condition $L_{ch} \lesssim L_r$ one needs high electron currents. However, the maximal current that can be used in this process is strongly restricted because of the effects of damaging the crystal as well as increasing the beam divergence and the strong bremsstrahlung background. As we saw in Section 8.2 the regime of wave amplification when the electron beam is modulated —"macroscopic transition current" differs from zero — may operate without any initial seeding power, and radiation intensity in this regime reaches a significant value even for small interaction lengths. In the considered case initially electrons should be in the coherent superposition state of transverse levels and the maximal intensity that can be extracted here is

$$W \sim N_0 \hbar\omega' \mathrm{v}_z \left(\frac{L_r}{L_{ch}}\right)^2,$$

which for allowable electron currents at the frequency $\hbar\omega' \sim 1$ keV is of order of 1 kW/cm^2.

8.8 Compton FEL on the Channeling Particle Beam

Consider the scheme of X-ray coherent radiation generation by means of mildly relativistic high-density channeled particle beam and strong counter-propagating pump laser field. In this case the necessity of inverse population of transverse levels for lasing vanishes and as we will see the exponential growth rate of the considered process is resonantly enhanced by several orders with respect to the Compton FEL. We will assume that the pump laser field is not too strong (the Rabi frequency is small compared with resonance detuning) and, consequently, the population of transverse excited state remains small. The main terms responsible for the wave amplification in this case are $\rho_{00}(p_z, p_z + \hbar k' + \hbar k, t)$ and $\rho_{01}(p_z, p_z + \hbar k')$. Hence, from the set of Eqs. (8.176)–(8.179) in the first order by the fields when

$$\rho_{ij}(p_z, p'_z) = \rho_{ij}^{(0)}(p_z, p'_z) + \rho_{ij}^{(1)}(p_z, p'_z)$$

and keeping only the resonant terms we will obtain

$$\frac{\partial \rho_{00}^{(1)}(p_z, p_z + \hbar k' + \hbar k, t)}{\partial t} = \frac{e\beta_\perp}{2\hbar c}\Big[A_e \rho_{10}^{(0)}(p_z + \hbar k', p_z + \hbar k' + \hbar k, t)$$

$$\times e^{i\Gamma(p_z, p_z + \hbar k', \omega')t} + A_0 \rho_{01}^{(0)}(p_z, p_z + \hbar k', t)$$

$$\times e^{-i\Gamma(p_z + \hbar k' + \hbar k, p_z + \hbar k', \omega)t}\Big], \tag{8.187}$$

$$\frac{\partial \rho_{01}^{(1)}(p_z, p_z + \hbar k')}{\partial t} = -\frac{eA_0\beta_\perp}{2\hbar c}\rho_{00}^{(1)}(p_z, p_z + \hbar k' + \hbar k)e^{i\Gamma(p'_z + \hbar k, p'_z, \omega)t}, \tag{8.188}$$

and

$$\frac{\partial \rho_{01}^{(0)}(p_z, p'_z)}{\partial t} = -\frac{e\beta_\perp}{2\hbar c}\Big[A_0 \rho_{00}^{(0)}(p_z, p'_z + \hbar k)e^{i\Gamma(p'_z + \hbar k, p'_z, \omega)t}$$

$$+ A_e \rho_{00}^{(0)}(p_z, p'_z - \hbar k')e^{i\Gamma(p'_z - \hbar k', p'_z, \omega')t}\Big]. \tag{8.189}$$

The Maxwell equation (8.180) for this process is

$$\frac{\partial A_e}{\partial t} = \frac{4\pi ce}{\omega'}\beta_{\perp}$$

$$\times \sum_{p_z} \left[\rho_{01}^{(0)}(p_z, p_z + \hbar k') + \rho_{01}^{(1)}(p_z, p_z + \hbar k')\right] e^{-i\Gamma(p_z, p_z + \hbar k', \omega')t}. \quad (8.190)$$

Here we will consider the probe wave amplification in time at which the spatial dependence of the quantities will be neglected. It is also assumed that the initial electron beam is uniform and, consequently,

$$\rho_{00}^{(0)}(p_z, p_z') = 2\pi\hbar F_0\left(\frac{p_z + p_z'}{2}\right)\delta_{p_z p_z'}, \quad (8.191)$$

where $F(p_z)$ is the classical momentum distribution function of electrons.

Taking into account Eq. (8.191), the solution of Eq. (8.189) for the first-order nondiagonal elements of the electrons' density matrix is

$$\rho_{01}^{(0)}(p_z, p_z + \hbar k') = i\frac{\pi\beta_{\perp}}{c} eA_e F_0(p_z) \frac{e^{i\Gamma(p_z, p_z + \hbar k', \omega')t}}{\Gamma(p_z, p_z + \hbar k', \omega')}, \quad (8.192)$$

$$\rho_{10}^{(0)}(p_z + \hbar k', p_z + \hbar k' + \hbar k) = -i\frac{\pi\beta_{\perp}}{c} eA_0 F_0(p_z + \hbar k' + \hbar k)$$

$$\times \frac{e^{-i\Gamma(p_z + \hbar k' + \hbar k, p_z + \hbar k', \omega)t}}{\Gamma(p_z + \hbar k' + \hbar k, p_z + \hbar k', \omega)}. \quad (8.193)$$

Substituting Eqs. (8.192) and (8.193) into Eqs. (8.187) and (8.190) and taking into account that

$$\Gamma(p_z, p_z + \hbar k', \omega') - \Gamma(p_z + \hbar k' + \hbar k, p_z + \hbar k', \omega) = \Gamma_{0p_z}$$

(see the definition (8.167)), where

$$\Gamma_{0p_z} = \frac{\mathcal{E}_0(p_z) - \mathcal{E}_0(p_z + \hbar k' + \hbar k) + \hbar\omega' - \hbar\omega}{\hbar} \quad (8.194)$$

is the resonance detuning for the Compton scattering, we obtain the self-consistent set of equations which determines the evolution and dynamics of the considered FEL:

$$\frac{dA_e}{dt} = i\Delta A_e + \frac{4\pi c}{\omega'} \int e^{-i\Gamma(p_z, p_z + \hbar k', \omega')t} J(p_z, t) dp_z, \quad (8.195)$$

$$\frac{dJ}{dt} = -\frac{A_0 e^2 \beta_\perp^2}{2\hbar c} e^{i\Gamma(p_z+\hbar k'+\hbar k, p_z+\hbar k', \omega)t} \delta n(p_z, t), \tag{8.196}$$

$$\frac{d\delta n}{dt} = i\frac{A_0 A_e e^2 \beta_\perp^2}{4\hbar^2 c^2} e^{i\Gamma_{0p_z} t} \left[\frac{F_0(p_z)}{\Gamma(p_z, p_z + \hbar k', \omega')} \right.$$

$$\left. - \frac{F_0(p_z + \hbar k' + \hbar k)}{\Gamma(p_z + \hbar k' + \hbar k, p_z + \hbar k', \omega)} \right]. \tag{8.197}$$

Here for convenience we have introduced new quantities

$$\delta n(p_z, t) \equiv \frac{1}{2\pi\hbar} \rho_{00}^{(1)}(p_z, p_z + \hbar k' + \hbar k, t),$$

$$J(p_z, t) \equiv \frac{e\beta_\perp}{2\pi\hbar} \rho_{01}^{(1)}(p_z, p_z + \hbar k', t),$$

and the summation is replaced by integration. Then

$$\Delta = \frac{2\pi e^2 \beta_\perp^2}{\hbar\omega'} \int dp_z \frac{F_0(p_z)}{\Gamma(p_z, p_z + \hbar k', \omega')} \tag{8.198}$$

is the frequency shift due to the particle beam polarization (induced dipole moment).

Performing Laplace transformation on Eqs. (8.195), (8.196), and (8.197) we arrive at the following characteristic equation:

$$q - i\Delta = -i\frac{\pi e^4 \beta_\perp^4 A_0^2}{2\hbar^3 c^2 \omega'} \int \frac{dp_z}{(q + i\Gamma_{0p_z})(q + i\Gamma_{p_z, p_z+\hbar k', \omega'})}$$

$$\times \left[\frac{F_0(p_z)}{\Gamma_{p_z, p_z+\hbar k', \omega'}} - \frac{F_0(p_z + \hbar k' + \hbar k)}{\Gamma_{p_z+\hbar k'+\hbar k, p_z+\hbar k', \omega}} \right]. \tag{8.199}$$

This is a transcendental equation that allows one to determine the small-signal gain in various regimes. For the cold electron beam (8.95), taking into account the condition $|q| >> |\Gamma_{0p_z}|$, $|\Delta|$ (high gain regime) and neglecting the quantum recoil, from Eq. (8.199) one can obtain the exponential growth rate:

$$G = \frac{\sqrt{3}}{2} \left[\frac{4\pi r_e}{\lambda_c} \frac{mc}{\hbar \Omega_{01} \gamma} \frac{\xi_0^2}{\delta^2} \beta_\perp^4 N_0 \right]^{1/3}. \tag{8.200}$$

Here $\lambda_c = \hbar/mc$ is the particle Compton wavelength, r_e is the electron classical radius, and

$$\delta = \frac{|\omega + \mathrm{v}_z k - \Omega_{01}|}{\Omega_{01}}$$

is the relative detuning of the resonance.

Equation (8.200) defines the exponential growth rate of X-rays in the crystal at "Compton" scattering of a strong pump laser radiation on the channeled particle beam at the resonance. Instead of ξ_0^2 in the Compton effect on the free electrons the effective interaction parameter in the channeling process is determined by the resonance parameter ξ_0^2/δ^2. For the high-gain regime it is necessary that $GL_r/c > 1$, where L_r is the relaxation length in the crystal.

The obtained results are also applicable for positron beams channeled in the zeolite crystals containing hollow channels with the diameter $R \sim 10 \div 100$ Å. In this case, main time channeled particles move in the hollow channel and atomic electrons are disposed in the thin layer of the internal surface of the channel and the scattering processes are suppressed and the relaxation time is much larger than in the monocrystals ($L_r \sim 0.1$ cm). Besides, if $\lambda < R$ (λ is the wavelength of amplifying radiation) the X-ray absorption and scattering process is also suppressed, which in turn reduces the threshold currents and the considered setup will be more preferable. In this case, the potential of the channel can be approximated by the potential $U(\rho) = 0$, if $\rho < R$; and ∞, if $\rho \geq R$. Then the resonance can be achieved by the infrared pump lasers as $\hbar\Omega_{01} \sim 0.1$ eV and one can consider the SASE regime as a small setup single-pass soft X-ray FEL.

8.9 Nonlinear Scheme of X-Ray Laser on the Ion and Pump Laser Beams

As an alternative version of FEL we will consider the problem of generation of coherent shortwave radiation by relativistic ion beams when due to the existence of bound states, the ion–photon interaction cross section resonantly increases with respect to the electron–photon scattering one. From this point of view stimulated radiation from relativistic ion beams is a synthesis of conventional quantum generators and FELs in the X-ray domain.

We consider as our model a relativistic beam of two level ions, co-propagating (Z axis) probe EM wave with a frequency ω and wave vector $\mathbf{k}$ and counterpropagating strong pump EM wave of frequency ω_0 and wave vector $\mathbf{k}_0$. The EM waves are treated as classical fields and the total electrical field is given by

$$\mathbf{E}(\mathbf{r}, t) = \frac{1}{2}\epsilon_0 E_0 e^{i\omega_0 t - i\mathbf{k}_0\mathbf{r}} + \frac{1}{2}\epsilon E_e(t, \mathbf{r}) e^{i\omega t - i\mathbf{k}\mathbf{r}} + \text{c.c.}. \tag{8.201}$$

The probe wave is characterized by slowly varying amplitude $E_e(t, \mathbf{r})$ and unit polarization vector ϵ, while a pump wave is characterized by a given amplitude E_0 and polarization vector ϵ_0 (both waves are linearly polarized). We assume that an internal ionic electron is nonrelativistic and the transition takes place from an S state to a P state. The Hamiltonian governing the evolution of the ion beam in the field (8.201) takes the following second quantized form in the resonant approximation:

$$\widehat{H} \simeq \sum_{\mathbf{p},s=1,2} \mathcal{E}_s(\mathbf{p})\widehat{a}^+_{s,\mathbf{p}}\, \widehat{a}_{s,\mathbf{p}} + \sum_{\mathbf{p}} \Big[\hbar\Omega_{0\mathbf{p}} e^{i\omega_0 t}\widehat{a}^+_{1,\mathbf{p}-\hbar\mathbf{k}_0}\, \widehat{a}_{2,\mathbf{p}}$$

$$+\hbar\Omega_{\mathbf{p}}(\mathbf{r},t)e^{i\omega t}\widehat{a}^+_{1,\mathbf{p}-\hbar\mathbf{k}}\, \widehat{a}_{2,\mathbf{p}} + \text{h.c.}\Big]. \tag{8.202}$$

Here

$$\mathcal{E}_s(\mathbf{p}) = \sqrt{c^2\mathbf{p}^2 + (m_i c^2 + w_s)^2}; \quad s = 1, 2 \tag{8.203}$$

is the total energy of the ion with the momentum $\mathbf{p}$ of the center-of-mass motion and w_1, w_2 are the binding energies of the internal ionic electron in the ground and excited states, respectively (m_i is the ion mass). Then $\widehat{a}^+_{s,\mathbf{p}}$, $\widehat{a}_{s,\mathbf{p}}$ denote ionic creation and annihilation operators for the internal states $s = 1, 2$ with center-of-mass momentum $\mathbf{p}$. These operators satisfy the usual either bosonic or fermionic type equal times commutation rules. The couplings

$$\Omega_{0\mathbf{p}} = \frac{E_0\epsilon_0\mathbf{d}_{12}}{2\hbar}\left(1 - \frac{\mathbf{v}\mathbf{k}_0}{\omega_0}\right), \tag{8.204}$$

$$\Omega_{\mathbf{p}}(\mathbf{r},t) = \frac{E_e(t,\mathbf{r})\epsilon\mathbf{d}_{12}}{2\hbar}\left(1 - \frac{\mathbf{v}\mathbf{k}}{\omega}\right) \tag{8.205}$$

take into account the dipole interaction as well as the interaction of magnetic moment $[\mathbf{d}_{12} \times \mathbf{v}]/c$ (because of moving electric dipole) with the magnetic field of the waves. In Eq. (8.205) $\mathbf{v} = \mathbf{p}/m_i\gamma$ is the ion velocity, γ is the Lorentz factor, and $\mathbf{d}_{12}$ is the ionic transition dipole moment.

We will use again the Heisenberg representation where evolution of the operators are given by Eq. (8.17) and expectation values are determined by the initial density matrix of the ion beam (see Eq. (8.18)). Then the Heisenberg equations should be supplemented by the Maxwell equation for slowly varying amplitude $E_e(t,\mathbf{r})$ analogously to Eq. (8.19). The resonant current for ion beam is defined by the nondiagonal element of the single-particle density matrix

$$\rho_{12\mathbf{p},\mathbf{p}+\hbar\mathbf{k}}(t) =< \widehat{a}^+_{2,\mathbf{p}+\hbar\mathbf{k}}\widehat{a}_{1,\mathbf{p}} > . \tag{8.206}$$

Hence, for the determination of the self-consistent field we need the evolution equation for the single particle density matrix

$$\rho_{ij\mathbf{p},\mathbf{p}'}(t) = < \widehat{a}^{+}_{j,\mathbf{p}'}\widehat{a}_{i,\mathbf{p}} > . \tag{8.207}$$

We will assume that initially ions are in the ground state and the pump laser field is not so strong or it is far off resonance and consequently, the excited state population remains small. In analogy with the previous section introducing the functions

$$\rho_{11\mathbf{p},\mathbf{p}+\hbar\mathbf{k}-\hbar\mathbf{k}_0}(t) = \rho^{(0)}_{11\mathbf{p},\mathbf{p}+\hbar\mathbf{k}-\hbar\mathbf{k}_0} + (2\pi\hbar)^3 e^{i(\omega-\omega_0)t}\delta n(\mathbf{p},t), \tag{8.208}$$

$$\rho_{12\mathbf{p},\mathbf{p}+\hbar\mathbf{k}}(t) = \rho^{(0)}_{12\mathbf{p},\mathbf{p}+\hbar\mathbf{k}} + (2\pi\hbar)^3 e^{i\omega t}J(\mathbf{p},t) \tag{8.209}$$

from the Heisenberg and Maxwell equations one can obtain the self-consistent set of equations which determines the evolution and dynamics of the considered system:

$$\frac{\partial E_e}{\partial t} + \frac{c^2\mathbf{k}}{\omega}\frac{\partial E_e}{\partial \mathbf{r}} - i\Delta E_e = 4\pi i\omega\epsilon\mathbf{d}^{*}_{12}\int\left(1-\frac{\mathbf{vk}}{\omega}\right)J(\mathbf{p},t)d\mathbf{p}, \tag{8.210}$$

$$\frac{\partial J}{\partial t} + \mathbf{v}_0\frac{\partial J}{\partial \mathbf{r}} + i\Gamma_1(\mathbf{p})J = i\Omega_{0\mathbf{p}}\delta n(\mathbf{p},t), \tag{8.211}$$

$$\frac{\partial \delta n}{\partial t} + \mathbf{v}_0\frac{\partial \delta n}{\partial \mathbf{r}} + i\Gamma_0(\mathbf{p})\delta n = i\Omega^{*}_{0\mathbf{p}}\Omega_{\mathbf{p}}$$
$$\times\left[\frac{F_0\left(\mathbf{p}+\hbar\mathbf{k}-\hbar\mathbf{k}_0\right)}{\Gamma_1(\mathbf{p})-\Gamma_0(\mathbf{p})} - \frac{F_0\left(\mathbf{p}\right)}{\Gamma_1(\mathbf{p})}\right]. \tag{8.212}$$

To take into account the pulse propagation effects we have replaced the time derivatives $\partial/\partial t \to \partial/\partial t + \mathbf{v}_0\partial/\partial\mathbf{r}$, where $\mathbf{v}_0$ is the mean velocity of the ion beam. Here it is assumed that the initial beam is uniform and consequently

$$\rho_{11\mathbf{p},\mathbf{p}'}(0) = (2\pi\hbar)^3 F_0\left(\frac{\mathbf{p}+\mathbf{p}'}{2}\right)\delta_{\mathbf{p},\mathbf{p}'}, \tag{8.213}$$

where $F_0\left(\mathbf{p}\right)$ is the ions' classical center-of-mass momentum distribution function and $\delta_{\mathbf{p},\mathbf{p}'}$ is the Kronecker symbol (summation is replaced by integration). Then

$$\Delta = \frac{2\pi\omega\left|\epsilon\mathbf{d}_{12}\right|^2}{\hbar}\int d\mathbf{p}\frac{\left(1-\frac{\mathbf{vk}}{\omega}\right)^2 F_0\left(\mathbf{p}\right)}{\Gamma_1(\mathbf{p})} \tag{8.214}$$

is the frequency shift because of the ion beam polarization (refractive index caused by ion beam) and

$$\varGamma_0(\mathbf{p}) = \frac{\mathcal{E}_1(\mathbf{p}) - \mathcal{E}_1(\mathbf{p} + \hbar\mathbf{k} - \hbar\mathbf{k}_0) + \hbar\omega - \hbar\omega_0}{\hbar} \tag{8.215}$$

is the resonance detuning for the Compton scattering of the strong wave on ions, while

$$\varGamma_1(\mathbf{p}) = \frac{\mathcal{E}_1(\mathbf{p}) + \hbar\omega - \mathcal{E}_2(\mathbf{p} + \hbar\mathbf{k})}{\hbar} \tag{8.216}$$

is the resonance detuning for absorption/emission of the probe wave's quanta.

To determine the conditions under which we will have collective instability and consequently the exponential growth of the probe wave, one should perform the same procedure as was made for the high-gain regime of amplification on an electron beam. We will assume again the steady-state operation at which one can drop all partial time derivatives in Eqs. (8.210), (8.211), and (8.212). Performing Laplace transformation (8.91) on Eqs. (8.210), (8.211), and (8.212) we arrive at the following characteristic equation for variable q:

$$q - i\varDelta = \int \frac{K(\mathbf{p})d\mathbf{p}}{(q + i\varGamma_0(\mathbf{p}))\,(q + i\varGamma_1(\mathbf{p}))}, \tag{8.217}$$

where

$$K(\mathbf{p}) = \frac{2\pi i\omega \left|\epsilon\mathbf{d}_{12}\right|^2 \left|\varOmega_{0\mathbf{p}}\right|^2}{\hbar \mathrm{v}_{0z}^2 c}\left(1 - \frac{\mathbf{vk}}{\omega}\right)^2$$

$$\times \left[\frac{F_0(\mathbf{p})}{\varGamma_1(\mathbf{p})} - \frac{F_0(\mathbf{p} + \hbar\mathbf{k} - \hbar\mathbf{k}_0)}{\varGamma_1(\mathbf{p}) - \varGamma_0(\mathbf{p})}\right]. \tag{8.218}$$

This is the transcendental equation which allows one to determine a small-signal gain in various regimes.

For initially cold ion beam

$$F_0(\mathbf{p}) = N_{i0}\delta(\mathbf{p} - \mathbf{p}_0)$$

(N_{i0} is the beam density) taking into account Eqs. (8.215), (8.216), as well as the conditions $|q| >> |\varGamma_0(\mathbf{p}_0)|$, $|\varDelta|$ (high-gain regime), and $|q| << |\varGamma_1(\mathbf{p}_0)|$ and neglecting the quantum recoil, from Eq. (8.217) one can obtain the exponential growth rate of the probe X-ray:

$$G = \frac{\sqrt{3}}{2}\left[\frac{\varOmega_r^2}{\delta^2}\frac{2\pi\omega^3 \left|\epsilon\mathbf{d}_{12}\right|^2}{\mathrm{v}_{0z}^2 \gamma_0^5 m_i c^3} N_{i0}\right]^{1/3}. \tag{8.219}$$

Here

$$\varOmega_r = \frac{E_0 \epsilon_0 \mathbf{d}_{12}}{2\hbar} \tag{8.220}$$

is the Rabi frequency associated with the pump wave,

$$\delta = \omega_{12} - \omega_0 \gamma_0 \left(1 + \frac{\mathrm{v}_{0z}}{c}\right) \tag{8.221}$$

is the resonance detuning, and

$$\omega_{12} = \frac{w_2 - w_1}{\hbar}$$

is the transition frequency for internal ionic electron.

Equation (8.219) defines the exponential growth rate of X-rays at the induced "Compton" scattering of a strong pump laser radiation on the ion beam, which is resonantly enhanced with respect to the Compton laser on free electrons.

8.10 Crystal Potential as a Pump Field for Generation of Coherent X-Ray

Consider now the possibility of coherent X-ray radiation generation by a fast, multiply charged, channeled ion beam in a crystal without a pump laser field. In the proposed process the X-ray transitions involving the K or L shell electrons in ions can be resonantly excited by the periodic crystal potential seen by fast channeled ions. The emission frequencies in this case are determined by the discrete spectrum of the electron states in ions and by the Doppler shift due to the ion center-of-mass motion. With respect to moving ions, the crystal electrostatic potential plays the role of an effective pumping field with the Rabi frequency corresponding to a high power "X-ray laser". By varying the crystal thickness, one can obtain diverse equivalent "X-ray pulses" leading to various coherent superposition states, from which one can obtain coherent X-ray radiation from the ion beam spontaneous superradiation.

Below we will consider superradiant coherent X-ray generation when an ion beam moves close to the crystal lattice axis. This radiation is predicted by the second quantized Maxwell and quantum equations governing the motion of an ion beam in a crystal.

For channeling an ion beam in a crystal, we assume that the incident angle of ions (with a charge number of the nucleus Z_i) with respect to a crystalline axis (OZ) is smaller than the Lindhard angle. Then the potential of the atomic chain, which governs the ion motion, can be represented in the form

$$V(z, r_\perp) = \sum_n V_n(r_\perp) \exp\left[i \frac{2\pi n}{d} z\right], \tag{8.222}$$

where d is the crystal lattice period along the channel axis, $V_n(r_\perp)$ is defined by the single atomic potential of the crystal, which is given by the screening Coulomb potential with the radius of screening R and a charge number of the nucleus Z_c that has the form

$$V_n(r_\perp) = \frac{2eZ_c}{d} K_0\left(r_\perp q_n\right),$$

$$q_n = \sqrt{\frac{1}{R^2} + \left(\frac{2\pi n}{d}\right)^2}, \qquad (8.223)$$

where K_0 is a modified Bessel function.

The potential (8.222) acts on the internal electron as well as on the ion center-of-mass motion, providing channeling. The center of mass of the ion represents slow oscillations in the transversal direction ($\mathbf{r}_\perp$) and free motion (on average) along the crystalline axis. For the ionic electron the atomic chain potential acts as an exciting field. The latter is obvious in the rest frame of reference of the ion (neglecting transversal oscillations) where there is an oscillating time/space electromagnetic field with a fundamental frequency $2\pi\gamma \mathrm{v}_z/d$ (γ is the Lorentz factor, v_z is the ion longitudinal velocity). If one of the harmonics (n) of this frequency is close to the frequency ω_{12} associated with the energy difference of the ionic electron levels

$$\frac{2\pi n\gamma \mathrm{v}_z}{d} \simeq \omega_{12}, \qquad (8.224)$$

we can expect resonant excitation of ions. The latter represents the conservation law for the total energy (neglecting quantum recoil) in the laboratory frame of reference.

As the physical picture of the considered process is more evident in the frame of reference connected with the ion beam and the problem becomes nonrelativistic in this frame, then it is more convenient to pass to the rest frame of the ion beam (moving with the mean velocity v_0 of the beam). If the resonance condition (8.224) holds, we can keep only the resonant harmonic in the potential (8.222) and the Hamiltonian describing the quantum kinetics of the channeled ion beam takes the following second quantized form in the resonant approximation:

$$\widehat{H}_{ic} \simeq \sum_{\mathbf{p},s=1,2} \mathcal{E}_s(\mathbf{p})\widehat{a}^+_{s,\mathbf{p}}\, \widehat{a}_{s,\mathbf{p}} + \sum_{\mathbf{p}} \left[\hbar\Omega_c e^{i\omega_c t}\widehat{a}^+_{1,\mathbf{p}-\hbar\mathbf{g}_n}\, \widehat{a}_{2,\mathbf{p}} + \mathrm{h.c.}\right]. \qquad (8.225)$$

Here we have introduced the lattice vector

$$\mathbf{g}_n = (0, 0, -\frac{2\pi n\gamma_0}{d}),$$

where $\gamma_0 = (1 - \mathrm{v}_0^2/c^2)^{-1/2}$, and $\widehat{a}^+_{s,\mathbf{p}}$, $\widehat{a}_{s,\mathbf{p}}$ denote ionic creation and annihilation operators for the states $s = 1, 2$ with center-of-mass momentum $\mathbf{p}$ and energy

$$\mathcal{E}_s(\mathbf{p}) = \frac{\mathbf{p}^2}{2m_i} + w_s$$

(w_s is the binding energy of the ionic electron). These operators satisfy either the usual bosonic or fermionic type equal time commutation rules. The coupling is

$$\Omega_c = \frac{2eZ_c\gamma_0}{\hbar d} \left\{-ig_n f_z K_0 \left(\overline{r}_\perp q_n\right) + \frac{\mathbf{f}\mathbf{r}_\perp}{r_\perp} q_n K_1 \left(\overline{r}_\perp q_n\right)\right\}, \tag{8.226}$$

where $\mathbf{f}$ is the ionic transition dipole moment, which represents the Rabi frequency, with the assumption that the crystal potential acts as a quasi-monochromatic wave with the frequency

$$\omega_c = \mathrm{v}_0 g_n; \quad g_n = \frac{2\pi n \gamma_0}{d}. \tag{8.227}$$

In Eq.(8.226) we have neglected the ion transverse oscillations, since they are much slower than the frequency of collisions of ions with the atoms of the crystalline axis. Here $\overline{r}_\perp$ is the ion mean transverse displacement.

The full Hamiltonian describing also the radiation processes will be

$$\widehat{H} = \widehat{H}_{ic} + \sum_{\mathbf{k},\mu=1,2} \hbar\omega \widehat{c}^+_{\mathbf{k},\mu} \widehat{c}_{\mathbf{k},\mu}$$

$$+ \sum_{\mathbf{p},\mathbf{k},\mu} \left[\hbar \Omega_{\mathbf{k},\mu} \widehat{a}^+_{1,\mathbf{p}-\hbar\mathbf{k}}\, \widehat{a}_{2,\mathbf{p}} \widehat{c}_{\mathbf{k},\mu} + \text{h.c.}\right], \tag{8.228}$$

where the second term is the Hamiltonian of the photon field with the creation and annihilation operators $\widehat{c}^+_{\mathbf{k},\mu}$, $\widehat{c}_{\mathbf{k},\mu}$ of photons with momentum $\hbar\mathbf{k}$ and linear polarization ϵ_μ ($\mu = 1, 2$). The last term is the Hamiltonian of interaction of the ions with the photon field and

$$\Omega_{\mathbf{k},\mu} = \sqrt{2\pi\hbar\omega}\,(\epsilon_\mu \mathbf{f}) \tag{8.229}$$

is the Rabi frequency for the quantized photon field (the quantization volume is taken to be $V = 1$).

If the effective Rabi frequency is large enough and the crystal length is short enough, the spontaneous emission and the relaxation processes may be neglected during the time of interaction of ions with the crystal. In this case, the Heisenberg equation (8.17) for the operators $\widehat{a}_{s,\mathbf{p}}$ may be solved analytically. This gives the following solution:

$$\widehat{a}_{1,\mathbf{p}} = e^{-\frac{i}{\hbar}\mathcal{E}_1(\mathbf{p})t}\, e^{-i\frac{1}{2}\delta_{\mathrm{v}_z}\tau} \left\{\cos\Omega\tau + i\frac{\delta_{\mathrm{v}_z}}{2\Omega}\sin\Omega\tau\right\}\widehat{a}^{(0)}_{1,\mathbf{p}},$$

$$\widehat{a}_{2,\mathbf{p}} = -ie^{-\frac{i}{\hbar}\mathcal{E}_2(\mathbf{p})t}e^{i\frac{1}{2}\delta_{\mathrm{v}_z}\tau}\sin\Omega\tau\frac{\Omega_c}{\Omega}\widehat{a}^{(0)}_{1,\mathbf{p}-\hbar\mathbf{g}_n}. \quad (8.230)$$

Here $\widehat{a}^{(0)}_{1,\mathbf{p}}$ is the initial operator, τ is the ion interaction time with the crystal,

$$\delta_{\mathrm{v}_z} = \omega_{12} - \omega_c - g_n\mathrm{v}_z \quad (8.231)$$

is the resonance detuning, and

$$\Omega = \sqrt{|\Omega_c|^2 + \frac{\delta^2_{\mathrm{v}_z}}{4}} \quad (8.232)$$

is the effective Rabi frequency. We assume that initially ions are in the ground state, so that in Eq. (8.230) we have not written the terms with the operator $\widehat{a}^{(0)}_{2,\mathbf{p}}$. As we see, the population of electrons oscillates coherently between the states depending on the crystal length $L_c \simeq \mathrm{v}_z\tau$. If $|\Omega_c| >> |\delta_{\mathrm{v}_z}|$ and the crystal length corresponds to "pulse area" $|\Omega_c|\,\tau = j\pi/4$ ($j = 1, 2, ...$), the ion beam will then have the maximal polarization (macroscopic dipole moment).

To investigate the properties of ion beam radiation (in free space) we come back to the full Hamiltonian and perturbatively calculate the photonic operators $\widehat{c}_{\mathbf{k},\mu}(t)$:

$$\widehat{c}_{\mathbf{k},\mu}(t) = -i\pi\hbar\Omega_{\mathbf{k},\mu}\sum_{\mathbf{p}}\widehat{a}^{+}_{1,\mathbf{p}-\hbar\mathbf{k}}\,\widehat{a}_{2,\mathbf{p}}$$

$$\times\delta(\hbar\omega + \mathcal{E}_1(\mathbf{p}-\hbar\mathbf{k}) - \mathcal{E}_2(\mathbf{p})). \quad (8.233)$$

The output spectrum consists of coherent and incoherent radiation. The coherent superradiation is defined by the mean value of the photonic operators $< \widehat{c}_{\mathbf{k}\mu}(t) >$; i.e., it is proportional to the Fourier transform of the mean ion polarization $< \widehat{a}^{+}_{1,\mathbf{p}-\hbar\mathbf{k}}\,\widehat{a}_{2,\mathbf{p}} >$. To determine the intensity of coherent radiation we will assume that the mean number of photons is much smaller than the total number of ions: $N_{ph} << N_i$. In accordance with this assumption, one can neglect the retro radiation effects. Otherwise, ions would respond collectively, and as is known the N-particle spontaneous emission rate might be much larger than a single-particle spontaneous emission rate, consequently the considered equations for the photons and ions operators should be solved self-consistently.

From Eq. (8.233) we obtain the following equation for the total number of emitted photons with momentum $\hbar\mathbf{k}$ and polarization μ per unit time:

$$\frac{\partial N_{\mathbf{k}\mu}^{(coh)}}{\partial t} = 2\pi\hbar \left|\Omega_{\mathbf{k},\mu}\right|^2 \sum_{\mathbf{p}_1,\mathbf{p}} Re\left\{\rho_{12\mathbf{p}_1-\hbar\mathbf{k},\mathbf{p}_1}\right.$$

$$\left. \times\, \rho_{21\mathbf{p},\mathbf{p}-\hbar\mathbf{k}}\delta\left(\hbar\omega + \mathcal{E}_1(\mathbf{p}-\hbar\mathbf{k}) - \mathcal{E}_2(\mathbf{p})\right)\right\}, \tag{8.234}$$

where $\rho_{12\mathbf{p},\mathbf{p}'}(t) =< \widehat{a}_{2,\mathbf{p}'}^{+}\widehat{a}_{1,\mathbf{p}} >$ is the nondiagonal element of the single-particle density matrix defined by the operators (8.230). By summing over photon polarization and integrating over frequency one can obtain the following expression for the angular distribution of superradiant power per unit solid angle (dO):

$$\frac{dI_{coh}}{dO} = N_i^2 I_1(\widehat{\mathbf{k}}) \left|G\left(\widehat{\mathbf{k}}\frac{\omega_{12}}{c} - \mathbf{g}_n\right)\right|^2$$

$$\times \left|\int \exp\left(i\frac{\widehat{\mathbf{k}}\mathbf{v}}{c}\omega_{12}t\right) P(\mathrm{v}_z)F(\mathbf{v})d\mathbf{v}\right|^2, \tag{8.235}$$

where

$$G(\mathbf{q}) = \frac{1}{N_i}\int n(\mathbf{r})e^{i\mathbf{qr}}d\mathbf{r} \tag{8.236}$$

is the beam form-factor with $n(\mathbf{r})$ being the ion beam density function, $F(\mathbf{v})$ is the velocity distribution function of ions, $I_1(\widehat{\mathbf{k}})$ is the single ion radiation power with the unit vector $\widehat{\mathbf{k}}$ in the radiation direction, and

$$P(\mathrm{v}_z) = \frac{\Omega_c}{\Omega}\sin\Omega\tau\left\{\cos\Omega\tau + i\frac{\delta_{\mathrm{v}_z}}{2\Omega}\sin\Omega\tau\right\}\exp\left(-ig_n\mathrm{v}_z\tau\right). \tag{8.237}$$

For the beam spatial and velocity distributions we will assume Gaussian functions with isotropic transverse distributions. Then, from Eq.(8.235) for the differential power of the ion beam superradiation we obtain

$$\frac{dI_{coh}}{dO} = N_i^2 I_1(\widehat{\mathbf{k}}) \exp\left[-\frac{\delta\mathrm{v}_\perp^2}{c^2}\omega_{12}^2 t^2\sin^2\vartheta\right]\left|P(t,\vartheta)\right|^2$$

$$\times \exp\left[-\frac{l_\perp^2\omega_{12}^2}{c^2}\sin^2\vartheta - l_z^2 g_n^2(1+\frac{\omega_{12}}{cg_n}\cos\vartheta)^2\right], \tag{8.238}$$

where

$$P(t,\vartheta) = \int P(\mathrm{v}_z)\exp\left[i\frac{\mathrm{v}_z\cos\vartheta}{c}\omega_{12}t - \frac{\mathrm{v}_z^2}{2\delta\mathrm{v}_z^2}\right]\frac{d\mathrm{v}_z}{\sqrt{2\pi}\delta\mathrm{v}_z}. \tag{8.239}$$

Here $l_\perp$, l_z are the transverse and longitudinal bunch sizes of the beam with the transverse and longitudinal velocity spreads $\delta\mathrm{v}_\perp$, $\delta\mathrm{v}_z$. As is seen from Eq.(8.238), if the observed wavelengths are much smaller than the transverse size of an ion beam, the superradiation from the ion beam will occur primarily along the Z axis and will cover only a tiny solid angle, which will be defined by the transverse size of the ion beam

$$\Delta O \simeq \pi\frac{c^2}{l_\perp^2\omega_{12}^2}. \tag{8.240}$$

The superradiant pulse duration depends on velocity spreads of the beam and will be defined by the function $P(t,\vartheta)$. The analysis of Eq.(8.238) shows the existence of two superradiant regimes of X-ray generation. For the first regime when the phase matching condition holds

$$\omega_{12} = cg_n, \tag{8.241}$$

the superradiation from the ion beam may occur primarily in the backward direction and the longitudinal bunch size l_z of the ion beam should not be smaller than the wavelength of superradiation. On the other hand, for the resonant excitation the condition $|\delta_0| << \omega_{12}$ should be fulfilled. Then taking into account the phase matching condition (8.241), for the detuning (8.231) we have

$$\delta_0 \simeq \omega_{12}(1 - \frac{\mathrm{v}_0}{c}). \tag{8.242}$$

The latter means that for the backward superradiation it is necessary for a relativistic ion beam to satisfy the resonance condition $\delta_0 \simeq \omega_{12}/2\gamma_0^2 << \omega_{12}$.

For the mean power of backward superradiation from Eq.(8.238) one can obtain the following approximate formula:

$$I_{mean} \simeq N^2 I_1 \left|P(0,\pi)\right|^2 \Delta O. \tag{8.243}$$

In the opposite case when the resonance condition holds: $\omega_{12} = \omega_c = \mathrm{v}_0 g_n$, one can easily fulfill the condition for maximal dipole moment $|\Omega_c| >> \delta_0$ for the light ion beams ($Z_i < 10$, $\gamma_0 \simeq 1$), but since the phase matching condition (8.241) is violated: $\omega_{12} < cg_n$, the superradiation will take place if the longitudinal bunch size of the ion beam is smaller than the crystal lattice period d.

Bibliography

V.L. Ginzburg, Dokl. Akad. Nauk SSSR **56**, 145 (1947) [in Russian]
R.H. Pantell, G. Soncini, H.E. Puthoff, IEEE J. Quantum Electron. **4**, 905 (1968)
J. M. J. Madey, J. Appl. Phys. **42**, 1906 (1971)
M.L. Ter-Mikaelian: High-Energy Electromagnetic Processes in Condensed Media (Wiley–Interscience, New York 1972)
L.R. Elias et al., Phys. Rev. Lett. **36**, 771 (1976)
D.A. Deacon et al., Phys. Rev. Lett. **38**, 892 (1977)
P. Sprangle, C.M. Tang, W.M. Manheimer, Phys. Rev. Lett. **43**, 1932 (1979)
A.M. Kondratenko, E.L. Saldin, Part. Accel. **10**, 207 (1980)
G. Dattoli, A. Renieri, Nuovo Cimento B **59**, 1 (1980)
C.M. Tang, P. Sprangle, J. Appl. Phys. **53**, 831 (1981)
H. Haus, IEEE J. Quantum Electron. **QE-17**, 1427 (1981)
Free-Electron Generators of Coherent Radiation, Physics of Quantum Electronics, vol 5,7-9, ed by S. F. Jacobs, H. S. Pilloff, M. Sargent III, M.O. Scully, R. Spitzer (Addison-Wesley, Reading, MA 1982)
P. Sprangle, C.M. Tang, I. Bernstein, Phys. Rev. A **28**, 2300 (1983)
P. Dobiasch, P. Meystre, M.O. Scully, IEEE J. Quantum Electron. **19**, 1812 (1983)
R. Bonifacio, C. Pellegrini, L.M. Narducci, Opt. Commun. **50**, 373 (1984)
T.C. Marshall: Free Electron Lasers (MacMillan, New York 1985)
E. Jerby, A. Gover, IEEE J. Quantum Electron. **QE-21**, 1041 (1985)
J.B. Murphy, C. Pellegrini, Nucl. Instrum. Methods A **237**, 159 (1985)
Kwang-Je Kim, Phys. Rev. Lett. **57**, 1871 (1986)
K.J. Kim, Nucl. Instrum. Methods A **250**, 396 (1986)
J.M. Wang, L.H. Yu, Nucl. Instrum. Methods A **250**, 484 (1986)
S. Krinsky, L.H. Yu, Phys. Rev. A **35**, 3406 (1987)
J. Gea-Banacloche, G.T. Moore, R.R. Schlichter et al., IEEE J. Quantum Electron. **23**, 1558 (1987)
A. Friedman et al., Rev. Mod. Phys. **60**, 471 (1988)
G. Dattoli et al., Phys. Rev. A **37**, 4334 (1988)
A. Yariv: Quantum Electronics, 3rd edn (John Wiley and Sons, New York 1989)
H.K. Avetissian, A.K. Avetissian, K.Z. Hatsagortsian, Phys. Lett. A **137**, 463 (1989)
C.A. Brau: Free-Electron Lasers (Academic Press, New York 1990)
P. Luchini, H. Motz: Undulators and Free-Electron Lasers (Oxford Science Publications, Oxford 1990)
W.B. Colson, C. Pellegrini, A. Renieri: Laser Handbook, Vol.6 (North-Holland, Amsterdam 1990)
Proceedings of the Annual International Free Electron Laser Conferences published in Nucl. Instrum. Methods, Vol. A528, A507, A483, A475, A445, A407, A358, A341, A331, A318, A304

M. Xie, D.A.G. Deacon, J.M.J. Madey, Phys. Rev. A **41**, 1662 (1990)
G. Dattoli, A. Renieri, A. Torre: Lectures on the Free Electron Laser Theory and Related Topics (World Scientific, London 1993)
P. G. O'Shea et al., Phys. Rev. Lett. **71**, 3661 (1993)
G. Kurizki, M. O. Scully, C. Keitel, Phys. Rev. Lett. **70**, 1433 (1993)
E.M. Belenov et al., Zh. Éksp. Teor. Fiz. **105**, 808 (1994)
R. Bonifacio et al., Phys. Rev. Lett. **73**, 70 (1994)
H.P. Freund, Phys. Rev. E **52**, 5401 (1995)
H.P. Freund, T.M. Antonsen, Jr.: Principles of Free-Electron Lasers (Chapman and Hall, London 1996)
H.K. Avetissian et al., Phys. Rev. A **56**, 4121 (1997)
DESY Report No. 1997-048, ed by R. Brinkmann, G. Materlik, J. Rossbach, A. Wagner
J. Arthur et al., LCLS-Design Study Report No. SLAC-R-521, 1998
K.Z. Hatsagortsian, A.L. Khachatryan, Opt. Commun. **146**, 114 (1998)
E.L. Saldin, E.A. Schneidmiller, M.V. Yurkov: The Physics of Free Electron Lasers (Springer, Berlin 2000)
S.V. Milton et al., Phys. Rev. Lett. **85**, 988 (2000)
J. Andruszkow et al., Phys. Rev. Lett. **85**, 3825 (2000)
P.G. O'Shea, H. P. Freund, Science **292**, 1853 (2001)
S.V. Milton et al., Science **292**, 2037 (2001)
H.K. Avetissian, G.F. Mkrtchian, Phys. Rev. E **65**, 046505 (2002)
H.K. Avetissian, G.F. Mkrtchian, Nucl. Instrum. Methods A **483**, 548 (2002)
V. Ayvazyan et al., Phys. Rev. Lett. **88**, 104802 (2002)
V. Ayvazyan et al., Eur. Phys. J. D **20**, 149 (2002)
H.K. Avetissian, A.L. Khachatryan, G.F. Mkrtchian, Nucl. Instrum. Methods A 507, 31 (2003)
H.K. Avetissian, G.F. Mkrtchian, Nucl. Instrum. Methods A **507**, 479 (2003)
H.K. Avetissian, G.F. Mkrtchian, Nucl. Instrum. Methods A **528**, 530 (2004)
H.K. Avetissian, G.F. Mkrtchian, Nucl. Instrum. Methods A **528**, 534 (2004)
C.A. Brau, Phys. Rev. ST Accel. Beams **7**, 020701 (2004)
H.K. Avetissian, G.F. Mkrtchian: Quantum SASE Regime of X-ray FEL. 27th International FEL Conference, Stanford University, USA (2005)

9 Electron–Positron Pair Production in Superstrong Laser Fields

Considering the interaction of charged particles with strong radiation fields in vacuum we looked at the non-quantum electrodynamic (QED) properties of electromagnetic vacuum. At such consideration, vacuum stipulates only the classical dispersion properties of EM waves propagating with the speed of light c. However, the latter is valid for radiation fields that are not superstrong ($\xi_0 < 1$), otherwise the excitation of QED vacuum and production of electron–positron pairs becomes possible.

As follows from the physical meaning of the wave intensity parameter ξ_0, at values of $\xi_0 > 1$ the energy acquired by an electron over a wavelength of a coherent radiation field exceeds the electron rest energy mc^2. On the other hand, the energetic width of the vacuum gap or the threshold value for the electron–positron pair production is $2mc^2$. This means that electrons of the Dirac vacuum acquiring the energy $\mathcal{E} > 2mc^2$ at the interaction with the wave field of intensity $\xi_0 > 1$ will pass from negative energy states to positive ones (excitation of the Dirac vacuum) and electron–positron pair production becomes a fact (with the presence of a third body for the satisfaction of the conservation laws for this process).

The production of electron–positron pairs by plane EM waves of relativistic intensities ($\xi_0 >> 1$) is essentially a multiphoton process, which principally differs from the known "Klein paradox" — production of electron–positron pairs in stationary and homogeneous electric field proceeding over the electron Compton wavelength. The latter corresponds to the tunnel effect through the effective energetic barrier of finite width formed from the vacuum gap of infinite width by the presence of a uniform electric field (Schwinger mechanism). The physical mechanisms are similar to two different limits of Above Threshold Ionization of atoms in strong radiation fields — multiphoton and tunnel ionization.

This chapter considers the excitation of the Dirac vacuum in superstrong EM fields and the electron–positron pair production process in the presence of a diverse type third body.

9.1 Vacuum in Superstrong Electromagnetic Fields. Klein Paradox

It has long been well known that in the background of a stationary and homogeneous electric field the QED vacuum is unstable and electron–positron (e^-, e^+) pair production from the vacuum occurs (this mechanism is often referred to as the Schwinger mechanism). However, a measurable rate for pair production requires extraordinarily strong electric field strengths comparable to the critical vacuum field strength

$$E_c = \frac{m^2 c^3}{e\hbar}, \tag{9.1}$$

the work of which on an electron over the Compton wavelength $\lambda_c = \hbar/mc$ equals the electron rest energy. As we will see the probability of this process reaches optimal values when

$$\zeta = \frac{E_0}{E_c} \gtrsim 1, \tag{9.2}$$

where E_0 is the magnitude of a uniform electric field strength.

Fortunately, it seems possible to produce EM fields with electric field strengths of the order of the Schwinger critical field in the focus of expected X-ray FEL and consideration of this problem is theoretically important, since it requires one to go beyond perturbation theory, and its experimental observation would verify the validity of theory in the domain of strong fields.

To solve the problem of e^-, e^+ pair production in the given electric field we shall make use of the Dirac model — all vacuum negative energy states are filled with electrons and e^-, e^+ pair production by the electric field occurs when the vacuum electrons with initial negative energies $\mathcal{E}_0 < 0$ due to "acceleration" pass to the final states with positive energies $\mathcal{E} > 0$. To distinguish the free particle states we will switch on and switch off the electric field elaborating on a model which retains the main features of the spatially uniform electric field and allows one to obtain an exact solution for the Dirac equation and final expressions for the pair production rate in closed form. Thus, we will assume an electric field of the form

$$\mathbf{E}(t) = \frac{E_0}{\cosh^2\left(\frac{t}{T}\right)} \widehat{\mathbf{z}}, \tag{9.3}$$

where T is the characteristic period of the field and $\widehat{\mathbf{z}}$ is the unit vector along the field strength. The vector potential corresponding to this field may be written as

$$\mathbf{A}(t) = -c \int_{-\infty}^{t} \mathbf{E}(t)dt = -cE_0T\hat{\mathbf{z}}\left[\tanh\left(\frac{t}{T}\right) + 1\right]. \tag{9.4}$$

We will solve the Dirac equation in the spinor representation (see Eqs. (1.77), (1.78)). Since the interaction Hamiltonian does not depend on the space coordinates, the generalized momentum $\mathbf{p}_0$ is conserved. Hence, the solution of Eq. (1.77) may be represented in the form

$$\Psi(\mathbf{r},t) = \Psi_{\mathbf{p}_0}(t)\, e^{\frac{i}{\hbar}\mathbf{p}_0\mathbf{r}}, \tag{9.5}$$

and from Eq. (1.77) for the function $\Psi_{\mathbf{p}_0}(t)$ we obtain the following equation:

$$i\hbar\frac{d\Psi_{\mathbf{p}_0}}{dt} = \left[c\alpha\left(\mathbf{p}_0 + \frac{e}{c}\mathbf{A}(t)\right) + mc^2\beta\right]\Psi_{\mathbf{p}_0}. \tag{9.6}$$

In this section the electron charge will be assumed to be $-e$. Since $\mathbf{A}(-\infty) = 0$ the solution of Eq. (9.6) at $t \to -\infty$ should be superposition of the free particle solutions $\psi^{(\varkappa)}_{\mathbf{p}_0,\sigma}$ with negative ($\varkappa = -1$) and positive ($\varkappa = 1$) energies and polarizations $\sigma = \pm\frac{1}{2}$ (spin projections $S_z = \pm\frac{1}{2}$ in the rest frame of the particle):

$$\psi^{(\varkappa)}_{\mathbf{p}_0,1/2} = \sqrt{\frac{1}{2\mathcal{E}_0\left(\mathcal{E}_0 - \varkappa c p_{0z}\right)}}\begin{pmatrix} \varkappa mc^2 w^{(1/2)} \\ \left(\mathcal{E}_0 - \varkappa c\sigma\mathbf{p}_0\right) w^{(1/2)} \end{pmatrix} e^{-\frac{i}{\hbar}\varkappa\mathcal{E}_0 t}, \tag{9.7}$$

$$\psi^{(\varkappa)}_{\mathbf{p}_0,-1/2} = \sqrt{\frac{1}{2\mathcal{E}_0\left(\mathcal{E}_0 + \varkappa c p_{0z}\right)}}\begin{pmatrix} \left(\mathcal{E}_0 + \varkappa c\sigma\mathbf{p}_0\right) w^{(-1/2)} \\ \varkappa mc^2 w^{(-1/2)} \end{pmatrix} e^{-\frac{i}{\hbar}\varkappa\mathcal{E}_0 t}, \tag{9.8}$$

where $\mathcal{E}_0 = \sqrt{c^2\mathbf{p}_0^2 + m^2c^4}$, σ are Pauli matrices, and the spinors $w^{(\pm 1/2)}$ are

$$w^{(1/2)} = \begin{pmatrix} 1 \\ 0 \end{pmatrix}; \qquad w^{(-1/2)} = \begin{pmatrix} 0 \\ 1 \end{pmatrix}.$$

At $t \to \infty$, the electric field $\mathbf{E}(\infty) = 0$ but

$$\mathbf{A}(\infty) = -2cE_0T\hat{\mathbf{z}}, \tag{9.9}$$

and the solution of Eq. (9.6) at $t \to \infty$ should be superposition of the free particle solutions (9.7), (9.8) where the "final momentum"

$$\mathbf{p} = \mathbf{p}_0 - e\int_{-\infty}^{\infty} \mathbf{E}(t)dt = \mathbf{p}_0 + \frac{e}{c}\mathbf{A}(\infty) \tag{9.10}$$

stands for $\mathbf{p}_0$. Equation (9.6) in the quadratic form (see Eqs. (1.82), (1.83)) for the bispinor components

$$\Psi_{\mathbf{p}_0}(t) = \begin{pmatrix} f_1 \\ f_2 \\ f_3 \\ f_4 \end{pmatrix} \tag{9.11}$$

gives the following set of equations:

$$\left\{\hbar^2\frac{d^2}{dt^2} + \mathcal{E}_0^2 + e^2A^2(t) + 2ecp_{0z}A(t) \mp iec\hbar E(t)\right\} f_{1,2} = 0, \tag{9.12}$$

$$\left\{\hbar^2\frac{d^2}{dt^2} + \mathcal{E}_0^2 + e^2A^2(t) + 2ecp_{0z}A(t) \pm iec\hbar E(t)\right\} f_{3,4} = 0. \tag{9.13}$$

Thus, solving the equation

$$\left\{\hbar^2\frac{d^2}{dt^2} + \mathcal{E}_0^2 + e^2A^2(t) + 2ecp_{0z}A(t) - \delta iec\hbar E(t)\right\} \Phi = 0 \tag{9.14}$$

with $\delta = \pm 1$ one can construct the whole bispinor (9.11). Passing in Eq. (9.14) to the new variable

$$z = -e^{2\frac{t}{T}},$$

and seeking the solution in the form

$$\Phi(t) = (-z)^{i\frac{\mathcal{E}_0 T}{2\hbar}}(1-z)^{i\delta\frac{eE_0T^2c}{\hbar}}F(z), \tag{9.15}$$

we obtain the equation for hypergeometric function $F(\alpha, \beta, \gamma, z)$:

$$z(1-z)F'' + (\gamma - (\alpha + \beta + 1)z)F' - \alpha\beta F = 0. \tag{9.16}$$

The parameters α, β, γ are defined as follows:

$$\alpha(\mathcal{E}_0, \delta) = i\frac{\mathcal{E}_0 + \mathcal{E} + 2i\delta eE_0cT}{2\hbar}T,$$

$$\beta(\mathcal{E}_0,\delta) = i\frac{\mathcal{E}_0 - \mathcal{E} + 2i\delta eE_0cT}{2\hbar}T, \tag{9.17}$$

$$\gamma(\mathcal{E}_0) = 1 + i\frac{\mathcal{E}_0}{\hbar}T,$$

where according to Eqs. (9.10) and (9.9)

$$\mathcal{E} = \sqrt{c^2\left(\mathbf{p}_0 - 2eE_0T\widehat{\mathbf{z}}\right)^2 + m^2c^4}.$$

The general solution for hypergeometric equation (9.16) is

$$F(z) = F(\alpha,\beta,\gamma,z) + z^{1-\gamma}F(\alpha-\gamma+1,\beta-\gamma+1,2-\gamma,z). \tag{9.18}$$

Taking into account the relations

$$\alpha(\mathcal{E}_0,\delta) - \gamma(\mathcal{E}_0) + 1 = \alpha(-\mathcal{E}_0,\delta),$$

$$\beta(\mathcal{E}_0,\delta) - \gamma(\mathcal{E}_0) + 1 = \beta(-\mathcal{E}_0,\delta),$$

$$2-\gamma = \gamma(-\mathcal{E}_0),$$

$$i\frac{\mathcal{E}_0}{2\hbar}T + 1 - \gamma(\mathcal{E}_0) = -i\frac{\mathcal{E}_0}{2\hbar}T,$$

the general solution for bispinor $\Psi_{\mathbf{p}_0}(t)$ can be written as follows:

$$\Psi_{\mathbf{p}_0}(t) = \begin{pmatrix} A_1\Phi(\mathcal{E}_0,1;z) + A_2\Phi(-\mathcal{E}_0,1;z) \\ B_1\Phi(\mathcal{E}_0,-1;z) + B_2\Phi(-\mathcal{E}_0,-1;z) \\ C_1\Phi(\mathcal{E}_0,-1;z) + C_2\Phi(-\mathcal{E}_0,-1;z) \\ D_1\Phi(\mathcal{E}_0,1;z) + D_2\Phi(-\mathcal{E}_0,1;z) \end{pmatrix}, \tag{9.19}$$

where

$$\Phi(\mathcal{E}_0,\delta;z) = (-z)^{i\frac{\mathcal{E}_0}{2\hbar}T}(1-z)^{i\delta\frac{eE_0c}{\hbar}T^2}$$

$$\times F(\alpha(\mathcal{E}_0,\delta),\beta(\mathcal{E}_0,\delta),\gamma(\mathcal{E}_0),z), \tag{9.20}$$

and the coefficients $A_{1,2}$, $B_{1,2}$, $C_{1,2}$, $D_{1,2}$ should be defined from the initial condition.

To determine the probability of e^-, e^+ pair production we use the initial condition: at $t \to -\infty$ when $\mathbf{A}(-\infty) = 0$ this wave function must turn into the free Dirac equation solution with negative energy in accordance with the Dirac model. Then taking into account that at

$$t \to -\infty; \quad z \to 0,$$

$$\Phi\left(\mathcal{E}_0, \delta; z \to 0\right) = (-z)^{i\frac{\mathcal{E}_0}{2\hbar}T} = e^{\frac{i}{\hbar}\mathcal{E}_0 t},$$

we obtain

$$\Psi^{(-1)}_{\mathbf{p}_0,1/2} = \sqrt{\frac{1}{2\mathcal{E}_0\left(\mathcal{E}_0 + cp_{0z}\right)}} \begin{pmatrix} -mc^2\Phi\left(\mathcal{E}_0, 1; z\right) \\ 0 \\ \left(\mathcal{E}_0 + cp_{0z}\right)\Phi\left(\mathcal{E}_0, -1; z\right) \\ \left(cp_{0x} + icp_{0y}\right)\Phi\left(\mathcal{E}_0, 1; z\right) \end{pmatrix}, \tag{9.21}$$

$$\Psi^{(-1)}_{\mathbf{p}_0,-1/2} = \sqrt{\frac{1}{2\mathcal{E}_0\left(\mathcal{E}_0 - cp_{0z}\right)}} \begin{pmatrix} \left(-cp_{0x} + icp_{0y}\right)\Phi\left(\mathcal{E}_0, 1; z\right) \\ \left(\mathcal{E}_0 + cp_{0z}\right)\Phi\left(\mathcal{E}_0, -1; z\right) \\ 0 \\ -mc^2\Phi\left(\mathcal{E}_0, 1; z\right) \end{pmatrix}. \tag{9.22}$$

After the interaction at $t \to +\infty; \quad z \to -\infty$ these wave functions become the superposition of the free Dirac equation solutions. To determine the asymptotes of these functions we will use the following property of the hypergeometric function:

$$F\left(\alpha, \beta, \gamma, z\right) = \frac{\Gamma\left(\gamma\right)\Gamma\left(\beta - \alpha\right)}{\Gamma\left(\beta\right)\Gamma\left(\gamma - \alpha\right)}\left(-z\right)^{-\alpha} F\left(\alpha, \alpha + 1 - \gamma, \alpha + 1 - \beta, \frac{1}{z}\right)$$

$$+\frac{\Gamma\left(\gamma\right)\Gamma\left(\alpha - \beta\right)}{\Gamma\left(\alpha\right)\Gamma\left(\gamma - \beta\right)}\left(-z\right)^{-\beta} F\left(\beta, \beta + 1 - \gamma, \beta + 1 - \alpha, \frac{1}{z}\right). \tag{9.23}$$

Hence, for the function Φ we obtain

$$\Phi\left(\mathcal{E}_0, \delta; z \to -\infty\right) = e^{-\frac{i}{\hbar}\mathcal{E}t} \frac{\Gamma\left(\gamma\left(\mathcal{E}_0\right)\right)\Gamma\left(\beta\left(\mathcal{E}_0, \delta\right) - \alpha\left(\mathcal{E}_0, \delta\right)\right)}{\Gamma\left(\beta\left(\mathcal{E}_0, \delta\right)\right)\Gamma\left(\gamma\left(\mathcal{E}_0\right) - \alpha\left(\mathcal{E}_0, \delta\right)\right)}$$

$$+e^{\frac{i}{\hbar}\mathcal{E}t}\frac{\Gamma\left(\gamma\left(\mathcal{E}_0\right)\right)\Gamma\left(\alpha\left(\mathcal{E}_0,\delta\right)-\beta\left(\mathcal{E}_0,\delta\right)\right)}{\Gamma\left(\alpha\left(\mathcal{E}_0,\delta\right)\right)\Gamma\left(\gamma\left(\mathcal{E}_0\right)-\beta\left(\mathcal{E}_0,\delta\right)\right)}. \tag{9.24}$$

Taking into account the relations

$$\frac{\mathcal{E}-\mathcal{E}_0+2eE_0cT}{\mathcal{E}_0-\mathcal{E}+2eE_0cT}=\frac{\mathcal{E}-cp_z}{\mathcal{E}_0+cp_{0z}},$$

$$p_{0z}-p_z=2eE_0T,$$

for the bispinor wave function (9.21) we obtain

$$\Psi^{(-1)}_{\mathbf{p}_0,1/2}\left(t\to+\infty\right)=C\left(\mathcal{E}\right)\psi^{(1)}_{\mathbf{p},1/2}+C\left(-\mathcal{E}\right)\psi^{(-1)}_{\mathbf{p},1/2}, \tag{9.25}$$

where

$$C\left(\mathcal{E}\right)=\sqrt{\frac{\mathcal{E}\mathcal{E}_0}{\left(\mathcal{E}_0-\mathcal{E}+2eE_0cT\right)\left(\mathcal{E}-\mathcal{E}_0+2eE_0cT\right)}}$$

$$\times\frac{2\Gamma\left(i\frac{\mathcal{E}_0}{\hbar}T\right)\Gamma\left(-i\frac{\mathcal{E}}{\hbar}T\right)}{\Gamma\left(i\frac{\mathcal{E}_0-\mathcal{E}+2eE_0cT}{2\hbar}T\right)\Gamma\left(i\frac{\mathcal{E}_0-\mathcal{E}-2eE_0cT}{2\hbar}T\right)}. \tag{9.26}$$

The probability of the e^-, e^+ pair production summed over the spin states is

$$W(\mathcal{E})=2\left|C\left(\mathcal{E}\right)\right|^2. \tag{9.27}$$

Taking into account that

$$\left|\Gamma\left(iy\right)\right|^2=\frac{\pi}{y\sin\pi iy},$$

for the probability (9.27) we obtain

$$W(\mathcal{E})=2\frac{\cosh\left(\pi\frac{2eE_0cT^2}{\hbar}\right)-\cosh\left(\pi\frac{\mathcal{E}-\mathcal{E}_0}{\hbar}T\right)}{\cosh\left(\pi\frac{\mathcal{E}+\mathcal{E}_0}{\hbar}T\right)-\cosh\left(\pi\frac{\mathcal{E}-\mathcal{E}_0}{\hbar}T\right)}. \tag{9.28}$$

The number of created e^-, e^+ pairs per unit space volume is

$$N=\int W(\mathcal{E})\frac{d\mathbf{p}_0}{\left(2\pi\hbar\right)^3},$$

which with Eq. (9.28) is written as

$$N = \frac{2}{(2\pi\hbar)^3}\int \frac{\cosh\left(\pi\frac{2eE_0cT^2}{\hbar}\right) - \cosh\left(\pi\frac{\mathcal{E}-\mathcal{E}_0}{\hbar}T\right)}{\cosh\left(\pi\frac{\mathcal{E}+\mathcal{E}_0}{\hbar}T\right) - \cosh\left(\pi\frac{\mathcal{E}-\mathcal{E}_0}{\hbar}T\right)} dp_{0z}dp_{0x}dp_{0y}. \quad (9.29)$$

The probability (9.28) has a maximum at $p_{0z} = eE_0T$ (the electrons and positrons are created with the same energy, i.e., $p_z = -eE_0T$). In the limit $T \to \infty$ the electric field (9.3) tends to a constant one: $\mathbf{E}(t) \to E_0\widehat{\mathbf{z}}$ and from Eq. (9.28) one can obtain the probability of the e^-, e^+ pair production in the static, spatially uniform electric field. In this case in the integral (9.29) over p_{0z} the main contribution gives the maximum point with the width $\delta p_{0z} \approx eE_0T$. Hence, at

$$(ceE_0T)^2 >> m^2c^4 + c^2p_{0\perp}^2; \qquad p_{0\perp} = \sqrt{p_{0x}^2 + p_{0y}^2},$$

we have

$$\mathcal{E}_0 \approx \mathcal{E} \approx ceE_0T + \frac{m^2c^4 + c^2p_{0\perp}^2}{2ceE_0T},$$

and for the number of e^-, e^+ pairs created per unit time and unit space volume we obtain

$$\frac{N}{T} \approx \frac{2}{(2\pi\hbar)^3}eE_0\int \exp\left[-\pi\frac{m^2c^4 + c^2p_{0\perp}^2}{ceE_0\hbar}\right] dp_{0x}dp_{0y}. \quad (9.30)$$

Integrating in Eq. (9.30) over transversal momentum we obtain the Schwinger formula:

$$\frac{N_{Sch}}{T} = \frac{e^2E_0^2}{4\pi^3\hbar^2c}\exp\left[-\pi\frac{m^2c^3}{e\hbar E_0}\right], \quad (9.31)$$

or in the terms of critical field

$$\frac{N_{Sch}}{T} = \frac{\zeta^2}{4\pi^3\lambda_c^3}\frac{mc^2}{\hbar}\exp\left[-\frac{\pi}{\zeta}\right]. \quad (9.32)$$

If $\zeta << 1$ the probability of pair production is exponentially suppressed and reaches the optimal values when $\zeta \gtrsim 1$ at which

$$\frac{N_{Sch}}{T} \gtrsim 10^{49}\mathrm{cm}^{-3}\mathrm{c}^{-1}.$$

9.2 Electron–Positron Pair Production by Superstrong Laser Field and γ-Quantum

For the electron–positron pair production by superstrong laser fields of relativistic intensities as a third body for the satisfaction of conservation laws in physically more interesting cases can serve a γ-quantum or a nucleus/ion.

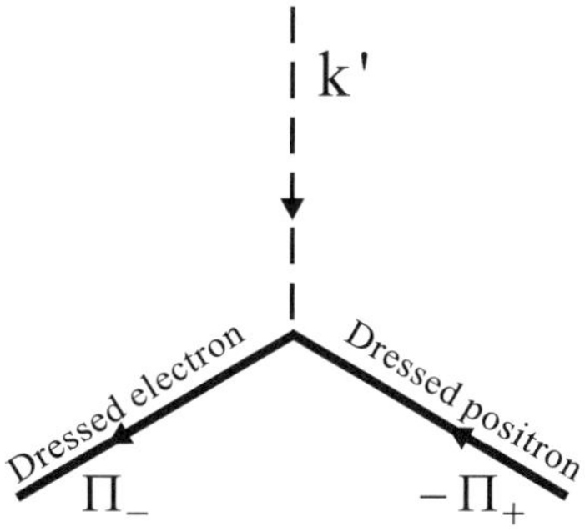

Fig. 9.1. Feynman diagram for electron–positron pair production by laser field and γ-quantum.

The e^-,e^+ pair production process by a plane monochromatic radiation field and a γ-quantum in the scope of QED is described by the first order Feynman diagram (Fig. 9.1) where wave functions (1.94) correspond to electron/positron lines. As in QED the production of electron and positron with quasimomentums Π_- and Π_+ respectively is interpreted as a transition of an electron from the vacuum state "$-\Pi_+$" to state Π_-. The Feynman diagram is topologically equivalent to that of the Compton effect. Hence, the S-matrix amplitude of this process can be obtained from the Compton-effect S-matrix amplitude (1.114) by the substitutions: $\epsilon_\mu^* \to \epsilon_\mu$, $k' \to -k'$, $\Pi \to -\Pi_+$, $\Pi' \to \Pi_-$:

$$S_{fi} = -i\,(2\pi\hbar)^4 \sqrt{\frac{\pi\alpha_0}{2\omega' c\Pi_{0+}\Pi_{0-}V^3}}\,\overline{u}_{\sigma'}(p_-)$$

$$\times \widehat{M}_{fi}^{(Compton)}\,(\epsilon^* \to \epsilon, k' \to -k', \Pi \to -\Pi_+, \Pi' \to \Pi_-)\,u_\sigma(-p_+). \quad (9.33)$$

We will assume that the γ-quantum is nonpolarized and corresponding summation over the electron and positron polarizations will be made. Taking into account that at the summation over the positron polarizations one should replace $u(-p_+)\overline{u}(-p_+)$ by $c^2(\widehat{p}_+ - mc)$ one can see that

$$\frac{1}{2}\sum_{\sigma',\sigma,\epsilon} |S_{fi}|^2$$

$$= -\frac{1}{2} \sum_{\sigma',\sigma,\epsilon,\epsilon} |S_{fi}|^2_{(Compton)} \left(k' \to -k', \Pi \to -\Pi_+, \Pi' \to \Pi_-\right). \qquad (9.34)$$

For the differential probability of e^-, e^+ pair production per unit time we have

$$dW = \frac{1}{2\Delta t} \sum_{\sigma',\sigma,\epsilon} |S_{fi}|^2 V \frac{d\mathbf{\Pi}_-}{(2\pi\hbar)^3} V \frac{d\mathbf{\Pi}_+}{(2\pi\hbar)^3}. \qquad (9.35)$$

Hence, using Eqs. (1.114) for the Compton effect and taking into account relation (9.34) for the differential probability (9.35) we obtain

$$dW = \sum_{s>s_m}^{\infty} W^{(s)} \delta\left(\Pi_- + \Pi_+ - \hbar k' - s\hbar k\right) d\mathbf{\Pi}_- d\mathbf{\Pi}_+, \qquad (9.36)$$

where

$$W^{(s)} = \frac{\alpha_0 m^2 c^6}{2\pi\omega' \hbar^2 \Pi_{0+} \Pi_{0-}} \left[|G_s|^2 - \left(1 - \frac{\hbar^2 (kk')^2}{2(p_+k)(p_-k)}\right) \right.$$

$$\times \left(\frac{(1+g^2)\xi_0^2}{4} \left(|G_{s-1}|^2 + |G_{s+1}|^2 - 2|G_s|^2 \right) \right.$$

$$\left.\left. + \frac{(1-g^2)\xi_0^2}{4} Re\left[2G^*_{s-1}G_{s+1} - G^*_s\left(G_{s-2} + G_{s+2}\right)\right] \right) \right]. \qquad (9.37)$$

The arguments of the functions $G_s(\alpha, \beta, \varphi)$ in this case are

$$\alpha = \frac{eA_0}{\hbar c} \left[\left(\frac{\mathbf{e}_1\mathbf{p}_-}{p_-k} - \frac{\mathbf{e}_1\mathbf{p}_+}{p_+k} \right)^2 + g^2 \left(\frac{\mathbf{e}_2\mathbf{p}_-}{p_-k} - \frac{\mathbf{e}_2\mathbf{p}_+}{p_+k} \right)^2 \right]^{1/2}, \qquad (9.38)$$

$$\beta = -\frac{e^2 A_0^2}{8\hbar c^2} (1-g^2) \left(\frac{1}{p_+k} + \frac{1}{p_-k} \right), \qquad (9.39)$$

$$\tan\varphi = \frac{g\left(\frac{\mathbf{e}_2\mathbf{p}_-}{p_-k} - \frac{\mathbf{e}_2\mathbf{p}_+}{p_+k} \right)}{\left(\frac{\mathbf{e}_1\mathbf{p}_-}{p_-k} - \frac{\mathbf{e}_1\mathbf{p}_+}{p_+k} \right)}. \qquad (9.40)$$

Since the pair production is a threshold effect, the number of photons absorbed from the strong wave must exceed the threshold value

$$s_m = \frac{2m^{*2}c^2}{\hbar^2 (k'k)}, \tag{9.41}$$

which follows from the conservation law of this process expressed by the δ-function in Eq. (9.36) and the dispersion law for quasimomentum (1.96). Note that in Eq. (9.41) the effective mass appears which depends on the laser intensity. If $s_m > 1$ (for low photon energies), production of the electron–positron pair may only proceed by nonlinear channels (even for $\xi_0 << 1$). Besides, this process does not have a classical limit and the quantum recoil is always essential.

For the concreteness we will investigate the case of circular polarization of the incident wave ($g = \pm 1$). In this case $|G_s|^2 = J_s^2(\alpha)$ and from Eq. (9.37) for the partial probabilities we have

$$W^{(s)} = \frac{e^2 m^2 c^5}{2\pi\omega'\hbar^3 \Pi_{0+}\Pi_{0-}} \left[J_s^2(\alpha) - \xi_0^2 \left(1 - \frac{\hbar^2 (kk')^2}{2(p_+k)(p_-k)} \right) \right.$$

$$\left. \times \left(\left(\frac{s^2}{\alpha^2} - 1 \right) J_s^2(\alpha) + J_s'^2(\alpha) \right) \right]. \tag{9.42}$$

Taking into account the conservation laws, as well as the relations $p_-k = \Pi_-k$ and $p_+k = \Pi_+k$, the argument of the Bessel function can be written as

$$\alpha = \xi_0 \frac{mc^2}{\hbar\omega} \left| \left[\mathbf{k} \left(\frac{\mathbf{p}_-}{p_-k} - \frac{\mathbf{p}_+}{p_+k} \right) \right] \right|$$

$$= \xi_0 \frac{mc}{\hbar} \left[2s\hbar \left(\frac{1}{\Pi_-k} + \frac{1}{\Pi_+k} \right) - m_*^2 c^2 \left(\frac{1}{\Pi_-k} + \frac{1}{\Pi_+k} \right)^2 \right]^{1/2}. \tag{9.43}$$

For a weak EM wave: $\xi_0 << 1$ and $s_m < 1$ (linear theory) the argument of the Bessel function $\alpha << 1$ and the main contribution to the probability of the pair production is the one-photon process. In this case $J_1^2(\alpha_1) \simeq \alpha_1^2/4$, $J_1'^2(\alpha_1) \simeq 1/4$, $\Pi_{0+} \simeq \mathcal{E}_+$, $\Pi_{0-} \simeq \mathcal{E}_-$ and taking into account that

$$1 - \frac{(kk')^2}{2(p_+k)(p_-k)} = -\frac{1}{2} \left[\frac{p_-k}{p_+k} + \frac{p_+k}{p_-k} \right],$$

we obtain the G. Breit, A. Wheeler formula:

$$W^{(1)} = \frac{e^2 m^2 c^5}{8\pi\omega'\hbar^3 \mathcal{E}_+\mathcal{E}_-} \xi_0^2 \left[2 \left(\frac{m^2c^2}{\hbar p_-k} + \frac{m^2c^2}{\hbar p_+k} \right) \right.$$

$$-\left(\frac{m^2c^2}{\hbar p_-k}+\frac{m^2c^2}{\hbar p_+k}\right)^2+\left[\frac{p_-k}{p_+k}+\frac{p_+k}{p_-k}\right]\Bigg]. \tag{9.44}$$

For a strong EM wave it is more convenient to choose the quantum recoil parameter as an integration variable:

$$\rho=\frac{\hbar^2\left(kk'\right)^2}{2\left(p_+k\right)\left(p_-k\right)}=\frac{\hbar^2\left(kk'\right)^2}{2\left(\Pi_+k\right)\left(\Pi_-k\right)}. \tag{9.45}$$

Taking into account the azimuthal symmetry with respect to the wave propagation direction one can make the following replacement:

$$\delta\left(\Pi_-+\Pi_+-\hbar k'-s\hbar k\right)\frac{d\mathbf{\Pi}_-d\mathbf{\Pi}_+}{\Pi_{0+}\Pi_{0-}}=>\frac{2\pi}{c^2}\frac{1}{\rho\sqrt{\rho^2-2\rho}}d\rho, \tag{9.46}$$

and we obtain

$$W=\frac{e^2m^2c^3}{\omega'\hbar^3}\sum_{s>s_m}^{\infty}\int_2^{2s/s_m}\Bigg[J_s^2(\alpha_s(\rho))+\xi_0^2(\rho-1)$$

$$\times\left(\left(\frac{s^2}{\alpha_s^2(\rho)}-1\right)J_s^2(\alpha_s(\rho))+J_s'^2(\alpha_s(\rho))\right)\Bigg]\frac{d\rho}{\rho\sqrt{\rho^2-2\rho}}, \tag{9.47}$$

where the argument of the Bessel function is

$$\alpha_s(\rho)=\frac{\xi_0}{\sqrt{1+\xi_0^2}}s_m\left[\frac{2s}{s_m}\rho-\rho^2\right]^{1/2}. \tag{9.48}$$

The latter reaches its maximal value

$$\alpha_{s\max}=\frac{\xi_0}{\sqrt{1+\xi_0^2}}s \tag{9.49}$$

at $\rho=s/s_m$. This value is in the integration range when $s>2s_m$. If $s_m>>1$, which is possible for not so hard γ-quantum, and at $\xi_0\gg1$ one can approximate the Bessel function by the Airy one (see Eq. (1.69) for Compton effect) and for the probability of the pair production we obtain

$$W\simeq\frac{e^2m^2c^3}{\omega'\hbar^3}\sum_{s>s_m}^{\infty}\int_2^{2s/s_m}\left\{\left[1+\xi_0^2(\rho-1)\left(\frac{s^2}{\alpha^2(\rho)}-1\right)\right]\left(\frac{2}{s}\right)^{2/3}Ai^2(Z)\right.$$

$$+\xi_0^2 (\rho - 1)\, Ai'^2 (Z) \left(\frac{2}{s}\right)^{4/3} \Bigg\} \frac{d\rho}{\rho\sqrt{\rho^2 - 2\rho}}, \tag{9.50}$$

where

$$Z = \frac{1}{1+\xi_0^2} \left(\frac{s}{2}\right)^{2/3} \left(1 + \xi_0^2 \left(1 - \frac{s_m}{s}\rho\right)^2\right). \tag{9.51}$$

As far as the Airy function exponentially decreases with increasing of the argument one can conclude that the optimal parameters for the pair production process are determined from the condition $Z_{\min} \sim 1$, where

$$Z_{\min} = \left(\frac{s}{2}\right)^{2/3} \left(1 - \frac{\alpha^2_{s\,\max}}{s^2}\right) \simeq \left(\frac{s}{2\xi_0^3}\right)^{2/3},$$

which gives

$$2\xi_0^3 \gtrsim s_m.$$

For $\xi_0 >> 1$, $s_m \simeq 2m^2c^2\xi_0^2/(\hbar^2 k' k)$ we obtain

$$\zeta = \frac{\hbar^2 k' k}{m^2 c^2} \xi_0 \gtrsim 1. \tag{9.52}$$

The latter means that in the rest frame of created electron the electric field strength of the EM wave exceeds the critical vacuum field (9.1). Hence, ζ is the quantum parameter of interaction in the scale of the critical vacuum field.

For $Z_{\min} >> 1$ or $\zeta << 1$ (so called tunneling regime of the pair production process) one can use the following asymptotic formula for the Airy function:

$$Ai(Z) \simeq \frac{1}{2\sqrt{\pi}} Z^{-1/4} \exp\left(-\frac{2Z^{3/2}}{3}\right).$$

Hence, the probability of the electron–positron pair production

$$W \propto \exp\left(-\frac{4}{3\zeta}\right) \tag{9.53}$$

is exponentially suppressed.

For the moderate relativistic intensities $\xi_0 \sim 1$ to show the dependence of the probability on the wave intensity and quantum parameter of interaction ζ the normalized probability

$$\widetilde{W} = \frac{\omega' \hbar^3}{e^2 m^2 c^3} W \tag{9.54}$$

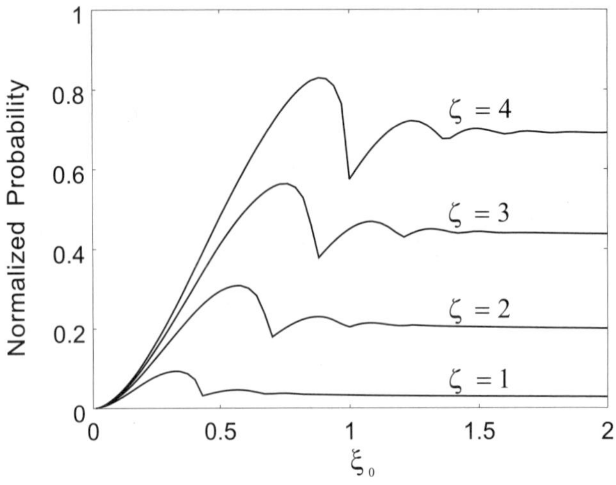

Fig. 9.2. The normalized probability $\widetilde{W} = \hbar^3\omega' W/(e^2m^2c^3)$ as a function of relativistic parameter of intensity ξ_0 for various ζ.

is displayed in Fig. 9.2 as a function of ξ_0 for various ζ.

9.3 Pair Production via Superstrong Laser Beam Scattering on a Nucleus

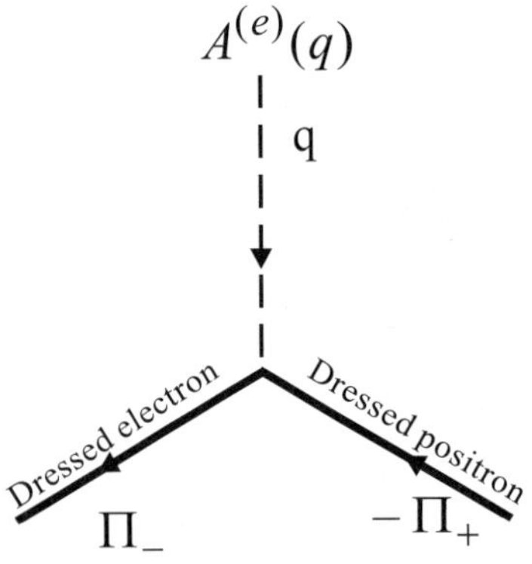

Fig. 9.3. Feynman diagram for electron–positron pair production via laser beam scattering on a nucleus.

The electron–positron pair production via superstrong laser beam scattering on a nucleus can be described again by the first-order Feynman diagram (Fig. 9.3) where wave functions (1.94) correspond to electron/positron lines.

The Feynman diagram is topologically equivalent to that of the stimulated bremsstrahlung (SB) effect. As in the previous section the S-matrix amplitude of this process can be obtained from the S-matrix amplitude of SB (1.128) by the substitutions: $\Pi \rightarrow -\Pi_+$, $\Pi' \rightarrow \Pi_-$:

$$S_{fi} = \frac{-i\pi e}{Vc\sqrt{\Pi_{0+}\Pi_{0-}}}\overline{u}_{\sigma'}(p_-)\widehat{M}_{fi}^{(SB)}\left(\Pi \rightarrow -\Pi_+, \Pi' \rightarrow \Pi_-\right)u_\sigma(-p_+). \tag{9.55}$$

Making the summation over the electron and positron polarizations one can see that

$$\sum_{\sigma',\sigma}\left|S_{fi}\right|^2 = \sum_{\sigma',\sigma}\left|S_{fi}\right|^2_{SB}\left(\Pi \rightarrow -\Pi_+, \Pi' \rightarrow \Pi_-\right). \tag{9.56}$$

The differential probability of e^-, e^+ pair production per unit time is written as

$$dW = \frac{1}{\Delta t}\sum_{\sigma',\sigma}\left|S_{fi}\right|^2 V\frac{d\mathbf{\Pi}_-}{(2\pi\hbar)^3}V\frac{d\mathbf{\Pi}_+}{(2\pi\hbar)^3}. \tag{9.57}$$

Hence, using Eq. (1.129) for the SB process and taking into account Eq. (9.56) for the differential probability of pair production per unit time we obtain

$$dW = \sum_{s>s_m}^{\infty} W^{(s)}\delta\left(\Pi_{0+} + \Pi_{0-} - s\hbar\omega\right)d\mathbf{\Pi}_-d\mathbf{\Pi}_+, \tag{9.58}$$

where

$$W^{(s)} = \frac{4\pi}{\Pi_{0+}\Pi_{0-}}\frac{e^2\left|\varphi\left(\mathbf{q}_s\right)\right|^2}{(2\pi\hbar)^6\hbar}\left\{\frac{\hbar^2\mathbf{q}_s^2c^2}{4}\left|B_s\right|^2 + \frac{e^2\hbar^2\left[\mathbf{kq}_s\right]^2}{4(kp_-)(kp_+)}\right.$$

$$\left.\times\left[\left|\mathbf{B}_{1s}\right|^2 - ReB_{2s}B_s^*\right] - \left|\mathcal{E}_+B_s + \frac{e\left(\mathbf{p}_+\mathbf{B}_{1s}\right)\omega}{(kp_+)c} + \frac{e^2\omega}{2c^2(kp_+)}B_{2s}\right|^2\right\}, \tag{9.59}$$

and

$$\hbar\mathbf{q}_s = \mathbf{\Pi}_- + \mathbf{\Pi}_+ - s\hbar\mathbf{k}.$$

The threshold value of the photon number for this process is defined as follows:

$$s_m = \frac{2m^*c^2}{\hbar\omega}. \tag{9.60}$$

The arguments α, β, φ of the functions B_s, $\mathbf{B}_{1s}$, B_{2s} are defined according to Eqs. (9.38)–(9.40).

In the case of circular polarization of an incident strong wave ($g = 1$) we have

$$G_s(\alpha, 0, \varphi) = (-1)^s J_s(\alpha) e^{is\varphi}.$$

Taking into account the azimuthal symmetry with respect to the wave propagation direction one can make the following replacement:

$$\delta\left(\Pi_{0+} + \Pi_{0-} - s\hbar\omega\right) d\mathbf{\Pi}_- d\mathbf{\Pi}_+ \to 2\pi m^* \frac{\Pi_{0-}\left|\mathbf{\Pi}_-\right| \Pi_{0+}\left|\mathbf{\Pi}_+\right|}{c^2}$$

$$\times \sin\theta_+ \sin\theta_- d\theta_- d\theta_+ d\phi d\gamma_+, \qquad (9.61)$$

where $\gamma_+ = \Pi_{0+}/(m^* c^2)$, θ_+, θ_- are the scattering angles of positron and electron with respect to the EM wave propagation direction and ϕ is the angle between the planes formed by $\mathbf{\Pi}_-$, $\mathbf{k}$ and $\mathbf{\Pi}_+$, $\mathbf{k}$. Hence, for the differential probability of e^-, e^+ pair production per unit time we have

$$dW = \frac{2\pi^2 \alpha_0 m^*}{(2\pi\hbar)^6 c} \sum_{s>s_m}^{\infty} \left|\mathbf{\Pi}_-\right| \left|\mathbf{\Pi}_+\right| \left|\varphi(\mathbf{q}_s)\right|^2$$

$$\times \left\{ \left[\hbar^2 \mathbf{q}_s^2 c^2 - 4\left(\Pi_{0+} - \frac{s\hbar\omega}{(kp_+)} \frac{\varkappa\left[\mathbf{kp}_+\right]}{\varkappa^2}\right)^2 \right] J_s^2(\alpha_s) \right.$$

$$+ \frac{\hbar^2 e^2 A_0^2}{(kp_-)(kp_+)} \left[\mathbf{kq}_s\right]^2 \left[\left(\frac{s^2}{\alpha_s^2} - 1 \right) J_s^2(\alpha_s) + J_s'^2(\alpha_s) \right]$$

$$\left. - \frac{4e^2 A_0^2}{(kp_+)^2} \frac{\left[\varkappa\left[\mathbf{kp}_+\right]\right]^2}{\varkappa^2} J_s'^2(\alpha_s) \right\} \sin\theta_+ \sin\theta_- d\theta_- d\theta_+ d\phi d\gamma_+. \qquad (9.62)$$

In this equation the electron quasienergy and quasimomentum are defined via Π_{0+} according to conservation law and

$$\varkappa = \frac{\left[\mathbf{kp}_+\right]}{p_+ k} - \frac{\left[\mathbf{kp}_-\right]}{p_- k}. \qquad (9.63)$$

The Bessel function argument in Eq. (9.62)

$$\alpha_s = \frac{eA_0}{\hbar\omega} \left|\varkappa\right|$$

can be represented in the form

$$\alpha_s = \frac{\xi_0 s_m}{2\sqrt{1+\xi_0^2}} \left[\frac{\beta_+^2 \sin^2\theta_+}{(1-\beta_+\cos\theta_+)^2} + \frac{\beta_-^2 \sin^2\theta_-}{(1-\beta_-\cos\theta_-)^2} \right.$$

$$\left. -2\frac{\beta_-\beta_+ \sin\theta_+ \sin\theta_- \cos\phi}{(1-\beta_+\cos\theta_+)(1-\beta_-\cos\theta_-)} \right]^{1/2}, \tag{9.64}$$

where

$$\beta_\pm = \frac{c\,|\mathbf{\Pi}_\pm|}{\Pi_{0\pm}}; \quad \Pi_{0-} = s\hbar\omega - \Pi_{0+}.$$

In this particular case we utilize Eq. (9.62) in order to obtain the electron–positron pair production probability on the Coulomb potential for which the Fourier transform is

$$\varphi(\mathbf{q}_s) = \frac{4\pi Z_a e}{\mathbf{q}_s^2}. \tag{9.65}$$

Then taking into account Eq. (9.65) for the differential probability of e^-, e^+ pair production by a strong plane monochromatic wave per unit time at the scattering on the Coulomb field we will have

$$dW = \alpha_0^2 \frac{Z_a^2 m^*}{2\pi^2\hbar} \sum_{s>s_m}^{\infty} \frac{|\mathbf{\Pi}_-|\,|\mathbf{\Pi}_+|}{\hbar^4 \mathbf{q}_s^4}$$

$$\left\{ \left[\hbar^2\mathbf{q}_s^2 c^2 - \frac{4}{\varkappa^4}\left(\varkappa\left(\frac{\Pi_{0-}\,[\mathbf{k\Pi}_+]}{\Pi_+ k} + \frac{\Pi_{0+}\,[\mathbf{k\Pi}_-]}{\Pi_- k}\right)\right)^2 \right] J_s^2(\alpha_s) \right.$$

$$-\frac{4c^2\Lambda_0^2}{\varkappa^2}\left(\frac{[[\mathbf{k\Pi}_-]\,[\mathbf{k\Pi}_+]]}{(k\Pi_-)\,(k\Pi_+)}\right)^2 J_s'^2(\alpha_s) + \frac{c^2\Lambda_0^2}{(k\Pi_-)\,(k\Pi_+)}\left[\mathbf{k}\left(\mathbf{\Pi}_- + \mathbf{\Pi}_+\right)\right]^2$$

$$\left. \times \left[\left(\frac{s^2}{\alpha_s^2} - 1\right) J_s^2(\alpha_s) + J_s'^2(\alpha_s) \right] \right\} \sin\theta_+ \sin\theta_- d\phi d\theta_- d\theta_+ d\gamma_+. \tag{9.66}$$

For a weak EM wave the main contribution in this process is the one-photon process. Dividing the differential probability (9.66) by the initial flux density

$$J = \frac{1}{\hbar\omega}\frac{c}{4\pi}E_0^2$$

we obtain the H.A. Bethe, W. Heitler formula:

$$d\sigma = \alpha_0^3 \frac{Z_a^2}{2\pi} \frac{|\mathbf{p}_-|\,|\mathbf{p}_+|}{\hbar^4 \mathbf{q}_1^4} \frac{1}{\hbar\omega^3}$$

$$\times \left\{ \hbar^2 \mathbf{q}_1^2 c^2 \left(\frac{[\mathbf{k}\mathbf{p}_+]}{p_+k} - \frac{[\mathbf{k}\mathbf{p}_-]}{p_-k} \right)^2 - 4 \left(\frac{\mathcal{E}_-\,[\mathbf{k}\mathbf{p}_+]}{p_+k} + \frac{\mathcal{E}_+\,[\mathbf{k}\mathbf{p}_-]}{p_-k} \right)^2 \right.$$

$$\left. + \frac{2\hbar^2\omega^2}{(kp_-)\,(kp_+)} \left[\mathbf{k}\,(\mathbf{p}_- + \mathbf{p}_+)\right]^2 \right\} \sin\theta_+ \sin\theta_- d\phi d\theta_- d\theta_+ d\mathcal{E}_+. \tag{9.67}$$

In general the expression for the differential probability of e^-, e^+ pair production by strong radiation field (9.66) is very complicated (one should perform four-dimensional integration and summation over photon numbers) but without integration one can make conclusions about optimal values of laser parameters for the measurable pair production probability using the properties of the Bessel function. The Bessel function argument in Eq. (9.66) $\alpha_s(\gamma_+, \theta_+, \theta_-, \phi)$ as a function of θ_+, θ_-, ϕ reaches its maximal value at

$$\cos\theta_+ = \beta_+, \quad \cos\theta_- = \beta_-, \quad \cos\phi = -1,$$

and is equal to

$$\overline{\alpha}_s\,(\gamma_+) = \frac{\xi_0 s_m}{2\sqrt{1+\xi_0^2}} \left(\sqrt{\gamma_+^2 - 1} + \sqrt{\left(\frac{2s}{s_m} - \gamma_+\right)^2 - 1} \right). \tag{9.68}$$

The latter is always small compared with the Bessel function index. Indeed, as follows from the conservation law

$$1 \leq \gamma_+ \leq \frac{2s}{s_m} - 1,$$

and in this range $\overline{\alpha}_s\,(\gamma_+)$ reaches its maximal value

$$\overline{\alpha}_{s\,\max} = \frac{\xi_0}{\sqrt{1+\xi_0^2}} \sqrt{s^2 - s_m^2} < s \tag{9.69}$$

at the $\gamma_+ = s/s_m$. Hence, for $\xi_0 \gg 1$ and $s_m >> 1$ the main contribution to the differential probability will give the number of photons $s >> s_m$ and as in the previous section one can approximate the Bessel function by the Airy one (1.69). The Airy function argument for $\alpha \simeq \overline{\alpha}_{s\,\max}$ will be

$$Z(s) \simeq \frac{1}{2^{2/3}\xi_0^2} s^{2/3} \left(1 + \xi_0^2 \frac{s_m^2}{s^2}\right). \tag{9.70}$$

As the Airy function exponentially decreases with increasing of the argument one can conclude that the optimal parameters for the pair production process are determined from the condition $Z_{\min} \sim 1$, $Z_{\min}$ being the minimum value of $Z(s)$. The latter corresponds to the number of photons $s = \sqrt{2}\xi_0 s_m$ at which

$$Z_{\min} = Z\left(\sqrt{2}\xi_0 s_m\right) = 3\left(\frac{E_c}{2E_0}\right)^{2/3}, \tag{9.71}$$

where E_c is the vacuum critical field strength (9.1). Hence, at $\xi \geq 1$ the probability reaches optimal values when $\zeta \equiv E_c/E_0 \geq 1$ (at $\xi_0 << 1$ quantum effects are optimal when $\zeta \sim \xi_0$, which corresponds to linear theory, that is, the perturbation theory of QED). When $\zeta << 1$ according to Eq. (9.53) the probability is exponentially suppressed:

$$W \propto \exp(-2\sqrt{3}/\zeta), \tag{9.72}$$

as in the Schwinger mechanism for e^-, e^+ pair production in the uniform electrostatic field, where $W \propto \exp(-\pi/\zeta)$. For the available superstrong optical lasers $\zeta \sim 10^{-4}$, which practically does not allow for measurable pair creation probability. As was argued, one can achieve $\zeta \sim 10^{-1}$ at the focus of expected X-ray FEL facilities, which will allow for measurable pair creation probability by the Schwinger mechanism.

Note that in the considered process of pair production on a nucleus one can achieve the condition $\zeta \geq 1$ (even $\zeta >> 1$) in the scheme of counterpropagating nucleus beam and X-ray FEL. Then, in the rest frame of the nucleus we will have $\zeta \simeq 2\zeta_L \gamma_L$, where γ_L is the Lorenz factor of nucleus and ζ_L is the field parameter in the laboratory frame. Since ξ_0 is the Lorenz invariant, then if $\xi_0 \geq 1$ and $\gamma_L > E_c/2E_0$ in the laboratory frame, the probability of multiphoton e^-, e^+ pair production reaches its optimal value.

9.4 Nonlinear e^-, e^+ Pair Production in Plasma by Strong EM Wave

As was shown in Chapter 6 for electron–positron pair production by a γ-quantum or a plane monochromatic EM wave, a macroscopic medium with a refractive index $n_0(\omega_0) < 1$ may serve as a third body for the satisfaction of conservation laws. In such a plasmalike medium the multiphoton production of e^-, e^+ pairs by a strong laser radiation field is possible at ordinary densities of plasma, in contrast to single-photon production $\gamma \to e^- + e^+$, which is only accessible in a superdense plasma with the electron density $\rho \gtrsim 3 \cdot 10^{34} \mathrm{cm}^{-3}$.

In laser fields with $\xi_0 \sim 1$ when the energy of the interaction of an electron (of the Dirac vacuum) with the field over a wavelength becomes comparable to the electron rest energy ($eE_0\lambda_0 \sim mc^2$) the multiphoton pair-production process goes in through nonlinear channels. At such intensities, in general, the dispersion law of a plasma becomes nonlinear, too; i.e., the refractive index depends on the wave intensity: $n_0 = n_0(\omega_0, \xi_0^2)$. As is known, because of the intensity effect, the transparency range of a plasma widens and the dispersion law $n_0(\omega_0, \xi_0^2) < 1$, which is necessary for the production of e^-, e^+ pairs, holds all the more. But the intensities required for the appearance of a real nonlinearity in dispersion become essential when $\xi_0 >> 1$. Hence, in considering fields $\xi_0 \sim 1$ the dispersion law of a plasma can be regarded as linear $(n_0^2(\omega_0) = 1 - 4\pi\rho e^2/m\omega_0^2)$.

Let a plane transverse linearly polarized EM wave with frequency ω_0 and vector potential

$$\mathbf{A}(\mathbf{r}, t) = \mathbf{A}_0 \cos(\omega_0 t - \mathbf{k}_0 \mathbf{r}); \quad |\mathbf{k}_0| = n_0 \frac{\omega_0}{c} \tag{9.73}$$

propagate in a plasma. The multiphoton degree s for the e^-, e^+ pair production in the light fields is defined by the condition (reaction threshold)

$$s\hbar\omega_0 \geqslant \frac{2mc^2}{\sqrt{1 - n_0^2}}. \tag{9.74}$$

To determine the multiphoton probabilities of this process it is convenient to solve the problem in the center-of-mass frame of the produced pair (C frame), in which the wave vector of the photons is $\mathbf{k}' = 0$ (the refractive index of the plasma in this frame is $n' = 0$). The velocity of the C frame with respect to the laboratory frame (L frame) is $\mathrm{v} = cn_0$. The traveling EM wave is transformed in the C frame into a varying electric field (the magnetic field $H' = 0$) with a vector potential

$$\mathbf{A}'(t') = \frac{\mathbf{A}_0}{2}[\exp(i\omega' t') + \exp(-i\omega' t')], \quad \omega' = \omega_0 \sqrt{1 - n_0^2}. \tag{9.75}$$

It is easily noted that with Eq. (9.75) taken into account the reaction threshold condition (9.74) is obtained from the laws of the conservation of energy $\mathcal{E}'_- + \mathcal{E}'_+ = s\hbar\omega'$ and momentum $\mathbf{p}'_- + \mathbf{p}'_+ = s\hbar\mathbf{k}' = 0$ in the C frame ($\mathcal{E}'_-$, $\mathbf{p}'_-$ and $\mathcal{E}'_+$, $\mathbf{p}'_+$ are the energy and momentum of the electron and positron, respectively, in the C frame).

To solve the problem of s-photon production of an e^-, e^+ pair in the given radiation field (9.73), we shall make use of the Dirac model (all vacuum negative-energy states are filled with electrons and the interaction of the external field proceeds only with this vacuum: on the other hand, the interaction with the plasma electrons reduces to a refraction of the wave only).

The Dirac equation in the field (9.75) has the form

$$i\hbar\frac{\partial\Psi}{\partial t} = \left[c\widehat{\alpha}\left(\mathbf{p}' - e\mathbf{A}'\left(t'\right)\right) + \widehat{\beta}mc^2\right]\Psi, \tag{9.76}$$

where the Dirac matrices $\widehat{\alpha}$, $\widehat{\beta}$ will be chosen in the standard representation, with σ the Pauli matrices. Since in the C frame the interaction Hamiltonian does not depend on the space coordinates, the solution of Eq. (9.76) can be represented in the form of a linear combination of free solutions of the Dirac equation with amplitudes $a_i(t')$ depending only on time:

$$\Psi_{\mathbf{p}'}\left(\mathbf{r}', t'\right) = \sum_{i=1}^{4} a_i(t')\Psi_i^{(0)}\left(\mathbf{r}', t'\right). \tag{9.77}$$

Here

$$\Psi_{1,2}^{(0)}\left(\mathbf{r}', t'\right) = \sqrt{\frac{\mathcal{E}' + mc^2}{2\mathcal{E}'}}\begin{pmatrix} \varphi_{1,2} \\ \frac{c\sigma\mathbf{p}'}{\mathcal{E}'+mc^2}\varphi_{1,2} \end{pmatrix} e^{\frac{i}{\hbar}\left(\mathbf{p}'\mathbf{r}' - \mathcal{E}'t'\right)},$$

$$\Psi_{3,4}^{(0)}\left(\mathbf{r}', t'\right) = \sqrt{\frac{\mathcal{E}' + mc^2}{2\mathcal{E}'}}\begin{pmatrix} \frac{-c\sigma\mathbf{p}'}{\mathcal{E}'+mc^2}\chi_{3,4} \\ \chi_{3,4} \end{pmatrix} e^{\frac{i}{\hbar}\left(\mathbf{p}'\mathbf{r}' + \mathcal{E}'t'\right)}, \tag{9.78}$$

where

$$\mathcal{E}' = \sqrt{c^2\mathbf{p}'^2 + m^2c^4}, \qquad \varphi_1 = \chi_3 = \begin{pmatrix}1\\0\end{pmatrix}, \qquad \varphi_2 = \chi_4 = \begin{pmatrix}0\\1\end{pmatrix}. \tag{9.79}$$

The solution of Eq. (9.76) in the form Eq. (9.77) corresponds to an expansion of the wave function in a complete set of orthonormal functions of the electrons (positrons) with specified momentum (with energies $\mathcal{E}' = \pm\sqrt{c^2\mathbf{p}'^2 + m^2c^4}$ and spin projections $S_z = \pm 1/2$). The latter are normalized to one particle per unit volume. According to the assumed model only the Dirac vacuum is present prior to the turning on of the field, i.e.,

$$\left|a_3(-\infty)\right|^2 = \left|a_4(-\infty)\right|^2 = 1, \qquad \left|a_1(-\infty)\right|^2 = \left|a_2(-\infty)\right|^2 \tag{9.80}$$

(the field is turned on adiabatically at $t = -\infty$). From the condition of conservation of the norm we have

$$\sum_{i=1}^{4}\left|a_i(t')\right|^2 = 2, \tag{9.81}$$

which expresses the equality of the number of created electrons and positrons, whose creation probability is, respectively, $\left|a_{1,2}(t')\right|^2$ and $1 - \left|a_{3,4}(t')\right|^2$.

Substituting Eq. (9.77) into Eq. (9.76), multiplying by the Hermitian conjugate functions $\Psi_i^{(0)^\dagger}(\mathbf{r}', t')$, and taking into account orthogonality of the eigenfunctions (9.78) and (9.79), we obtain a set of differential equations for the unknown functions $a_i(t')$. Since in the C frame there is symmetry with respect to the direction $\mathbf{A}'_0$ (the OY axis), we can take, without loss of generality, the vector $\mathbf{p}'$ to lie in the $x'y'$ plane ($p'_z = 0$). Further, having introduced, to simplify the notation, the new symbols

$$a_1(t') \equiv b_1(t'),$$

$$a_4(t') \equiv b_4(t')\left[1 - \frac{c^2 p_y'^2}{\mathcal{E}'^2}\right]^{-1/2}\left[\frac{c^2 p'_x p'_y}{\mathcal{E}'(\mathcal{E}' + mc^2)}\right.$$

$$\left. + i\left(1 - \frac{c^2 p_y'^2}{\mathcal{E}'(\mathcal{E}' + mc^2)}\right)\right], \tag{9.82}$$

we obtain for the amplitudes $b_1(t')$ and $b_4(t')$ ($|b_4(t')| = |a_4(t')|$) the following set of equations:

$$\frac{db_1(t')}{dt'} = i\frac{ecp'_y A'_y(t')}{\hbar\mathcal{E}'} b_1(t')$$

$$+ i\frac{eA'_y(t')}{\hbar}\sqrt{1 - \frac{c^2 p_y'^2}{\mathcal{E}'^2}}\, b_4(t') \exp\left(\frac{2i\mathcal{E}' t'}{\hbar}\right),$$

$$\frac{db_4(t')}{dt'} = -i\frac{ecp'_y A'_y(t')}{\hbar\mathcal{E}'} b_4(t')$$

$$+ i\frac{eA'_y(t')}{\hbar}\sqrt{1 - \frac{c^2 p_y'^2}{\mathcal{E}'^2}}\, b_1(t') \exp\left(-\frac{2i\mathcal{E}' t'}{\hbar}\right). \tag{9.83}$$

A similar set of equations is also obtained for the amplitudes $b_2(t')$ and $b_3(t')$. To solve the system (9.83), we make the transformations

$$b_1(t') = c_1(t') \exp\left[i\frac{ecp'_y}{\hbar\mathcal{E}'}\int_{-\infty}^{t'} A'_y(\eta)d\eta\right],$$

$$b_4(t') = c_4(t') \exp\left[-i\frac{ecp'_y}{\hbar\mathcal{E}'}\int_{-\infty}^{t'} A_y(\eta)d\eta\right], \tag{9.84}$$

where $c_1(t')$ and $c_4(t')$ satisfy the initial conditions, according to Eqs. (9.80) and (9.82), $|c_1(-\infty)| = 0$ and $|c_4(-\infty)| = 0$.

For the new amplitudes $c_1(t')$ and $c_4(t')$ from Eqs. (9.83), we obtain the set of equations

$$\frac{dc_1(t')}{dt'} = f(t')c_4(t'),$$

$$\frac{dc_4(t')}{dt'} = -f^*(t')c_1(t'), \tag{9.85}$$

where

$$f(t') = i\frac{e}{\hbar}A'_y(t')\sqrt{1 - \frac{c^2 p_y'^2}{\mathcal{E}^2}} \exp\left[\frac{2i}{\hbar}\mathcal{E}'t' - \frac{2iecp'_y}{\hbar\mathcal{E}'}\int_{-\infty}^{t'} A'_y(\eta)d\eta\right]. \tag{9.86}$$

We can obtain the solution of Eqs. (9.83), which satisfies the initial conditions of the problem (9.80), with the help of successive approximations, if

$$\left|\int_{-\infty}^{t'} f(\tau)d\tau\right| << 1. \tag{9.87}$$

Then, for the transition amplitude $c_1(t')$, we have

$$c_1(t') = \sum_{j=0}^{\infty} B_{2j+1}(t'), \tag{9.88}$$

where

$$B_{2j+1}(t') = (-1)^j \int_{-\infty}^{t'} f(\tau_1)d\tau_1 \int_{-\infty}^{\tau_1} f^*(\tau_2)d\tau_2 \int_{-\infty}^{\tau_2} f^*(\tau_3)d\tau_3 \cdots$$

$$\times \int_{-\infty}^{\tau_{2j-1}} f^*(\tau_{2j})d\tau_{2j} \int_{-\infty}^{\tau_{2j}} f^*(\tau_{2j+1})d\tau_{2j+1}. \tag{9.89}$$

We are interested in nonlinear pair production process in the strong wave field. For that let us calculate the first term of the sum (9.88):

$$B_1(t') = \int_{-\infty}^{t'} f(\tau_1) d\tau_1,$$

substituting the concrete form of the wave vector potential $A'_y(\eta)$ from Eq. (9.75) into Eq. (9.86) and carrying out the integration. Then for $B_1(t')$ we obtain

$$B_1(t') = \frac{\mathcal{E}'}{2cp'_y}\left(1 - \frac{c^2 p'^2_y}{\mathcal{E}'^2}\right)^{1/2} \sum_{l=-\infty}^{+\infty} \frac{l\hbar\omega'}{2\mathcal{E}' - l\hbar\omega'} J_l(\alpha) e^{\frac{i}{\hbar}(2\mathcal{E}' - l\hbar\omega')t'}, \quad (9.90)$$

where $J_s(z)$ is the Bessel function,

$$\alpha \equiv 2\xi_0 \frac{mc^2}{\mathcal{E}'} \frac{cp'_y}{\hbar\omega'}, \qquad \xi_0 = \frac{eE'_0}{mc\omega'}, \qquad E'_0 = \frac{\omega'}{c} A_0.$$

As ξ_0 is a relativistic invariant parameter, in Eqs. (9.90) $\xi_0 = eE_0/mc\omega_0$, where ω_0 and E_0 are the frequency and amplitude of the electric field of the wave in the L frame.

For the considered fields, when $\xi_0 \lesssim 1$, condition (9.87) always satisfies: $|B_1(t')| << 1$, but the latter is not enough, yet, in order to be confined to that term in determination of the amplitude $c_1(t')$. Because the resonant term $l = s = 2\mathcal{E}'/(\hbar\omega')$ $(s >> 1)$ gives a real contribution in the multiphoton pair production process and in Eq. (9.90), the maximal value of the Bessel function can be shifted from the resonant value. Since $s >> 1$, that shift will be as small and negligible as possible when the argument of the Bessel function is $\alpha \sim s >> 1$. Thus, the condition, when the pair production process will have an essential nonlinear character, is

$$\alpha = 2\xi_0 \frac{mc^2}{\mathcal{E}'} \frac{cp'_y}{\hbar\omega'} >> 1. \quad (9.91)$$

If condition (9.91) is satisfied, we can be restricted to the first term of the sum (9.88) for the amplitude $c_1(t')$:

$$c_1(t') = B_1(t'). \quad (9.92)$$

The obtained approximate solution of the Dirac equation is thus applicable with such intensities of EM wave, when conditions (9.87) and (9.91) are satisfied simultaneously:

$$\frac{1}{s} << \xi_0 \lesssim 1. \quad (9.93)$$

According to Eqs. (9.82) and (9.84), for the transition amplitude of the electron from the Dirac vacuum to the state with positive energy (in a definite spinor state) in the wave field we have

$$|a_1(t')|^2 = |b_1(t')|^2 = |c_1(t')|^2 .$$

To obtain the probability amplitude for the production of electrons and positrons after the wave has been turned off we introduce a small detuning of the resonance in Eq. (9.90), corresponding to an s-photon transition: $2\mathcal{E}' = s\hbar\omega' + \hbar\Gamma$ ($\Gamma << \omega'$).

The production probability of the e^-, e^+ pair, summed over the spin states, is determined by the quantity

$$|a_1(t')|^2 + |a_2(t')|^2 = 2\,|a_1(t')|^2 \equiv 2\,|C_1(t')|^2 .$$

The differential probability of the s-photon process per unit time and phase-space volume $d\mathbf{p}'/(2\pi\hbar)^3$ (the normalization volume $V = 1$) in the center-of-mass frame of the produced particles is given by

$$dw_s^C = \frac{dW_s^C(t')}{t'} = 2\lim_{t'\to\infty}\frac{|c_1(t')|^2}{t'}\frac{d\mathbf{p}'}{(2\pi\hbar)^3}. \tag{9.94}$$

Substituting Eq. (9.90) into Eq. (9.94) and making use of the definition of the δ-function in the form

$$\lim_{t'\to\infty}\frac{\sin^2\Gamma t'}{\pi\Gamma^2 t'} = \delta(\Gamma) = \hbar\delta\left(2\mathcal{E}' - s\hbar\omega'\right),$$

we obtain

$$dw_s^C = \frac{s^2\omega'^2\left(\mathcal{E}'^2 - c^2p_y'^2\right)}{16\pi^2\hbar^2c^2p_y'^2}J_s^2\left(\frac{2eA_0cp_y'}{\hbar\omega'\mathcal{E}'}\right)\delta\left(\mathcal{E}' - \frac{s\hbar\omega'}{2}\right)d\mathbf{p}'. \tag{9.95}$$

Integrating Eq. (9.95) over $d\mathbf{p}'$, we obtain the total probability of the s-photon e^-, e^+ pair production in a plasma by the strong EM wave:

$$w_s^C = \frac{\hbar s^5\omega'^5}{32\pi c^4 p'}\left\{\left[\frac{2\alpha_s^2}{4s^2-1} - 1\right]J_s^2(\alpha_s) + \frac{\alpha_s^2J_{s-1}^2(\alpha_s)}{2s(2s-1)}\right.$$

$$\left. + \frac{\alpha_s^2J_{s+1}^2(\alpha_s)}{2s(2s+1)} - \frac{4c^2p'^2}{s^2\hbar^2\omega'^2}\frac{\alpha_s^{2s}}{2^{2s}(2s+1)(s!)^2}\right.$$

$$\times {}_2F_3\left(s+\frac{1}{2}, s+\frac{1}{2}, s+1, 2s+1, s+\frac{3}{2}; -\alpha_s^2\right)\Bigg\}, \tag{9.96}$$

where ${}_2F_3\left(s+\frac{1}{2}, s+\frac{1}{2}, s+1, 2s+1, s+\frac{3}{2}; -\alpha_s^2\right)$ is the generalized hypergeometric function and

$$\alpha_s = \frac{2mc^2\xi_0}{\hbar\omega'}\left(1-\frac{4m^2c^4}{s^2\hbar^2\omega'^2}\right)^{1/2}.$$

As is seen from Eq. (9.95), the pair production probability decreases highly in the directions perpendicular to the field ($p'_y = 0$), and the obtained approximate nonlinear solution describes the process behavior well at the angles not too close to $\pi/2$. Thus, Eq. (9.96), which is a result of integration over all angles, does not contain a large error.

The quantity W_s is a relativistic invariant, and so Eq. (9.96) defines the pair production probability in the L frame as well. As for the angular distribution of the probability of s-photon pair production in the L frame, it can be obtained from the expression $dW_s^C(t')$ for the differential probability in the C frame by a Lorentz transformation. Here the quantity multiplying $d\mathbf{p}'$ is the expression of $dW_s^C(t')$ (see Eq. (9.94)) transforms like the time component of the current density four-vector of the electrons in the Dirac vacuum ($\mathcal{E}' < 0$). One must here take into account that the momentum of real electrons coincides with the momentum of the vacuum electron $\mathbf{p}'$, while the momentum of a positron equals $-\mathbf{p}'$ and the vacuum phase-space volume element $d\mathbf{p}'/(2\pi\hbar)^3$ (in unit volume) goes over correspondingly into the volume element in momentum space of electrons and positrons. Further, transforming the quantities in Eq. (9.95) from the C frame to the L frame, we obtain for the differential probability of s-photon pair production per unit time in the L frame:

$$dw_s^L = \frac{dW_s^L(t)}{t} = \frac{s^2\omega_0^2\left(1-n_0^2\right)\left(\mathcal{E}-n_0cp_x\right)}{16\pi^2\hbar^2c^2p_y^2\mathcal{E}}\left[\frac{\left(\mathcal{E}-n_0cp_x\right)^2}{1-n_0^2}-c^2p_y^2\right]$$

$$\times J_s^2\left(\frac{2eA_0cp_y}{\hbar\omega_0\left(\mathcal{E}-n_0cp_x\right)}\right)\delta\left(\mathcal{E}-n_0cp_x-\frac{s\hbar\omega_0\left(1-n_0^2\right)}{2}\right)d\mathbf{p}', \tag{9.97}$$

where $\mathcal{E}$ and $\mathbf{p}$ are the energy and momentum of the produced electron or positron. Integrating Eq. (9.97) over the electron (positron) energy, we obtain the angular distribution of the probability of the s-photon production of electrons (positrons) per solid angle element, $do = \sin\vartheta d\vartheta d\varphi$ (the azimuthal asymmetry of the probability in the L frame is due to the linear polarization of the wave: in the case of circular polarization the probability distribution has azimuthal symmetry):

$$dw_s^L = \sum_{\upsilon=1}^{2} \frac{s^3\omega_0^3\left(1-n_0^2\right)^2}{32\pi^2\hbar c^3\left(cp_\upsilon - n_0\mathcal{E}_\upsilon\cos\vartheta\right)\sin\vartheta\cos^2\varphi}$$

$$\times\left[\frac{s^2\hbar^2\omega_0^2\left(1-n_0^2\right)}{4} - c^2p_\upsilon^2\sin^2\vartheta\cos^2\varphi\right]$$

$$\times J_s^2\left[\frac{4mc^3\xi_0 p_\upsilon\sin\vartheta\cos\varphi}{s\hbar^2\omega_0^2\left(1-n_0^2\right)}\right]d\vartheta d\varphi, \tag{9.98}$$

where

$$p_{1,2} = \frac{1}{2c\left(1-n_0^2\cos^2\vartheta\right)}\left\{sn_0\hbar\omega_0\left(1-n_0^2\right)\cos\vartheta\right.$$

$$\left.\pm\left[s^2\hbar^2\omega_0^2\left(1-n_0^2\right)^2 - 4m^2c^4\left(1-n_0^2\cos^2\vartheta\right)\right]^{1/2}\right\},$$

$$\mathcal{E}_{1,2} = \frac{1}{2\left(1-n_0^2\cos^2\vartheta\right)}\left\{s\hbar\omega_0\left(1-n_0^2\right)\right.$$

$$\left.\pm n_0\cos\vartheta\left[s^2\hbar^2\omega_0^2\left(1-n_0^2\right)^2 - 4m^2c^4\left(1-n_0^2\cos^2\vartheta\right)\right]^{1/2}\right\}. \tag{9.99}$$

The angle φ varies from 0 to 2π, while ϑ varies from 0 to $\vartheta_{\max}$, which is determined from the energy and momentum conservation laws (9.99). Further, depending on the value of the plasma refractive index n_0, the electron (positron) production at the given angle is possible for a particular momentum or for one of two momenta with different magnitude. For values

$$n_0 < \sqrt{1-\frac{2mc^2}{s\hbar\omega_0}}$$

(in this case the threshold condition (9.74) for the process is certainly satisfied), we should take in Eqs. (9.99) only the upper sign, corresponding to the fact that in the probability (9.98) only $\nu = 1$ (p_1) remains and $\vartheta_{\max} = \pi$; i.e., particles are produced in all directions for the given angle ϑ with definite momentum. In the opposite case we must also take into account the reaction threshold condition in the region of values of the index of refraction,

$$\sqrt{1-\frac{2mc^2}{s\hbar\omega_0}} < n_0 < \sqrt{1-\frac{4m^2c^4}{s^2\hbar^2\omega_0^2}},$$

and an electron (positron) is produced in a given direction with one of two different values of momentum p_1 and p_2 in a cone, opened forward, whose opening angle is

$$\vartheta_{\max} = \arcsin\left\{\left[\left(1-n_0^2\right)\left(s^2\hbar^2\omega_0^2\left(1-n_0^2\right)-4m^2c^4\right)\right]^{1/2}/2mc^2 n_0\right\}.$$

The problem of e^-, e^+ pair production by the photon field is solved in the C frame and the probability expressions (9.94)–(9.96) in that frame are adduced with express purpose. This is of independent physical interest, since Eqs. (9.94)–(9.96) describe the process of pair production in vacuum by a uniform periodic electric field (electric undulator)

$$\mathbf{E}(t) = \mathbf{E}_0 \cos\omega_0 t, \tag{9.100}$$

with the reaction threshold (see Eq. (9.74) when $n' = 0$)

$$s\hbar\omega_0 \geqslant 2mc^2. \tag{9.101}$$

By integrating over the electron (positron) energy, we obtain the angular distribution of the nonlinear production of electrons (positrons) in the periodic electric field (in contrast to the pair production by the photon field (9.98), here there is azimuthal symmetry):

$$dw_s = \frac{s^3\omega_0^3}{32\pi\hbar c^3}\frac{4m^2c^4\cos^2\vartheta + \hbar^2 s^2\omega_0^2\sin^2\vartheta}{\left(\hbar^2 s^2\omega_0^2 - 4m^2c^4\right)^{1/2}\cos^2\vartheta}$$

$$\times J_s^2\left[\frac{2ceE_0\left(\hbar^2 s^2\omega_0^2 - 4m^2c^4\right)^{1/2}\cos\vartheta}{s\hbar^2\omega_0^3}\right]\sin\vartheta d\vartheta, \tag{9.102}$$

where ϑ is the angle between the directions of the momentum of produced electrons (positrons) and the electric field.

Finally, we consider the case of weak fields, $eA/\left(\hbar\omega_0\right) << 1$ ($\xi_0 << 1/s$), when perturbation theory is applicable. In this case, as was noted above, we cannot be confined to the first term of the sum (9.88), since every term $B_{2l+1}(t')$ of the sum at $\alpha << 1$ (see Eq. (9.90) for the expression of α) includes a resonant multiplier $\sim \xi_0^s$ (at $2l+1 \leqslant s$) in the lowest order of perturbation theory. Then from Eq. (9.88) we obtain the formula of perturbation theory for the pair production probability in the C frame, which has a more compact analytical form (here we could get free of the sum of unwieldy products):

$$dw_s^C = 2\pi\hbar\Phi^2\delta\left(2\mathcal{E}' - s\hbar\omega'\right)\frac{d\mathbf{p}'}{(2\pi\hbar)^3}, \tag{9.103}$$

where

$$\Phi = \beta \left(\frac{\alpha}{2}\right)^s \omega' \left[\frac{1}{(s-1)!} + \sum_{K=1}^{[(s-1)/2]} \sum_{S_1=1}^{s-2K} \right.$$

$$\ldots \sum_{S_j=1}^{s-1-(S_1+\ldots+S_{j-1})-2K+j} \ldots \sum_{S_{2K}=1}^{s-1-(S_1+\ldots+S_{2K-1})} \tag{9.104}$$

$$\left\{ \frac{(-1)^{S_2+S_4+\ldots+S_{2K}}}{(s-S_1)(S_1+S_2)\ldots[s-(S_1+S_2+\ldots+S_{2K-1})](S_1+S_2+\ldots+S_{2K})} \right.$$

$$\left. \left. \times \frac{\beta^{2K}}{(S_1-1)!(S_2-1)!\ldots(S_{2K}-1)![s-1-(S_1+S_2+\ldots+S_{2K})]!} \right\} \right].$$

Here $s \geqslant 3$, and parameters

$$\beta = \frac{\mathcal{E}'}{2cp'_y}\left(1 - \frac{c^2 p'^2_y}{\mathcal{E}'^2}\right)^{1/2}, \qquad \alpha = s\xi_0 \frac{mc^3 p'_y}{\mathcal{E}'^2}; \qquad \xi_0 << \frac{1}{s}.$$

9.5 Pair Production by Superstrong EM Waves in Vacuum

As we saw in the previous section the conservation laws for the pair production in the field of a plane monochromatic wave can be satisfied in a plasmalike medium where EM waves propagate with a phase velocity larger than the speed of light in vacuum. In this case

$$\frac{\omega^2}{c^2} - \mathbf{k}^2 > 0, \tag{9.105}$$

which means that we have a "photon with nonzero rest mass" providing the creation of the particles with the rest masses. The satisfaction of conservation laws for the e^-, e^+ pair production process in the EM field is equivalent to the satisfaction of the condition

$$\mathbf{E}^2 - \mathbf{H}^2 > 0, \tag{9.106}$$

where **E**, **H** are the electric and magnetic strengths of the field. The latter is obvious in the frame of reference where there is only an electric field that provides the pair creation (in the opposite case we would have only a magnetic field that cannot produce a pair). The condition (9.106) can be satisfied in the stationary maxima of a standing wave being formed by two counterpropagating waves (opposite laser beams) of the same frequencies. It can also be satisfied in the field of a plane monochromatic wave in a wiggler. Thus, these processes of multiphoton pair production via nonlinear channels in vacuum by superstrong laser fields are of special interest.

Let plane transverse linearly polarized EM waves with frequency ω and amplitude of vector potential $\mathbf{A}_0$

$$\mathbf{A}_1 = \mathbf{A}_0 \cos(\omega t - \mathbf{kr}), \quad \mathbf{A}_2 = \mathbf{A}_0 \cos(\omega t + \mathbf{kr}), \tag{9.107}$$

propagate in opposite directions in vacuum. To solve the problem of s-photon production of an e^-, e^+ pair in the given radiation fields (9.107) we shall make use of the Dirac model for electron–positron vacuum. The Dirac equation in the field (9.107) has the form

$$i\hbar\frac{\partial \Psi}{\partial t} = \left[c\widehat{\alpha}(\widehat{\mathbf{p}} - \frac{e}{c}\mathbf{A}_0 \cos(\omega t - \mathbf{kr}) - \frac{e}{c}\mathbf{A}_0 \cos(\omega t + \mathbf{kr})) + \widehat{\beta} mc^2\right]\Psi. \tag{9.108}$$

Then we have stationary maxima of a standing wave and Eq. (9.108) may be rewritten in the form

$$i\hbar\frac{\partial \Psi}{\partial t} = \left[c\widehat{\alpha}(\widehat{\mathbf{p}} - 2\frac{e}{c}\mathbf{A}_0 \cos \mathbf{kr} \cos \omega t) + \widehat{\beta} mc^2\right]\Psi. \tag{9.109}$$

According to the Dirac model the electron–positron pair production by the EM wave field occurs when the vacuum electrons with initial negative energies $\mathcal{E}_0 < 0$ due to s-photon absorption pass to the final states with positive energies $\mathcal{E} = \mathcal{E}_0 + s\hbar\omega > 0$. Since we study the case of superstrong laser fields in which the pairs are essentially produced at the length $l << \lambda$ (λ is the wavelength of laser radiation) and on the other hand the Hamiltonian of the interaction $H_{int} \sim \mathbf{p}(\mathbf{A}_1 + \mathbf{A}_2)$, then the significant contribution in the process of e^-, e^+ pair creation will be conditioned by the areas of stationary maxima in the direction along the electric field strength of the standing wave. Consequently, we can neglect the inhomogeneity of the field in the considered problem, i.e., Eq. (9.109) will reduce to the following equation:

$$i\hbar\frac{\partial \Psi}{\partial t} = \left[c\widehat{\alpha}(\widehat{\mathbf{p}} - 2\frac{e}{c}\mathbf{A}_0 \cos \omega t) + \widehat{\beta} mc^2\right]\Psi. \tag{9.110}$$

In this approximation the magnetic fields of the counterpropagating waves cancel each other. In the case of e^-, e^+ pair production in a plasma we had a similar equation in the center-of-mass frame of created particles (9.76). Thus, we will follow the approach developed in the previous section. Since the interaction Hamiltonian does not depend on the space coordinates, the solution of Eq. (9.110) can be represented in the form of a linear combination of free solutions of the Dirac equation with amplitudes $a_i(t)$ depending only on time (9.77). The application of the unitarian transformations (9.82) and (9.84) yields the set of equations

$$\frac{dc_1(t)}{dt} = f(t)c_4(t), \tag{9.111}$$

$$\frac{dc_4(t)}{dt} = -f^*(t)c_1(t). \tag{9.112}$$

Here the function $f(t)$ (see Eq. (9.86)) is expanded into series

$$f(t) = i \sum_{s'=-\infty}^{\infty} f_{s'} \exp\left[\frac{i}{\hbar}(2\mathcal{E} - s'\hbar\omega)t\right], \tag{9.113}$$

where

$$f_{s'} = \frac{\mathcal{E}}{2cp_y}\left(1 - \frac{c^2p_y^2}{\mathcal{E}^2}\right)^{\frac{1}{2}} s'\omega J_{s'}\left(4\xi_0 \frac{mc^2}{\mathcal{E}}\frac{p_y c}{\hbar\omega}\right), \tag{9.114}$$

and J_s is the ordinary Bessel function. The new amplitudes $c_1(t)$ and $c_4(t)$ satisfy the initial conditions

$$|c_1(-\infty)| = 0, |c_4(-\infty)| = 1.$$

Because of space homogeneity the generalized momentum of a particle is conserved so that the real transitions in the field occur from a $-\mathcal{E}$ negative energy level to positive $+\mathcal{E}$ energy level (in the assumed approximation) and, consequently, the multiphoton probabilities of e^-, e^+ pair production will have maximal values for the resonant transitions $2\mathcal{E} \simeq s\hbar\omega$. The latter just is the conservation law of the pair production process at which both electrons and positrons will be created back-to-back according to zero total momentum: $\mathbf{p}_{e^-} + \mathbf{p}_{e^+} = 0$, since the considered field is only time dependent. Thus, we can utilize the resonant approximation, as in a two-level atomic system in the monochromatic wave field.

The probabilities of multiphoton e^-, e^+ pair production will have maximal values for the resonant transitions

$$2\mathcal{E} - s\hbar\omega \simeq 0. \tag{9.115}$$

In this case the function $f(t)$ can be represented in the following form:

$$f(t) = F_s + \Phi(t), \tag{9.116}$$

where

$$F_s = if_s e^{i\delta_s t} \tag{9.117}$$

is the slowly varying function on the scale of the wave period and

$$\Phi(t) = ie^{i\delta_s t} \sum_{s' \neq s, s'=-\infty}^{\infty} f_{s'} e^{i(s-s')\omega t} \tag{9.118}$$

is the rapidly oscillating function. Here we have introduced resonance detuning

$$\hbar\delta_s = 2\mathcal{E} - s\hbar\omega. \tag{9.119}$$

As a consequence of this separation the probability amplitudes can be represented in the form

$$c_1(t) = c_1^{(s)}(t) + \beta_1(t), \tag{9.120}$$

$$c_4(t) = c_4^{(s)}(t) + \beta_4(t), \tag{9.121}$$

where $c_1^{(s)}(t)$ and $c_4^{(s)}(t)$ are the slowly varying amplitudes corresponding to $c_1(t)$ and $c_4(t)$. The functions $\beta_1(t)$ and $\beta_4(t)$ are rapidly oscillating functions. Substituting Eqs. (9.120), (9.121) into Eqs. (9.111), (9.112) and separating slow and rapid oscillations, taking into account Eq.(9.116), we will obtain the following set of equations for the slowly varying amplitudes $c_{1,4}^{(s)}(t)$:

$$\frac{dc_1^{(s)}}{dt} = F_s c_4^{(s)} + \overline{\Phi(t)\,\beta_4(t)}, \tag{9.122}$$

$$\frac{dc_4^{(s)}}{dt} = -F_s c_1^{(s)} - \overline{\Phi^*(t)\,\beta_1(t)}, \tag{9.123}$$

and for the rapidly oscillating functions $\beta_{1,4}$:

$$\frac{d\beta_1}{dt} = \Phi(t)\, c_4^{(s)}, \tag{9.124}$$

$$\frac{d\beta_4}{dt} = -\Phi^*(t)\, c_1^{(s)}. \tag{9.125}$$

In Eqs. (9.122) and (9.123) the bar denotes averaging over time much larger than wave period. In the set of Eqs. (9.124) and (9.125) we have neglected the terms $\sim F_s\, \beta_{1,4}(t)$ due to the rapid oscillations

$$|F_s\beta_\eta(t)| << \left|\frac{d\beta_1}{dt}\right|. \tag{9.126}$$

Solving the set of Eqs. (9.124) and (9.125), taking into account that $c_{1,4}^{(s)}$ are slowly varying functions, we obtain

$$\beta_1 = c_4^{(s)} \int_0^t \Phi\left(t'\right) dt',$$

$$\beta_4 = -c_1^{(s)} \int_0^t \Phi^*\left(t'\right) dt'.$$

Then substituting $\beta_{1,4}(t)$ into Eqs. (9.122) and (9.123), we will have the following equations for the functions $c_{1,4}^{(s)}$:

$$\frac{dc_1^{(s)}}{dt} = F_s c_4^{(s)} - i\frac{\delta_f}{2}c_1^{(s)}, \tag{9.127}$$

$$\frac{dc_4^{(s)}}{dt} = -F_s c_1^{(s)} + i\frac{\delta_f}{2}c_4^{(s)}, \tag{9.128}$$

where

$$\delta_f = \overline{-2i\Phi\left(t\right) \int_0^t \Phi^*\left(t'\right) dt'} = \frac{2}{\omega} \sum_{s' \neq s, s'=-\infty}^{\infty} \frac{\left|f_{s'}\right|^2}{s - s'}. \tag{9.129}$$

The set of Eqs. (9.127) and (9.128) can be solved in the general case of arbitrary wave envelope $A_0(t)$ only numerically. But it admits an exact solution for a monochromatic wave describing "Rabi oscillations" of the Dirac vacuum. In this case the set of Eqs. (9.127) and (9.128) for the phase transformed amplitudes $c_1^{(s)} \exp\left(-i\delta_s t/2\right)$ and $c_4^{(s)} \exp\left(i\delta_s t/2\right)$ is a set of ordinary linear differential equations with fixed coefficients. The general solution of the latter is given by a superposition of two linearly independent solutions which with the initial condition is

$$c_1^{(s)}(t) = i\frac{|f_s|}{\Omega_s}e^{i\frac{\delta_s}{2}t} \sin\left(\Omega_s t\right), \tag{9.130}$$

$$c_4^{(s)} = e^{-i\frac{\delta_s}{2}t}\left[\cos(\Omega_s t) + \frac{i\Delta_s}{2\Omega_s}\sin(\Omega_s t)\right], \tag{9.131}$$

where

$$\Delta_s = \delta_f + \delta_s \tag{9.132}$$

is the resulting detuning and

$$\Omega_s = \sqrt{|f_s|^2 + \frac{\Delta_s^2}{4}} \tag{9.133}$$

is the "Rabi frequency" of the Dirac vacuum at the interaction with a periodic EM field. As is seen from Eq. (9.130) with this frequency the probability amplitude of e^-, e^+ pair production oscillates in the standing wave field during the whole interaction time similar to Rabi oscillations in two-level atomic systems. In this case the "Rabi frequency" has a nonlinear dependence on the amplitudes of the opposite EM wave fields. Considerable number of electron–positron pairs can be produced by a proper choice of intensity and duration of laser pulses.

The set of Eqs. (9.127) and (9.128) has been derived using the assumption that the amplitudes $c_{1,4}^{(s)}(t)$ are slowly varying functions on the scale of the EM wave period, i.e.,

$$\left|\frac{dc_{1,4}^{(s)}(t)}{dt}\right| << \left|c_{1,4}^{(s)}(t)\right|\omega. \tag{9.134}$$

These conditions with Eq.(9.126) define the condition of applicability of the applied resonant approximation which is equivalent to the condition

$$\Omega_s << \omega. \tag{9.135}$$

The probability of the s-photon e^-, e^+ pair production with the certain energy $\mathcal{E}$, summed over the spin states, is

$$W_s = 2\left|c_1^{(s)}(t)\right|^2 = \frac{2|f_s|^2}{\Omega_s^2}\sin^2(\Omega_s t). \tag{9.136}$$

Hence, from Eq. (9.114) we have

$$W_s = \frac{s^2\omega^2\left(p^2\sin^2\vartheta + m^2c^2\right)}{2p^2\cos^2\vartheta} J_s^2\left(4\xi_0\frac{mc^3p\cos\vartheta}{\hbar\omega\mathcal{E}}\right)\frac{\sin^2(\Omega_s t)}{\Omega_s^2}, \tag{9.137}$$

where ϑ is the angle between the directions of the momentum of produced electrons (positrons) and the amplitude of the total field electric strength.

Let us consider the case of short interaction time when

$$\Omega_s t << 1. \tag{9.138}$$

In this case we can determine a probability of multiphoton pair production per unit time according to the following definition of the Dirac δ-function:

$$\frac{\sin^2(\Omega_s t)}{\Omega_s^2} \to 2\pi\hbar t\delta(2\mathcal{E} - s\hbar\omega).$$

The differential probability of an s-photon e^-, e^+ pair production process per unit time and unit space volume, summed over the spin states, is given by the following formula:

$$dw_s = \frac{s^2\omega^2(p^2\sin^2\vartheta + m^2c^2)}{16\hbar^2\pi^2p^2\cos^2\vartheta}$$

$$\times J_s^2\left(4\xi_0\frac{mc^3p\cos\vartheta}{\hbar\omega\mathcal{E}}\right)\delta\left(\mathcal{E} - \frac{s\hbar\omega}{2}\right)d\mathbf{p}. \tag{9.139}$$

By integrating over the electron (positron) energy we obtain the angular distribution of the s-photon differential probability density of created electrons (positrons):

$$\frac{dw_s}{do} = \frac{s^3\omega^3}{64\pi^2\hbar c^3}\frac{4m^2c^4 + \hbar^2s^2\omega^2\tan^2\vartheta}{\left(\hbar^2s^2\omega^2 - 4m^2c^4\right)^{1/2}}$$

$$\times J_s^2\left(\frac{4ceE_0\left(\hbar^2s^2\omega^2 - 4m^2c^4\right)^{1/2}\cos\vartheta}{s\hbar^2\omega^3}\right), \tag{9.140}$$

where $do = \sin\vartheta d\vartheta d\varphi$ is the differential solid angle.

Analogously one can describe the multiphoton pair production process in a wiggler by a superstrong laser pulse of relativistic intensities. Thus, as we saw in Section 5.4 at the induced interaction of a charged particle with a plane EM wave in an undulator, or with the counterpropagating waves of different frequencies (Section 5.3) the two interference waves are formed which propagate with the phase velocities $\mathrm{v}_{ph} > c$ and $\mathrm{v}_{ph} < c$. According to the conditions (9.105) and (9.106) the wave propagating with the phase velocity $\mathrm{v}_{ph} > c$ will be responsible for the pair production process. By the appropriate transformations the processes of e^-, e^+ pair production in these EM field configurations can be reduced to the considered pair production process (as in the case of plasma) in this section. Namely, one should solve the problem in the center-of-mass frame of the produced pair moving with respect to the laboratory frame with the velocity $v = c^2/v_{ph}$.

Bibliography

J. Schwinger, Phys. Rev. **82**, 664 (1951)
V.P. Yakovlev, Zh. Eksp. Teor. Fiz. **49**, 318 (1965)
T. Erber, Rev. Mod. Phys. **38**, 626 (1966)
F.V. Bunkin, I.I. Tugov, Dokl. Akad. Nauk SSSR **187**, 541 (1969) [in Russian]
A.I. Nikishov, Zh. Eksp. Teor. Fiz. **57**, 1210 (1969)
N.B. Narojni, A.I. Nikishov, Yadernaya Fizika **11**, 1072 (1970) [in Russian]
A.I. Nikishov, Nucl. Phys. B **21**, 346 (1970)
V.S. Popov, Pis'ma Zh. Eksp. Teor. Fiz. **11**, 254 (1970)
E. Brezin, C. Itzykson, Phys. Rev. D **2**, 1191 (1970)
A.I. Nikishov, V.I. Ritus, Usp. Fiz. Nauk **100**, 724 (1970) [in Russian]
F.V. Bunkin, A.E. Kazakov, Dokl. Akad. Nauk SSSR **193**, 1274 (1970) [in Russian]
Ya.B. Zeldovich, V.S. Popov, Usp. Fiz. Nauk **105**, 403 (1971) [in Russian]
V.S. Popov, Pis'ma Zh. Eksp. Teor. Fiz. **13**, 261 (1971)
V.S. Popov, Zh. Eksp. Teor. Fiz. **61**, 1334 (1971)
A.I. Nikishov, Problemi Teoreticheskoi. Fiziki, Moscow (1972) [in Russian]
V.S. Popov, Zh. Eksp. Teor. Fiz. **62**, 1248 (1972)
V.S. Popov, Zh. Eksp. Teor. Fiz. **63**, 1586 (1972)
V.I. Ritus, Nucl. Phys. B **44**, 236 (1972)
N.B. Narojni, A.I. Nikishov, Zh. Eksp. Teor. Fiz. **63**, 1135 (1972)
A.A. Grib, V.M. Mostepanenko, V.M. Frolov, Teor. Math. Fiz. **13**, 377 (1972) [in Russian]
A.M. Perelomov, Phys. Lett. A **39**, 353 (1972)
H.L. Berkowitz, R. Rosen, Phys. Rev. D **5**, 1308 (1972)
N.B. Narojni, A.I. Nikishov, Zh. Eksp. Teor. Fiz. **63**, 862 (1973)
V.S. Popov, Yadernaya Fizika **19**, 1140 (1974) [in Russian]
A.I. Nikishov, V.I. Ritus, Tr. Fiz. Inst. Akad. Nauk SSSR **111**, 5 (1979)
V.N. Radionov, Zh. Eksp. Teor. Fiz. **78**, 105 (1980)
H.K. Avetissian, A.K. Avetissian, Kh.V. Sedrakian, Zh. Éksp. Teor. Fiz. **99**, 50 (1991)
H. K. Avetissian et al., Phys. Rev. D **54**, 5509 (1996)
D. Burke et al., Phys. Rev. Lett. **79**, 1626 (1997)
Y. Kluger, E. Mottola, Phys. Rev. D **58**, 125015 (1998)
J.C.R. Bloch et al., Phys. Rev. D **60**, 116011 (1999)
A. Ringwald, Phys. Lett. B **510**, 107 (2001)
V.S. Popov, JETP Lett. **74**, 133 (2001)
H.M. Fried et al., Phys. Rev. D **63**, 125001 (2001)
R. Alkofer et al., Phys. Rev. Lett. **87**, 193902 (2001)
H. K. Avetissian et al., Phys. Rev. E **66**, 016502 (2002)

C.D. Roberts et al., Phys. Rev. Lett. **89**, 153901 (2002)
A. Di Piazza, G. Calucci, Phys. Rev. D **65**, 125019 (2002)
H.K. Avetissian et al., Nucl. Instrum. Methods. A **507**, 582 (2003)
A. Di Piazza, Phys. Rev. D **70**, 053013 (2004)
P. Krekora, Q. Su, R. Grobe, Phys. Rev. Lett. **92**, 040406 (2004)
P. Krekora, Q. Su, R. Grobe, Phys. Rev. Lett. **93**, 043004 (2004)

Index

Springer Series in
OPTICAL SCIENCES

Volume 1

1 **Solid-State Laser Engineering**
By W. Koechner, 6th revised and updated ed. 2006, 450 figs., 45 tabs., XIV, approx 760 pages

Published titles since volume 80

85 **Sensing with Terahertz Radiation**
By D. Mittleman (Ed.), 2003, 207 figs., 14 tabs., XVI, 337 pages

86 **Progress in Nano-Electro-Optics I**
Basics and Theory of Near-Field Optics
By M. Ohtsu (Ed.), 2003, 118 figs., XIV, 161 pages

87 **Optical Imaging and Microscopy**
Techniques and Advanced Systems
By P. Török, F.-J. Kao (Eds.), 2003, 260 figs., XVII, 395 pages

88 **Optical Interference Coatings**
By N. Kaiser, H.K. Pulker (Eds.), 2003, 203 figs., 50 tabs., XVI, 504 pages

89 **Progress in Nano-Electro-Optics II**
Novel Devices and Atom Manipulation
By M. Ohtsu (Ed.), 2003, 115 figs., XIII, 188 pages

90/1 **Raman Amplifiers for Telecommunications 1**
Physical Principles
By M.N. Islam (Ed.), 2004, 488 figs., XXVIII, 328 pages

90/2 **Raman Amplifiers for Telecommunications 2**
Sub-Systems and Systems
By M.N. Islam (Ed.), 2004, 278 figs., XXVIII, 420 pages

91 **Optical Super Resolution**
By Z. Zalevsky, D. Mendlovic, 2004, 164 figs., XVIII, 232 pages

92 **UV-Visible Reflection Spectroscopy of Liquids**
By J.A. Räty, K.-E. Peiponen, T. Asakura, 2004, 131 figs., XII, 219 pages

93 **Fundamentals of Semiconductor Lasers**
By T. Numai, 2004, 166 figs., XII, 264 pages

94 **Photonic Crystals**
Physics, Fabrication and Applications
By K. Inoue, K. Ohtaka (Eds.), 2004, 209 figs., XV, 320 pages

95 **Ultrafast Optics IV**
Selected Contributions to the 4th International Conference on Ultrafast Optics, Vienna, Austria
By F. Krausz, G. Korn, P. Corkum, I.A. Walmsley (Eds.), 2004, 281 figs., XIV, 506 pages

96 **Progress in Nano-Electro Optics III**
Industrial Applications and Dynamics of the Nano-Optical System
By M. Ohtsu (Ed.), 2004, 186 figs., 8 tabs., XIV, 224 pages

97 **Microoptics**
From Technology to Applications
By J. Jahns, K.-H. Brenner, 2004, 303 figs., XI, 335 pages

98 **X-Ray Optics**
High-Energy-Resolution Applications
By Y. Shvyd'ko, 2004, 181 figs., XIV, 404 pages

99 **Mono-Cycle Photonics and Optical Scanning Tunneling Microscopy**
Route to Femtosecond Ångstrom Technology
By M. Yamashita, H. Shigekawa, R. Morita (Eds.) 2005, 241 figs., XX, 393 pages

100 **Quantum Interference and Coherence**
Theory and Experiments
By Z. Ficek and S. Swain, 2005, 178 figs., XV, 418 pages

101 **Polarization Optics in Telecommunications**
By J. Damask, 2005, 110 figs., XVI, 528 pages

102 **Lidar**
Range-Resolved Optical Remote Sensing of the Atmosphere
By C. Weitkamp (Ed.), 161 figs., XX, 416 pages

103 **Optical Fiber Fusion Splicing**
By A.D. Yablon, 2005, 137 figs., XIII, 306 pages

104 **Optoelectronics of Molecules and Polymers**
By A. Moliton, 2005, 229 figs., XXXI, 498 pages

105 **Solid-State Random Lasers**
By M. Noginov, 2005, 131 figs., XII, 238 pages

106 **Coherent Sources of XUV Radiation**
Soft X-Ray Lasers and High-Order Harmonic Generation
By P. Jaeglé, 2005, 150 figs., approx. 264 pages

107 **Optical Frequency-Modulated Continuous-Wave (FMCW) Interferometry**
By J. Zheng, 2005, 137 figs., XVIII, 254 pages

108 **Laser Resonators and Beam Propagation**
Fundamentals, Advanced Concepts and Applications
By N. Hodgson and H. Weber, 2005, 587 figs., XXV, 794 pages

109 **Progress in Nano-Electro Optics IV**
Characterization of Nano-Optical Materials and Optical Near-Field Interactions
By M. Ohtsu (Ed.), 2005, 123 figs., XIV, 206 pages

110 **Kramers-Kronig Relations in Optical Materials Research**
By V. Lucarini, J.J. Saarinen, K.-E. Peiponen, E.M. Vartiainen, 2005, 37 figs., X, 162 pages

111 **Semiconductor Lasers**
Stability, Instability and Chaos
By J. Ohtsubo, 2005, 169 figs., XII, 438 pages

112 **Photovoltaic Solar Energy Generation**
By A. Goetzberger and V.U. Hoffmann, 2005, 139 figs., XII, 234 pages

113 **Photorefractive Materials and Their Applications 1**
Basic Effects
By P. Günter and J.P. Huignard, 2005, 169 figs., approx. XII, 300 pages

114 **Photorefractive Materials and Their Applications 2**
Materials
By P. Günter and J.P. Huignard, 2005, 100 figs., approx. XII, 300 pages

115 **Photorefractive Materials and Their Applications 3**
Applications
By P. Günter and J.P. Huignard, 2005, 100 figs., approx. XII, 300 pages

116 **Spatial Filtering Velocimetry**
Fundamentals and Applications
By Y. Aizu and T. Asakura, 2006, 112 figs., approx. XII, 220 pages

117 **Progress in Nano-Electro-Optics V**
Nanophotonic Fabrications, Devices, Systems, and Their Theoretical Bases
By Motoichi Ohtsu, 2006, 122 figs., 3 tables, approx. 193 pages

118 **Mid-infrared Semiconductor Optoelectronics**
By A. Krier, 2006, 443 figs., approx. 776 pages

119 **Optical Interconnects**
The Silicon Approach
By L. Pavesi and G. Guillot, 2006, 265 figs., approx. 410 pages

120 **Relativistic Nonlinear Electrodynamics**
Interaction of Charged Particles with Strong and Super Strong Laser Fields
By H.K. Avetissian, 2006, 23 figs., XIII, approx. 320 pages